C0-ASG-770

ADVANCES IN X-RAY ANALYSIS

Volume 22

ADVANCES IN X-RAY ANALYSIS

Volume 22

Edited by

Gregory J. McCarthy

The Pennsylvania State University
University Park, Pennsylvania

and

Charles S. Barrett, Donald E. Leyden,
John B. Newkirk, and Clayton O. Ruud

University of Denver
Denver, Colorado

Sponsored by
JCPDS–International Centre for Diffraction Data
and
University of Denver Research Institute
and
Department of Chemistry
University of Denver

PLENUM PRESS • NEW YORK AND LONDON

The Library of Congress cataloged the first volume of this title as follows:

Conference on Application of X-ray Analysis.
Proceedings 6th- 1957- [Denver]

v. illus. 24-28 cm. annual.
No proceedings published for the first 5 conferences.
Vols. for 1958- called also: Advances in X-ray analysis, v. 2-
Proceedings for 1957 issued by the conference under an earlier name: Conference on Industrial Applications of X-ray Analysis. Other slight variations in name of conference.
Vol. for 1957 published by the University of Denver, Denver Research Institute, Metallurgy Division.
Vols. for 1958- distributed by Plenum Press, New York.
Conferences sponsored by University of Denver, Denver Research Institute.
1. X-rays–Industrial applications–Congresses. I. Denver. University. Denver Research Institute II. Title: Advances in X-ray analysis.
TA406.5.C6 58-35928

Library of Congress Catalog Card Number 58-35928
ISBN 0-306-40163-0

Proceedings of the Twenty-Seventh Annual Conference on Applications of X-Ray Analysis held in Denver, August 1–4, 1978

Plenum Press, New York
A Division of Plenum Publishing Corporation
227 West 17th Street, New York, N.Y. 10011

Foreword

Many of the science and engineering problems under investigation at industry, government and university laboratories come under the headings "energy, materials and resources." X-ray analysis plays a key role in these investigations. This is reflected in the content of the present volume of Advances in X-ray Analysis. Nearly half of the papers come under such headings as energy production and conversion, materials optimization and mineral characterization. The remainder continue the long tradition of this series in presenting the latest advances in apparatus and procedures for x-ray diffraction and fluorescence analyses.

In keeping with recent practice, this year's Denver Conference on Applications of X-ray Analysis emphasized x-ray diffraction and was co-sponsored by JCPDS, International Center for Diffraction Data.

The first group of papers in this volume were presented in a plenary session on unusual specimen preparation, handling and analyses by x-ray diffraction. In the lead paper, D. K. Smith and C. S. Barrett combine their seven decades of experience with surveys of diffractionists throughout the world to present a comprehensive review of non-routine specimen preparation in powder diffractometry. Next, M. J. Camp discusses the particular procedures and constraints under which foresnic laboratories employ diffraction methods.

F. A. Mauer and C. R. Robbins discuss a method for characterization of high temperature-pressure reactions *in situ* by energy dispersive x-ray diffraction that offers the promise of sorting out the mechanisms underlying materials compatibility and durability.

Special methods and examples of the characterization of radioactive materials by x-ray powder diffraction are included in an overview paper by David Schiferl and R. B. Roof.

With so many of the advances in electronic and optical materials coming in the field of thin films, it was quite appropriate to include two papers on the characterization of these materials by x-ray methods. A. R. Storm describes camera techniques being routinely employed in his laboratories and T. C. Huang and W. Parrish discuss thin film characterization by x-ray fluorescence and diffraction.

Numerous contributed papers complement these invited papers.

Gregory J. McCarthy
January 1979

Preface

This volume is the proceedings of the 1978 Denver Conference on the Applications of X-Ray Analysis, 27th in the series. The conference was sponsored by the University of Denver and the JCPDS-International Centre for Diffraction Data. Conference chairmen in residence were D. E. Leyden and C. O. Ruud, with C. S. Barrett and J. B. Newkirk as honorary chairmen. The invited conference chairman, G. J. McCarthy of the Pennsylvania State University, organized the Plenary Session entitled "Special Techniques in Powder Diffraction."

In addition to the plenary and regular contributed sessions, a special session on Progress in the Reduction of Matrix Effects in XRF was organized and chaired by P. A. Pella and K. F. J. Heinrich of the National Bureau of Standards.

A number of invited speakers were on the program and their names and titles of their papers are listed below.

D. K. Smith and C. S. Barrett, "Unusual Specimen Preparation Techniques for X-Ray Diffraction Powder Analysis."

Michael J. Camp, "Special Problems in Forensic Materials Analysis by XRD."

F. A. Mauer and C. R. Robbins, "X-Ray Powder Diffraction Measurements in Reactive Atmospheres at 1000°C and 7 MPa."

David Schiferl and R. B. Roof, "X-Ray Diffraction on Radioactive Materials."

A. R. Storm, "Thin Film Characterization of Thin Films by X-Ray Diffraction."

T. C. Huang and W. Parrish, "Characterization of Thin Films by X-Ray Fluorescence and Diffraction Analysis."

R. Jenkins, "Empirical Matrix Correction Methods."

D. N. Breiter, "Matrix Correction Methods Using Monochromatic X-Ray Excitation."

A. P. Quinn, "Quasi-Fundamental Correction Methods Using Broadband X-Ray Excitation."

R. P. Gardner, "Progress in Monte Carlo Simulation Methods."

K. K. Nielson, "Progress in X-Ray Fluorescence Correction Methods Using Scattered Radiation."

J. W. Criss, "Selecting and Combining Various Methods."

K. F. J. Heinrich, "Progress in the Reduction of Matrix Effects--A Discussion."

Two days of tutorial workshops were also presented on Monday and Tuesday preceding the conference. The workshops are described as follows.

Search/Match Techniques for Qualitative Powder Diffractometry

Sponsored by JCPDS-International Centre for Diffraction Data

Manual Techniques

Catharine Foris
Mark Holomany
Ron Jenkins
Theresa Maguire
Julian Messick
Ben Post

Computer Search Methods

Catharine Foris
Younghe Hahm
Mark Holomany
Cam Hubbard
Ron Jenkins
Theresa Maguire
Julian Messick
Ben Post
Walter Schreiner

Introduction to Mathematical Data Reduction Procedures for Use in Quantitative X-Ray Fluorescence Spectrometry

Chaired by P. A. Pella and K. F. J. Heinrich, National Bureau of Standards

Peter A. Pella, "Sample Handling and X-Ray Measurements."

Henry Chessin, "Physics of X-Ray Absorption and Scattering in Thick Samples."

Ron Jenkins, "Empirical Models for Correction of Interelement Effects."

Harold E. Marr, "Applications of Empirical Models."

Dennis N. Breiter, "Monochromatic Excitation Using Secondary Target Emitters (*Primarily* Energy-Dispersive)."

Alan P. Quinn, "Broadband Excitation Approximation, *e.g.* CORSET (Energy-Dispersive and Wavelength)."

Kirk K. Nielson, "Use of Compton Scattered Radiation for Correction of Matrix Effects."

John W. Criss, "Selecting and Combining Various Methods."

Co-chairmen for the conference sessions were:

C. S. Barrett
D. W. Beard
H. Chessin
J. W. Criss
P. Dismore
R. P. Gardner
K. F. J. Heinrich
M. R. James
D. E. Leyden
F. A. Mauer
G. J. McCarthy
M. Morris
J. B. Newkirk
P. A. Pella
P. K. Predecki
C. O. Ruud
D. M. Smith

Conference aids were:

Herb Acree
Larry Apple
Bill Bodnar
Jane Brawner
Barbara Cain
John Cronin
Al Reynolds
Mary Straw
Ali Zanjani

On behalf of the conference committee in residence and the attendees, I thank the persons listed above and the JCPDS-International Centre for Diffraction Data for making this an informative meeting.

Clayton O. Ruud

Unpublished Papers

The following papers were for oral presentation only, or not published for a variety of reasons.

"Thin Film Characterization by X-Ray Diffraction," Arthur R. Storm, Bell Telephone Laboratories, Murray Hill, NJ 07974.

"Coal Ash Analysis by XRF," Leroy Jacobs and Don Smith, Commercial Testing and Engineering Co., 14335 West 44th Ave., Golden, CO

80401; and Curtis Gold, Philips Electronic Instruments, Inc., 917 Acoma St., Denver, CO 80204.

"Unusual Techniques for X-Ray Diffractometry in the Transmission Mode of Operation," Marjorie D. Chris, Hercules Incorporated Research Center, Wilmington, DE 19899.

"UV/X-Ray Imaging for the Alcator Tokamak," R. Petrasso, P. Burstein, M. Gerassimenko, W. Hamilton, B.Krogstad, F. Seguin, M. Seiden and J. Ting," American Science and Engineering, Inc., 955 Massachusetts Ave., Cambridge, MA 02139; and R. Granetz and R. R. Parker, MIT, Cambridge, Massachusetts.

"Matrix Correction Methods Using Monochromatic X-Ray Excitation," Dennis Breiter, The Buhrke Company, 508 Washington Ave., Golden, CO 80401.

"Selecting and Combining Various Methods," John W. Criss, Code 6683, X-Ray Optics Branch, Naval Research Laboratory, Washington, DC 20375.

"Escape Peak Intensities in Si(Li) Detectors," P. Sioshansi, M. R. Khan, A. S. Lodhi, and A. Aframian, Nuclear Research Center, P.O. Box 3327, Tehran, Iran; and F. Jundt and A. Gallmann, Centre de Recherches Nucléaires et Université Louis Pasteur, Strasbourg, France.

"Cation-Exchange Filters for Sample Preconcentration in X-Ray Spectrometric Analysis," Howard M. Kingston and Peter A. Pella, National Bureau of Standards, Center for Analytical Chemistry, Washington, DC 20234.

"Determination of Trace Elements in Liquids by X-Ray Fluorescence Spectrometry Using Incoherent Scattered Radiation as an Internal Standard," Robert D. Giauque, R. B. Garrett and L. Y. Goda, Energy and Environment Division, Lawrence Berkeley Laboratory, Berkeley, CA 94720.

"Current Application of Residual Stress Analysis in the Aircraft Industry," R. E. Herfert, Aircraft Division, Northrop Corporation, 3901 W. Broadway, Hawthorne, CA 90250.

"Practical Accuracy Considerations in Use of X-Ray Diffraction Residual Stress Analyzers," B. Tooper, EMS Laboratories, 12517 Crenshaw, Hawthorne, CA 90250; and K. G. Anderson, Stress Measurements Inc., Office 604, 4940 Viking Dr., Minneapolis, MN 55435.

"Inconel 600 Tubing Surface Defect Characterization by X-Ray Line Broadening," Paul S. Prevey, Lambda Research Inc., 7213 Market

Place, Cincinnati, OH 45216; and J. M. Nielsen, Babcock and Wilcox, Alliance Research Center, Alliance, Ohio.

"Stability and Decay of Surface Residual Stresses and their Influence on the Fatigue Properties of Ti-6Al-4V," G. R. Leverant, Dept. of Materials Sciences, Southwest Research Institute, P.O. Drawer 28510, San Antonio, TX 78284.

"EXNAL: A Fundamental Parameters X-Ray Spectrum Analysis Program," R. Gunnink, N. Bonner, C. Bazan and D. C. Camp, Lawrence Livermore Laboratory, P.O. Box 808 (L-233), Livermore, CA 94550.

"Inaccuracies Encountered in Sulfur Analysis by Proton Induced X-Ray Emission," John F. Ryder, N. F. Mangelson, T. Majors, M. W. Hill, D. J. Eatough and L. D. Hansen, Brigham Young University, 226 ESC, Provo, UT 84602.

"Proton Induced X-Ray Emission Analysis of Thick Targets--A New Approach," Md. R. Khan and A. S. Lodhi, Atomic Energy Organization of Iran, Nuclear Research Centre, P.O. Box 3327, Tehran, Iran.

Contents

EVALUATION OF XRD PATTERNS

APPLICATIONS OF XRD

INSTRUMENTATION AND LASER SOURCES

MATHEMATICAL DATA ANALYSIS FOR XRF

XRF APPLICATIONS

XRF: INNOVATIONS

SPECIAL HANDLING PROBLEMS IN X-RAY DIFFRACTOMETRY

D. K. Smith* and C. S. Barrett°

Dept. of Geosciences, Pennsylvania State University, University Park, PA 16802*
Metallurgy and Materials Science, Denver Research Inst., University of Denver°

ABSTRACT

The success of a diffraction experiment often depends on the proper preparation of the sample under study. Many different methods of sample preparation have been devised for many different purposes. A survey was conducted to collect together many of these methods, and the results are reported in this paper.

INTRODUCTION

During the many years of short courses and workshops on x-ray diffraction, the topic of sample preparation has always received considerable attention. The reason for this attention is that the quality of the diffraction data is no better than the sample under study. A little special care in preparing the specimen can often mean the difference between interpretable data and unintelligible data. That extra effort at the beginning can often save hours at the end of the experiment.

There are many ways of preparing samples for the powder diffractometer, and good discussions are available in the standard textbooks (1-5). These discussions emphasize the packing of loose powders in a cavity for presentation to the diffractometer. Klug and Alexander (1) have described the parameters which must be considered when making up a specimen. Many other sample techniques are described in technical papers on reside in the heads of the experimenters. These techniques were usually developed for some special purpose or material. Some of these special tricks may never find their way into the literature. In an attempt to uncover some of these "tricks," a survey was conducted in which descriptions of such special methods were requested. The response was excellent, and the information supplied forms the main body of this paper. It will not be possible in the scope of this paper to identify all contributors. The authors would, however, like to personally thank each contributor. This paper will be restricted to methods applied to diffractometers.

A diffraction pattern contains three types of information: the position of a diffraction maximum, its intensity and the intensity distribution as a function of diffraction angle. The optimization of one of these aspects rarely leads to the optimization of the other properties. The ideal sample for a given experiment depends largely on the information desired. A sample which yields the most accurate d-values may not yield the most accurate intensities. A sample which is to be used only for identification of its constituents may be quite different from a sample to be used to prepare a standard reference pattern. A specimen to be used to measure strain will be different from a specimen to be used for quantitative analysis. Let us first consider a sample useful for reference data, then real samples and typical problems, and finally, special modifications for special purposes.

PREPARING A STANDARD REFERENCE PATTERN

Perhaps the most critical sample requirements are those which must be met when the data are to be reported as a standard reference pattern. The d-values must be accurate to the limit of the instrument, and the intensities must be representative of a truly random orientation of crystallites. It is not easy to achieve both these conditions in the same sample. In preparing the standard sample it is usually advisable to prepare different samples for determining the d-values and the intensities. The Joint Committee on Powder Diffraction Standards Associateship at the National Bureau Standards has developed a procedure which they follow in their production of reference patterns. The method is described by Morris et al. (6).

IDENTIFICATION

Perhaps the single most common use of powder diffraction methods is the identification of an unknown or the verification of a predicted phase. If the pattern is not too complicated or if the particles are not markedly shaped, rather simple methods may be used to prepare the sample. Because the d spacings are potentially the most accurate, being unaffected by orientation effects, it is important in any sample preparation technique to get the best d values possible. The use of an internal standard is advisable as it can compensate for several common systematic errors especially sample displacement. One must always remember that the position of measured diffraction peaks is very sensitive to the position of the sample with respect to the diffractometer. Contrary to common beliefs, it is not surface of the sample which is critical, rather it is the half-depth of penetration of the x-ray beam in the sample which more closely corresponds to the peak position. Neither powders in cavity mounts nor powders on the surface of microscope slides achieve the optimum sample position. Very light element compounds allow considerable penetration of the x-ray beam and concomitant peak displacement and broadening. Thick samples in cavity mounts may be subject to more displacement errors than thin smears on a glass slide.

The most common method of preparing a sample of loose powder is to pack it into a flat cavity mount. Mounts are made of aluminum or some other metal, Bakelite, glass or Lucite and in special cases quartz, LiF, or some other single crystal. The other common method of preparing a powder is the smear mount. A small amount of powder is slurried with a binder, and a drop of the slurry is spread on the surface of a microscope slide. Binders vary from a wetting agent such as acetone, alcohol, water, amyl acetate, oil, or ether; or a binder such as Duco in acetone or amyl acetate, collodion in amyl acetate or ether, Ambroid in nitrobenzene, mucilage in Karo, or rubber cement in toluene.

Both methods are subject to preferred orientation, and a variety of tricks have been used to minimize its effect. Using cavity mounts, the most common method is back loading (7) or side loading (8). A frosted or serrated glass surface, ceramic or cardboard flat is placed over the front of the mount and the sample is carefully added through the open back or a side port until the cavity is full. The back of the mount is covered, and the front piece is removed carefully so as not to disturb the surface. Front mounted samples may be retouched after packing by lightly cutting grooves with a sharp edge, rolling with a knurled surface or tamping with a serrated flat (9). This technique leaves a roughened surface which may slightly affect the pattern quality, and it is advisable to be able to adjust the sample height to compensate for the surface roughness (10).

An alternate method using a cavity mount is to mix the sample with an inert amorphous filler. Powdered glass, amorphous boron, cork, starch, gum arabic, tragacanth or gelatin have been used. This method results in increased penetration of the x-ray beam into the sample but less reflected intensity because of the absorption of the filler. Starch has a spherical habit, and after picking up shaped particles clinging to its surface, it packs randomly into the cavity. Irregular fragments such as glass behave similarly. Gum arabic may be mixed with the sample, set, then reground to form irregular fragments which pack randomly. Fillers must be chosen such that they have no diffraction lines which interfere with the sample.

To approach randomness on a microscope slide the sample may be mixed with a viscous binder such as collodion or petroleum jelly then spread on the microscope slide. Thin binders such as acetone allow particles to settle and shaped particles to align. An alternative method is to coat the microscope slide surface with a thin film of grease then dust the powder onto the surface through a fine mesh sieve. Once the powder has landed on the sticky surface, care must be taken not to touch or disturb the surface.

RANDOMIZING CRYSTALLITE ORIENTATIONS

Probably the most difficult problem in sample preparation is obtaining a truly random orientation of the crystallites. Those materials with equiaxed shapes and no cleavages that produce flat surfaces usually yield random samples using any method of sample preparation. However, if the particles have any crystallographically related shape, the shape will make achieving randomness difficult. The difficulty increases as the difference between the maximum and minimum particle dimensions increases. Fibrous shapes are generally more difficult to randomize than platy shapes.

Several methods including side or back loading, the use of viscous binders or dusting onto a surface have already been mentioned. Dispersion in a binder, which is allowed to set, yields a briquet which may be sectioned and polished. These methods help but do not always guarantee a sample of sufficient randomness for quantitative applications. The alternative approach is to cause the shaped particles to aggregate into spherical clumps which will pack randomly in a cavity. Several methods of spheridizing aggregates have been proposed. Mixing with starch, amorphous boron or silica gel gives a spherical core which attracts particles to its surface. Bloss et al. (11) describe lightly spraying a dispersed fine powder with an atomized binder such as a clear acrylic lacquer. The droplets pull particles into a spherical clump on contact. Several authors (12-15) describe the method of spray drying: atomizing a slurry of particles in a binder into the air and drying it before the spherical droplets fall into the collector. Various binders have been used including water, collodion in a fast evaporating solvent, rubber cement in toluene, and lacquer. Atomizers from perfume dispensers to paint sprayers have been used. Best results are obtained if the spray is directed upward and the droplets firm up at the peak of the fountain where they are most spherical. It is usually advisable to seive the resultant product to separate out the spherical particles from any complex aggregates and fine particles which did not spheridize.

One of the more interesting tricks to come to light in this survey is the device to randomize a sample devised by David Keating at Brookhaven National Laboratories. A flat sample spinner was modified with a thin foil cover on top and an air-cooled $BaTiO_3$ transducer underneath driven by the power supply of an ultrasonic cleaner. The sample was prepared as a slurry and vibrated during the data taking. The technique effectively minimized orientation effects.

Other methods which were mentioned were to prepare pellets by setting in a binder of epoxy or by pressing. Isostatic pressing of very fine powders usually leads to quite random mounts especially if the mount is sectioned and polished. Uniaxial pressing will usually lead to orientation effects, but standardized methods were effectively used that led to reproducibility of samples.

PREPARING ORIENTED SAMPLES

In many applications sufficiently quantitative results can be obtained if the sample can be prepared in a manner that produces intensity reproducibility even if some preferred orientation does exist. Standard quantity versus intensity curves must be prepared on reference samples prepared in the same manner. The method of Copeland and Bragg (16) has long been used in cement analyses and a procedure for α-quartz has been established by Bumsted (17) for respirable air quality control. Birks et al. (18) have developed a method for fibrous asbestos. Gaston Pouliot of the Ecole Polytechnique has described a very detailed specimen technique for comparing glacial clays by diffraction analysis. Its important features are the rigorous adherence to a specified stepwise procedure.

Clay minerals rarely show strong diffraction effects from planes other than the (00ℓ), so it is highly advantageous to prepare oriented samples for the diffractometer. The method of Gibbs (19), generally referred to as the smear mount, is probably the most commonly employed. A dispersed sample in distilled water is evaporated until the slurry is fairly thick. This slurry is then placed on a standard microscope slide and smeared over the surface using the edge of another slide or similar tool. Practice allows one to prepare a uniform smear with little effort. If the slurry is too thin, different size fractions may settle differently and the result may be an uneven distribution of phases. Alternatively, following Shaw (20) and Drever (21), a thin clay suspension from which all coarse particulates, greater than 2μm., have been removed by settling is rapidly filtered through a porous membrane. A vacuum system is fitted with small round filter pads (Millipore or Nuclepore), fitted glass or ceramic discs which can be removed and placed directly into a diffractometer sample holder. While the assembly is under vacuum, the suspension is poured through the funnel until a sufficient layer of clay particles is collected on the filter disc. The disc is removed and may be used wet, dried in an oven or desiccator, or exposed to special atmospheres. If a flexible membrane is used, the compact, while still wet, is placed clay side down on a frosted microscope slide or other support, and the membrane is carefully peeled away.

An extremely interesting method for orienting fibrous phases has been described by Birks et al. (18) and M. Fatimi et al. (20).

In this technique the fibers of asbestos are aligned nearly parallel to each other by an electrostatic process. The fibers are dispersed in water with a dispersing agent and a sonic generator. An aliquot is vacuum filtered onto a millipore filter, and the filter is then ashed. The residue is suspended in a weak solution of parlodion in amyl acetate and sonically dispersed again. A drop of the solution is then spread over a special grid of parallel wires which are alternately bussed, and 240 VAC are applied to the grid. When the solution has evaporated, a more concentrated solution of parlodion in amyl acetate is applied to the sample, the sample is allowed to dry again and then it is peeled from the grid. The fibers align perpendicular to the wires. A transmission diffractometer is used to measure the diffraction pattern perpendicular and parallel to the fiber directions.

REACTIVE SAMPLES AND SPECIAL ATMOSPHERES

The methods here-to-fore described have assumed that the sample was available or could be made as a powder in air. Many compounds are air reactive either with O_2, H_2O or CO_2. A researcher can arrange to visit a friend in a humid climate if the sample under study loses water easily or in an arid climate if it gains water, or choose a specific day or time of year when the conditions are appropriate. Alternatively, special samples or sample chambers may be used when appropriate travel cannot be arranged.

Many laboratories that (routinely) handle reactive samples have constructed elaborate dry boxes for preparing the samples and have even enclosed the diffractometer itself. A good dry box has been designed by Szymanski (23). More commonly, a sealable chamber is used which fits on the diffractometer after being loaded in the dry box. As usual the big problem with any device of this type is to have an impervious window which is still transparent to x-rays. Many different designs and materials have been used. Windows can be made from thin Be, both single and polycrystalline; Al foil; Mylar or other thin plastic film such as Saran Wrap; transformer winding paper; thin cleavage sheets of muscovite mica or graphite, or a film of glassy carbon. Most holders are sealed window devices which are loaded from the back or side. Mylar is most commonly used and can be sealed with a thin film of polyethylene glue from a hot glue gun. Epoxy or other cements, O-ring seals and double-sided tape are also effective. The quality of the seal necessary will be dictated by the reactivity of the sample. The severe limitation on the design of the cell other than size is usually the x-ray window in that it must be positioned and sealed in such a way that the beam path is uninterrupted over the desired diffraction range.

Catalysts are usually extremely reactive materials, and they often require maintenance of the sample in the reaction medium.

Several researchers have designed cells in which the catalyst can be maintained in a fluid or in an atmosphere of flowing gas. Some of these have continuously flowing systems from a reservoir into a diffraction cell. F. S. Molinaro of Northwestern University describes a glass enclosed reaction system for studying catalyst behavior in various gases. Part of the system consists of a reaction tube where the gas and catalyst can interact. This part can be easily heated. Then the system is tipped so that the catalyst is transferred to a thin hollow disc with a 0.1 mm thick mica window that fits in the diffractometer. The system never needs to be opened to provide the diffraction data. Following the experiment the apparatus may be flushed, cleaned and reused.

As an alternate to elaborate cells requiring the services of a machine shop or glass blower, many innovations have yielded sucessful experiments with reactive samples. Mylar, or other windows, may be cemented over the opening of a standard cavity mount; the mount is then back loaded in a dry box and sealed on the back side. A cell may be constructed with both front and back surfaces closed and the sample may be side loaded and sealed with cement or hot polyethylene from a glue gun. Many samples can be mixed directly with epoxy, Duco, petroleum jelly and then loaded in a cavity or on a microscope slide. Polished sections of metallographic mounts or other solid samples may be lightly sprayed with an acrylic lacquer or coated with collodion or other film. Alternatively a piece of transparent "Magic Mending" tape may be stretched over the surface. A powdered sample in a standard mount can be sprayed or coated similarly. Envelopes can also be made to surround the sample. Thin samples can be prepared by dusting a sample on the sticky side of transparent tape and applying a second piece of tape to seal the powder. The 3M Company produces tapes with a thin Mylar base and a cement coating. The cement coating can be removed from a small area on two pieces of tape with benzene on a cotton swab. The sample is then placed in the window region, and the two pieces of tape are pressed together. This envelope was used to enclose LiH for periods up to 24 hours. J. B. Dixon describes a simple permanent sealed mount by preparing an epoxy mixture of sample which is then spread on a piece of flat flexible polyethylene. A microscope slide is pressed into the exposed surface, and the whole is allowed to set. Then the polyethylene is carefully peeled away from the mount.

Simple modifications of the diffractometer safety chambers have allowed many studies to be made. Often, flowing dry N_2 gas is sufficient to minimize oxidation and hydration reactions. One or two gas ports are easy to install. Even controlled atmospheres, such as fixed relative humidities, are easy to implement. Do not use Ar gas as an inert atmosphere. This practice is common in reaction furnaces, but its x-ray absorption is too high for diffraction applications.

One problem many people commented on was the static electricity in a dry box. George Myer of Temple University indicated that an antistatic Po source worked very effectively to dissipate the charge.

Another technique for handling reactive powders especially when one wishes to recover the powder afterwards has been employed by many experimentalists. Glass capillaries normally used for Debye-Scherrer samples can be loaded and sealed in a dry box. A hot wire in the box is very effective for sealing. Several capillaries are then aligned in a parallel array either on a microscope slide or in a cavity, so their thickness can be compensated for by placing their centers on the plane of the diffractometer axis. If the capillaries are half full, the result can be a quite flat sample surface. Best results are usually obtained with the capillaries aligned parallel to the diffractometer axis.

Special techniques are required to handle radioactive specimens. This topic is being covered in a companion paper on this program by Schiferl and Roof (24) and will not be considered further.

MICROSAMPLES

An area which has received considerable attention in this survey is the handling of minute quantities primarily for identification. There are many who insist that the old camera techniques are more practical for microgram samples, but confirmed diffractometer users never give up. Their successes have been quite remarkable.

The major aim of handling microquantities is minimizing diffraction effects from everthing except the sample. Single crystal holders are the most effective sample supports. Cleaved crystals cut either on or off Bragg planes have been used. Fluorite, CaF_2; LiF; calcite, $CaCO_3$; and MgO can all be easily cleaved and used for sample supports. Crystals of Si and quartz can be cut as desired, lapped, and chemically polished to serve as sample suports as well. The HCl-HNO_3-$HC_2H_3O_2$-Br treatment for Si will result in damage free surfaces and a high polish. Quartz can best be polished mechanically by grinding on a flat surface, such as glass plate with 600 SiC followed by a 90 second etch in HF. This treatment effectively eliminates trouble from the damaged surface layers. A mechanical polish with SnO_2 or CeO_2 is almost as good but is much more work. A short 10 second etch is usually advisable following polishing. BT cut quartz oscillator plates make excellent substrates.

Care must be taken to crush the sample without losing any material and to carefully center the material in the x-ray beam. Samples are easily handled on glass slides under a binocular microscope. They can be crushed in a drop of acetone or between two glass slides. Grains may be transferred to the single crystal substrate by using a needle or wooden splinter wetted with water, mineral oil or ethyl alcohol.

For minute quantities it is not necessary to have a large single crystal as a support. A very convenient procedure is to fasten a small crystal firmly in the cavity of the usual Al sample mount after carefully determining with the aid of a fluorescent screen where the beam would hit the crystal.

An alternative method of mounting a micro sample after powdering on a microscope slide is to enclose the sample with a thin film of collodion in amyl acetate. After the film has dried, it may be peeled off and stretched across the opening in an Al mount. Care must be taken to keep the collodion film as thin as possible to minimize its contribution to the background.

For very minute quantities of material it is necessary to eliminate air scatter from the beam path. Evacuation of the beam path is advisable, but flowing He gas can be used. Modern safety chambers on the diffractometer usually require little modification to handle He, but be sure to keep the He away from the high voltage of the detector, as it has a low dielectric constant and can cause electrical breakdown.

Miscellaneous Techniques

This survey has lead to several methods which do not fall under the headings already used. Many laboratories have occasion to examine small to large irregularly shaped samples. Some of these have a flat surface, many do not. Quite large brackets and supports have been constructed in some situations. Researchers involved in stress analysis quite commonly encounter this problem and in some cases have taken the diffraction system to the object rather than bringing the part to the machine. Small samples may be supported in a variety of ways. If they are smaller than 2.5 cm., they can be held in the sample spinner assembly or in appropriately bent brackets with small blocks of wax or plasticene. The diffraction surface can be aligned using the leveling device for metallographic polished sections. A special bracket which slips into the diffractometer like the normal sample mount but has its reference surface inverted, i.e. coplanar with the diffractometer reference surface but facing downward, can support large flat surfaces. Usually, however, these samples can be held against the diffractometer reference surface directly with wax or wire.

Sample holders may be designed which can fill the needs of several systems. The whole sample assembly may be manipulated so as not to disturb the sample. An example is a modified sample holder from an automatic sample changer designed by C. A. Keisling of Worcester Polytechnic Institute. The holder is cutout to accept a silver microfilter which is pressed into position with a special jig so that the clay surface is in line with the surface of the mount. The assembly can be heated, treated in glycolated or controlled relative humidity atmospheres and re-analyzed.

Sometimes just getting the sample into some form which can be x-rayed is a feat. Samples with very thin coatings may often be used directly without crushing as the surface layers contribute strongly to a diffraction pattern. Crushing will expose too much substrate. J. D. Hanawalt of the University of Michigan describes the recovery of a thin film which formed on molten magnesium. Dropping small chips of dry ice into the molten metal produced bubbles which lifted off the skin as a thin film that could be then mounted intact on a microscope slide for identification. Zwell et al. (25) describe the use of extraction techniques to identify second phases in steels. The polished sample is lightly etched and then coated with a collodion film. After the film is dry, the specimen is deeply etched to remove the metal and leave the precipitates on the film which can be recovered and x-rayed. It should be stated here that proper chemical or physical treatment can often be used to advantage in concentrating phases in low concentration.

Samples that are studied by diffraction at sub-zero temperatures frequently are not in a state of equilibrium; the phases present may be metastable ones that are very sluggish in transforming to stable phases or stable mixtures of phases. It has been found that cold working a sample in this temperature range, for example by hammering, promotes transformations and aids the approach to equilibrium; however, the deformation often causes some broadening of peaks.

REFERENCES

1. Klug, H. P. and Alexander, L. E., X-ray Diffraction Procedures for Polycrystalline and Amorphous Material, J. Wiley and Sons (1974).

2. Cullity, B. D., Elements of X-ray Diffraction, Addison-Wesley Publishing Co. (1978).

3. Peiser, H. S., Rooksby, H. P. and Wilson, A. J. C., ed., X-ray Diffraction by Polycrystalline Materials. The Institute of Physcs (1955).

4. Barrett, C. S., and Massalski, T., The Structure of Metals, McGraw Hill, (1966).

5. Hutchison, C. S., Laboratory Handbook of Petrographic Techniques, J. Wiley and Sons (1974).

6. Morris, M. C., McMurdie, H. F., Evans, E. H., Paretzkin B., deGroot, J. H., Newberry, R., Hubbard, C. R. and Carmel, S. J., Standard X-ray Diffraction Powder Patterns National Bureau of Standards Monograph 25, Section 14 (1977).

7. Vassamillet, L. F., and King, H. W., in Advances in X-ray Analysis, Plenum Press, 6, 142 (1963).

8. Bystrom-Asklund, A. M. Sample Cups and a Technique for Sideward Packing of X-ray Diffractometer Specimens. Am. Mineral. 51, 1233 (1966).

9. Peters, Tj., A Simple Device to Avoid Orientation Effects in X-ray Diffractometer Samples. Norelco Reporter 17, 23 (1970).

10. Frevel, L. K., A Technique for Handling Preferred Orientation with an Adjustable Sample-Holder. J. Appl. Cryst. (in press).

11. Bloss, F. D., Frenzel, G., and Robinson, P. D., Reducing Orientation in Diffractometer Samples. Am. Mineral. 52, 1243 (1967).

12. Florke, O. W. and Saalfeld, H., Ein Verfahren zur Herstellung texturfreier Rongen-Pulverpraparate. Z. Kristallogr. 106, 460-466 (1955).

13. Jonas, E. C. and Kuykendall, J. R., Preparation of Montmorillonites for Random Powder Diffraction. Clay Mineral. 6, 232-235 (1966).

14. Hughes, R. and Bohor, B., Random Clay Powders Prepared by Spray Drying. Am. Mineral. 55, 1780 (1970).

15. Smith, S. T., Snyder, R. L. and Brownell, W. E., Elimination of Preferred Orientation by Spray Drying and Applications to Quantitative Analysis by X-ray Diffraction. Applications of X-ray Analysis (this conference).

16. Copeland, L. E. and Bragg, R. H., Preparation of Samples for the Geiger Counter Diffractometer ASTM Bulletin No. 228 (1958).

17. Bumstead, H. E., Determination of Alpha-Quartz in the Respirable Portion of Airborne Particulates by X-ray Diffraction. Am. Ind. Hyg. Assoc. J. 34, 150, (1973).

18. Birks, L. S., Fatemi, M., Gilfrich, J. V. and Johnson E. T., Quantative Analysis of Airborne Asbestos by X-ray Diffraction. Naval Research Laboratory NRL Report 7874 (1975).

19. Gibbs, R. J., Error due to Segregation in Quantitative Clay Mineral X-ray Diffraction Mounting Techniques. Am. Mineral. 50, 741-751 (1965).

20. Shaw, H. G., The Preparation of Oriented Clay Mineral Specimens for X-ray Diffraction Analysis by a Suction-onto ceramic Method. Clay Minerals 9, 349-350 (1972).

21. Drever, J. I., The Preparation of Oriented Clay Mineral Specimens for X-ray Diffraction Analysis by a Filter-Membrane Peel Method. Am. Mineral. 58, 553-4, (1973).

22. Fatemi, M. Johnson, E. T., Witlock, L. L., Birks, L. S., and Gilfrich, J. V., X-ray Analysis of Airborne Asbestos, Interim Report: Sample Preparation. Environmental Protection Agency EPA-600/2-77-062 (1976).

23. Szymanski, J. T., PhD Thesis, University of London (1963).

24. Schiferl, D. and Roof, R. B., X-ray Diffraction on Radioactive Materials, X-ray Analysis (this conference).

25. Zwell, L., Fasiska, E. J., and vonGremmingen F., X-ray Identification of Phases in Type 316 Austenitic Stainless Subjected to Creep Rupture. Trans. Met. Soc. AIME 224, 198-200 (1962).

SPECIAL PROBLEMS IN FORENSIC MATERIALS ANALYSIS BY XRD

Dr. Michael J. Camp

Northeastern University

Boston, MA 02115

There are three factors which may influence the method by which a piece of evidence is analyzed in a forensic laboratory. The first stems from the definition of the word forensic. It means "··· pertaining to, connected with, or used in the courts of law ···." (1) The major areas of forensic analysis are criminal, regulatory, and civil. The purpose of an analysis may be to aid in the prosecution or defense of criminal charges, to determine if a product meets its statutory requirements, to satisfy a consumer complaint, or to determine liability. For these reasons the legal aspects of an analysis may often put constraints upon the analyst. These may require special handling or documentation procedures.

The second factor arises from the physical nature of the sample itself. Such things as the amount available, its physical form, and heterogeneity may cause special problems.

The third factor is the nature of the results being sought. Forensic analysis does not always end with the identification of a single phase or mixture. In many cases the submitting agency wants to know what it is and where it came from. This latter question often influences the sample handling steps.

Any or all of these factors may be important during a given analysis. The legal aspects are always present. They are also the least understood by persons outside the forensic community (2,3,).

If you are an experienced diffractionist, i.e. you routinely run XRD analyses and interpret the results, there should be no arguments from the opposing council on your qualifications as an expert witness, your general competency to perform the analysis and your ability to form an opinion based upon that analysis. Questions may arise concerning documentation of your experimental method, handling of the sample, and the evidentiary chain. You may also expect rebuttal from an opposing expert. You must prepare for these questions before the analysis, not afterwards! A common event that causes much concern is that you have no idea of the importance of your analysis at the time you are doing it. Your results will have to be weighed along with all the other pieces of information. Six months to a year later is no time for second thoughts.

The best preparation is to do good work and to keep good notes. The major requirements are listed below. If at all possible, talk to the attorney for whom you are working. Follow his advice.

Documentation of Procedure. A description of the analytical procedure(s) should be written into a notebook as they are done. Don't forget the simple things such as a visual or microscopic examination or any wet chemistry or spot tests done as an aid to an instrumental technique. For the instrumental procedures record all of the settings of any variable parameter. The idea is to be able to have someone else rerun your analysis and to get the same results. A good rule of thumb is to record any information which would normally appear in the experimental section of a paper. If standards, controls or blanks are necessary, note which ones were examined. Literature references should be entered in full. The opposing council may be able to get a copy of your notes and examine you on them. So be neat and complete.

Sample Handling. Although this is part of the analytical procedure, it is singled out as the major area of attack. Your competency has been established, but there is always a chance to create doubt over the way the sample was handled. Of interest is what you have done to the sample before the test was run. Could it have been contaminated by other samples present in the lab or by unclean equipment? Did drying, grinding, or sieving alter the integrity of the specimen or effect the result? Documentation of these steps is very important, especially when you review your notes at a later date. The use of procedures accepted by your peers is the best way to counter questions of this nature.

Evidentiary Chain. This is an accounting of who has had the evidence in their possession or control. Each person forms a link. The chain begins when the sample first becomes of interest and ends when the sample is offered to the court. If one link can be broken, the court can rule the item to be inadmissible. The jury will never know of its existence. Your link is forged by documenting your possession. Record in your notebook when,where, and from whom you received the sample; how it was packaged; the date you began and ended your tests; and when and to whom you released it. Initial and date the original container. Reuse it if possible. If not, save it as it will have other people's markings on it. While in your possession, the sample should remain in a locked cabinet or safe. Reseal it when you are finished. You may keep it locked up or return it to the submitter. Get a receipt. It is best if you do the analysis yourself. If a technician does any part of the tests, advise the attorney involved. Courts vary as to the acceptance of a supervisor's testimony.

Rebuttal. It is hard to prepare for all of the possible rebuttal techniques. The opposition may request copies of your report, notes, or the remaining sample for their own analysis. Do not deliver anything until you see the court order and talk to your attorney. If at all possible, do not consume the entire sample. Leave enough for a separate analysis. The hardest cross-examination to prepare for is the ignorant questions that a well-meaning attorney can ask. These may have no relevance or they may be chemically absurd. You must be prepared to explain why there is no answer.

When sample size becomes a problem, one has to resort to microtechniques. A very useful method currently enjoying much popularity with forensic laboratories is the Gandolfi camera (4,5). A single particle of fifty micron size can yield a powder pattern suitable for identification. This small sample size allows the analyst to pick out individual particles from a complex evidence sample such as vacuum sweepings from a suspected drug laboratory, debris from an explosion scene, artifacts in soil, or the microcrystals formed during standard crystal tests.

The problem of heterogeneity can be solved in one of two ways. Either run microsamples of the individual components, or grind the sample and run as a mixture. Often both are useful when run in conjunction with the other. By subtracting the known individual phases from the mixture, previously unnoticed phases may be found. Alternatively the individual components may have commercial significance and useful intelligence.

The physical form of the sample may also cause some handling problems. Paint is one example. The crystalline components of a paint are the pigment (titanium dioxide, silicates, inorganic salts) plus various fillers and extenders (clays, talc, oxides). They are already finely ground, but they are dispersed in an organic film. It is not possible to separate the inorganics from the organics. Grinding the paint so it fits into a capillary is also impractical. Similar problems arise with inorganic extended synthetics.

A small piece of paint about 1x5 mm can be fastened to the outside of a regular capillary with a thick grease like Apiezon L.

The result is a powder film with a high background caused by the scattered radiation. However it is still readable. One has to sacrifice the intensity for the convenience of sample handling. Fortunately these samples do not require the identification of each component as much as a direct comparison of the total mixture.

In addition to chemical identification, forensic laboratories are also interested in intelligence which can be obtained from the knowledge of what particular phase is present. While wet chemistry will yield composition data, solid state methods are required to distinquish between polymorphs, hydrates, and some salt forms. In developing intelligence data on illicit drug labs, the salt form of the drug, the excipients used to cut the drug, and the presence of synthetic precursors need to be identified precisely. This information can be used to associate drug buys from different parts of the country.

Unusual phases or mixtures may have a specific commercial use. Once they are characterized, their origin may be found by consulting chemical supply houses or formularies. Two examples are sodium sesquicarbonate dihydrate (a) and sodium hydrogen sulfate (b):

$$Na_2CO_3 \cdot NaHCO_3 \cdot 2H_3O \qquad (a)$$

$$Na_3H(SO_4)_2 \qquad (b).$$

Both were received in criminal cases. Qualitative analysis was of no value since there are many many sodium sulfate and carbonate compounds known to exist. XRD identified the exact phase. The names were found in a chemical dictionary (6), and the authorities were able to associate each with a suspect. The former is used in the detergent, soap, and water softening industries; the latter is an industrial cleaner.

A third case involved a mixture. A white crystalline powder was submitted as a suspected drug. Screening tests failed to show the presence of any controlled substance or pharmaceutical. Under the microscope, three different crystal types were evident. They were separated by hand and a powder pattern was obtained from each. Sodium sulfite, hydroquinone, and potassium bromide were identified. Each has many industrial uses, but the common denominator was photographic supplies. The material was finally identified as a prepackaged film developer.

REFERENCES

1. The American College Dictionary, Random House, New York, (1959).

2. A.A. Moenssens, R.E. Moses and F.E. Inbau, Scientific Evidence in Criminal Cases, The Foundation Press, Mineola, NY, (1973).

3. J.R. Hanley and W.W. Schmidt, Legal Aspects of Criminal Evidence, McCutchan Publishing Corporation, Berkeley, CA, (1977).

4. D.V. Canfield and P.R.DeForest, "The Use of the Gandolfi Camera as a Screening and Conformation Tool in the Analysis of Explosive Residues," Journal of Forensic Sciences, 22, 337-747, (1977).

5. D.V. Canfield, J. Barrick, M.J. Camp and B.C. Giessen, "The Confirmation of Crystal Test Results by X-ray Diffraction," presented at the 29th Annual Meeting of the American Academy of Forensic Sciences, San Diego, CA, February, 1977.

6. Condensed Chemical Dictionary, 8th edition, Van Nostrand Reinhold, New York, (1971).

X-RAY POWDER DIFFRACTION MEASUREMENTS IN REACTIVE ATMOSPHERES AT 1000 °C AND 7 MPa (1000 PSIG)

F. A. Mauer and C. R. Robbins

National Bureau of Standards

Washington, D. C., 20234

ABSTRACT

Studies of changes in the flexural strength and phase composition of castable refractories are being carried out in simulated coal gasification reactor environments. This paper describes a pressure vessel in which x-ray powder diffraction measurements can be made by an energy dispersive method while the specimen is being heated to temperatures as high as 1000 °C in atmospheres containing H_2O, CO_2, CO, H_2, CH_4, NH_3, and H_2S at pressures up to 7 MPa (1000 psig).

INTRODUCTION

Castable refractories are relatively inexpensive materials that are used in applications where resistance to heat and abrasion, as well as chemical stability, are required. They have long been used, for example, in the petrochemical industry in high-wear areas of catalytic cracking units. Because of their favorable chemical and mechanical properties, they have replaced metals in many applications and have contributed to the reduction of operating costs in the petrochemical industry. Castable refractories are now being considered for similar applications in coal gasification plants. Environmental conditions will be even more severe than those in petrochemical plants, and chemical attack is likely to limit seriously their usefulness.

High pressure steam is one of the constituents of the gasifier atmosphere that is expected to degrade the mechanical properties of refractory liners. Wiederhorn et al. (1) have shown that in the case of a high-purity alumina castable refractory the flexural strength may be reduced by a factor of $\sim$5 and the erosion rate increased by a factor of $\sim$10 after hydrothermal exposure. The dramatic loss of strength and erosion resistance is associated in part with the dehydration of boehmite, a hydrated alumina which acts as a bonding phase in the matrix of the refractory. These results are based on measurements of flexural strength and x-ray analysis carried out at room temperature after hydrothermal exposure.

Because of the massive nature of the pressure vessels used for hydrothermal exposure, the cooling rate is limited. Additional reactions and transformations may occur after the specimen is treated at a given pressure and temperature while it is being brought to 25 °C and 1 atm. for testing. To eliminate any uncertainties resulting from such changes both the flexural strength measurements and the x-ray analysis are now being carried out *in situ*: that is, without removing the specimen from the pressure chamber used to simulate environmental conditions in the gasifier. The purpose of this paper is to describe the apparatus and methods used for the x-ray analysis.

The goals established for the x-ray analysis include the identification of crystalline phases, chemical reactions, and crystallographic transformations in specimen bars identical to those used for flexural strength tests. Measurements are to be carried out at pressures up to 7 MPa* and temperatures to 1000 °C in atmospheres containing H_2O, CO_2, CO, H_2, CH_4, NH_3, and H_2S.

In addition to the *in situ* measurements, powder diffraction measurements with a conventional diffractometer continue to be a major part of the program. They are used to check the results of *in situ* measurements as well as to characterize raw materials and specimens at various stages of fabrication and testing.

METHOD

Obtaining x-ray data from a specimen inside a pressure vessel, even at the modest pressure of 7 MPa, presents a difficult design problem if conventional diffraction methods are employed. In order to measure the necessary range of diffraction angles, x-ray windows that are either large in area or located very close to the specimen are needed. The choice of materials that are suitably

---

*The International Standard (SI) unit of pressure is the pascal, or newton per square meter. One mega pascal (MPa) = 10^6 pascals. The conversion to psig is given by p (psig) $\simeq$ p (MPa) x 145 - 14.7

transparent to the x-ray beam is very limited and does not include any high-strength materials that can be used with confidence to bridge a large gap in the pressure vessel wall. There is, however, one x-ray powder diffraction technique that requires only a single, narrow x-ray path through the pressure vessel. The entrance and exit windows can be as small as a quarter of an inch in diameter. This small diameter helps to make it possible to design for a desired pressure limit with a suitable safety factor while keeping the windows thin enough for adequate transmission of the x-ray beam. The technique referred to is the energy dispersive x-ray diffraction (EDXD) technique described by Giessen and Gordon (2) in 1968.

Unlike conventional diffraction methods, which use monochromatic radiation and a range of diffraction angles, the EDXD method employs white (continuum) radiation and a fixed diffraction angle, 2θ. The diffracted beam is analyzed with a solid state detector and multichannel pulse height analyzer to determine the energy distribution of the photons. Bragg's law in the form

$$d_{hk\ell} = 12.3981/(2E \sin\theta) \qquad (1)$$

is used to calculate interplanar spacings, $d_{hk\ell}$ (angstroms), from the energies, E (kiloelectronvolts), at which maxima in the number of counts are observed (3). For the work reported here, $\theta = 12.30°$, and the energy, E (keV), is related to the channel number, N, by the equation

$$E = .03502 \times N + .938. \qquad (2)$$

The constants, of course, must be determined for each detector system.

There are sacrifices that must be made as a trade off for the speed and convenience of the EDXD method. The most serious is the low resolution of the detector, which results in peaks that are more than an order of magnitude wider than in conventional diffractometry. The limited energy resolution of the solid state detector gives rise to symmetrical broadening, while specimen transparency contributes an additional asymetric component. Sparks and Gedcke (3) have analyzed errors associated with the EDXD method while Mantler and Parrish (4) have described a computer method of profile fitting to obtain corrected intensities and peak energies. In both of these cases, the standard Bragg-Brentano parafocusing geometry was used. Factors affecting the accuracy and resolution of measurements using intense, highly collimated synchrotron radiation and Debye-Scherrer geometry have been analyzed by Bordas et al. (5) and by Buras, et al. (6,7). In spite of excellent theoretical work and continued improvements in instrumentation the accuracy and resolution of the EDXD method continue to be poorer by about an order of magnitude than they are for conventional diffraction methods.

The EDXD patterns encountered in the present work, where several phases of low symmetry often co-exist, represent a greater degree of complexity than has been treated in previous studies by the EDXD method. There would be little hope for success if it were not for the fact that we are dealing with known structures and basing the interpretation on changes in a series of closely related patterns.

APPARATUS

Details of the construction of the pressure vessel are shown in Figure 1. The upper view is a section through the vessel in the plane of diffraction. Radiation from the x-ray source (A) passes through a 1/8 in. thick window of boron carbide (B) and, after being scattered by the specimen (C), passes through a second window (B) and slit (D) to enter the solid state detector (E). The bracket (F) which supports the specimen has slots milled in it for the x-ray beam. These limit the divergence of the incident and diffracted beams to 2°. An internal heater consisting of an alumina core (G) wound with approximately 6 ft. of 0.032 in. diameter 80% Pt-20% Rh wire (H) surrounds the specimen bar. A concentric alumina tube (I) is wrapped with platinum foil (J) to reduce heat transfer by radiation. The heater assembly is mounted in a U-shaped bracket of 0.031 in. thick Inconel sheet (M) which forms three sides of a box. The remaining three sides are formed by a cover (L) of 0.010 in. thick Inconel. The space between the heater and the box is packed with fibrous alumina insulation. Internal windows of 0.002 in. thick mica mounted in platinum disks (N) loosely cover the ends of the heater core to reduce convection currents while allowing the x-ray beam to pass. The importance of these disks should not be underestimated, as operating temperatures without them are limited to about 500 °C. Three thermocouples are provided. The first (O) is attached to the front edge of the specimen at the top. The second (P) is at the bottom of the vessel where it senses the temperature of the water that collects there. The third (not shown) is on the outside of the vessel in contact with the belt heater (not shown) that surrounds the vessel. Coned pressure fittings in ports (Q) and (R) are used to connect the gauge assembly and the water reservoir.

The pressure vessel, which is machined from 316L stainless steel, is 4 1/2 in. in o.d. at the height of the x-ray beam and has 1/2 in. thick walls. The flange is 6 in. in diameter. Twelve 3/8-16 high-tensile-strength bolts are used to attach the lid,

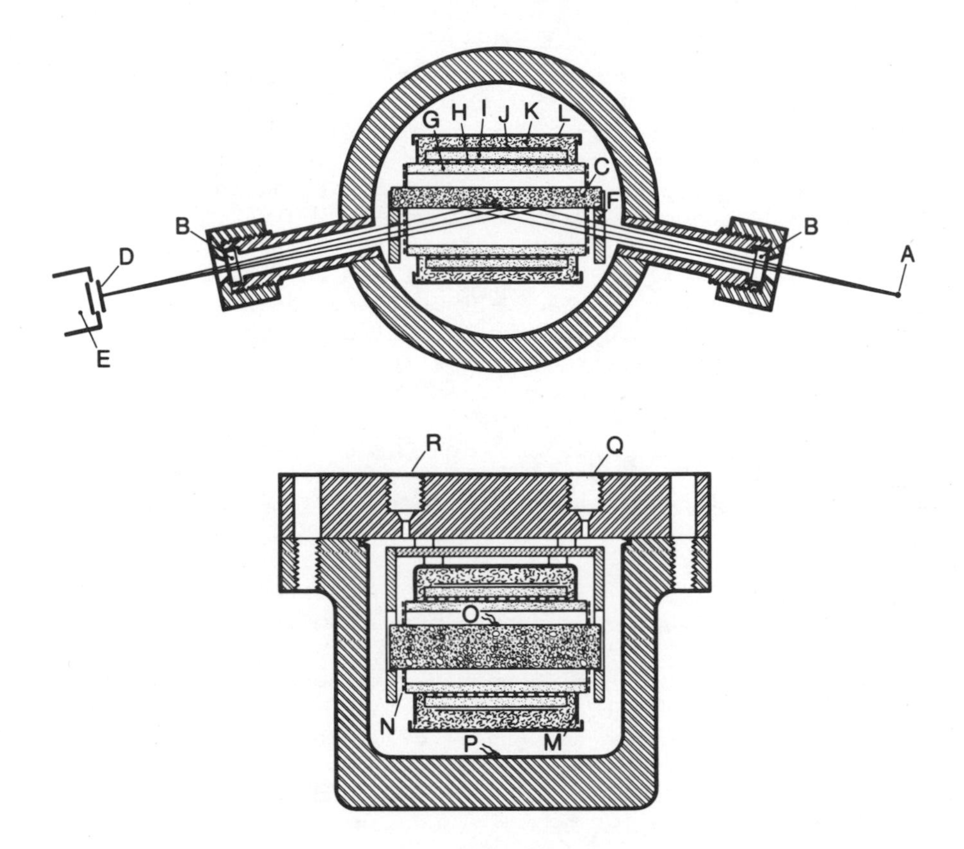

Figure 1. Assembly drawing of x-ray pressure vessel showing: (A) x-ray source, (B) B_4C windows, (C) specimen, (D) receiving slit, (E) solid state detector, (F) specimen mounting bracket and divergence slits, (G) heater core, (H) heater winding, (I) insulating cover, (J) heater sheath, (K) insulation, (L) heater case, (M) heater support bracket, (N) mica windows, (O) specimen thermocouple, (P) bottom thermocouple, (Q) port for pressure gauge assembly, and (R) port for water reservoir.

which is fitted with a gold-plated K-shaped flange seal made of Inconel X-750. The enclosed volume is approximately 435 ml. The side arms, which contain the x-ray windows, are also made of 316L stainless steel and are welded in place at an angle corresponding to $2\theta=24°$. The boron carbide windows are sealed with gold plated V-gaskets of Inconel X-750. As the windows appeared somewhat grainy even after they were polished, the surfaces in contact with the gaskets were sputtered with gold to fill any irregularities and provide a softer seat. Threaded caps compress the gaskets and support the load due to gas pressure.

In operation, the pressure vessel is mounted as shown in Figure 2. A base assembly supports both the pressure vessel and the solid state detector and provides the necessary adjustments for alignment. This base is pivoted at a point directly below the focal spot of the x-ray tube for ease in setting the take-off angle. Three leveling screws are used for setting the height and for making small adjustments to the angle of incidence. The base supports a heavy ring 8 1/4 in. in diameter in which the pressure vessel is clamped by four 3/8 in. radial set screws, the ends of which extend into holes in the six inch flange of the vessel. In addition to supporting the pressure vessel the mounting ring provides a means for counteracting the force applied in tightening the lid bolts. There are 12 3/8 in. holes drilled vertically through the ring. A 10 in. lever is attached by means of pins that fit any diametrical pair of holes. Equal and opposite forces applied to this lever and the wrench provide the necessary torque to tighten the bolts without disturbing the alignment of the instrument.

Since the whole vessel must often be operated at temperatures as high as 285 °C in order to maintain a steam pressure of 7 MPa, it is equipped with an external belt heater visible just below the x-ray ports in Figure 2. While heating the vessel, it is important to minimize the heat transferred to the base, the detector, and the x-ray tube. With this aim, the i.d. of the mounting ring was made 1/4 in. greater than the o.d. of the vessel flange. The four set screws comprise the only path for the conduction of heat from the vessel to the mount. A heavy copper sheet with a 1/8-in. aperture is mounted between the exit slit and the detector window to aid in the dissipation of heat that might otherwise be transferred to the detector. The only direct link between the vessel and the x-ray tube is a stainless steel beam tunnel that fits inside the safety shutter. Although this tunnel has thick walls (to reduce x-ray leakage) heat transfer to the x-ray tube has not been a problem.

External fittings visible in Figure 2 include the gauge block assembly and the water reservoir. The gauge block provides, in addition to the gauge, a pressure tubing connection for introducing gases, a needle valve, and a rupture disk. The water reservoir has a capacity of 40 ml. It is pressurized with nitrogen gas to ensure that water will flow into the pressure vessel when the needle valve is opened.

The lid of the pressure vessel is fitted with two pressure seal assemblies for electrical leads, one for the four thermocouple leads, and the other for a single power lead rated at 40 amperes. The vessel itself serves as the conductor for the grounded side of the low voltage power supply of the internal heater.

Figure 2. The pressure vessel and solid state detector mounted at the x-ray tube.

A safety shield of 1/4-in. thick steel is installed before the vessel is pressurized. A portion of this shield can be seen behind the pressure vessel in Figure 2. The enclosure is connected to an exhaust system in case of leakage of test atmospheres containing CO, H_2, or H_2S.

OPERATION

After installing the specimen bar with the thermocouple attached, an initial charge of 20 ml of water is added to the vessel before closing the lid. Approximately 15 ml would be required to pressurize the vessel with saturated steam at 7 MPa. Excess liquid remains at the bottom of the vessel and its temperature is controlled by means of the external heater to maintain the desired partial pressure of steam. Other components of the atmosphere are added as gases to obtain the desired total pressure. A reserve supply of water is kept in the 40 ml reservoir to replenish any used in hydrating the sample or lost during the run.

The wall temperature required to maintain a steam pressure of 7 MPa is approximately 285 °C. The external belt heater has a rating of 650 watts at 240 V ac, but the power input during steady-state operation does not exceed 400 watts.

Specimen temperature is separately controlled using the internal heater. The maximum current required for control at 1000 °C in steam is 14 amperes, corresponding to a power input of 390 watts. As the power input to the internal heater is increased, that to the external heater must be reduced to compensate. A small blower has been installed in case additional cooling is required. It is not generally needed.

Specimen temperature is usually increased in increments of about 50 °C. Diffraction patterns are recorded after 6 hours, and again after 24. A series of experiments covering the range to 1000 °C and 7 MPa requires from 3 to 6 weeks.

RESULTS

A recent series of experiments in which a bar of high alumina neat cement was heated in steam provides an excellent example of how *in situ* x-ray diffraction measurements can be used to study reactions affecting the mechanical properties of castable refractories. The composition in weight percent of the calcium aluminate cement used for the bar was 79% Al_2O_3, 18% CaO, 0.3% Fe_2O_3, 0.5% SiO_2, 0.5% Na_2O, and 0.4% MgO with 1.3% loss on ignition. The cement was mixed with distilled water and cast in the form of bars 15 mm wide x 75 mm long x 7.5 mm thick which were permitted to set for 24 hours in an environment of 100% relative humidity. Conventional powder diffraction patterns showed that the bars - which are to be used for flexural strength measurements as well as for the *in situ* x-ray measurements - consisted of CA and α-A (major phases) with small amounts of CAH_{10}, C_2AH_8, C_3AH_6, and AH_3 [in cement chemistry notation, C=CaO, A=Al_2O_3, and H=H_2O].

Energy dispersive diffraction patterns were prepared while the bar was being heated in steam. These patterns differ considerably from ones obtained by conventional methods, and special reference patterns are needed for use in identifying phases. The phases that will be encountered during the hydration of a calcium aluminate cement can be predicted on the basis of a pseudobinary phase diagram for the system CaO - Al_2O_3 - H_2O (8). They include, in addition to the CA and α-A originally found in the specimen, the hydrated phases C_3AH_6, $C_4A_3H_3$, AH (boehmite) and AH_3 (gibbsite). Samples representing each of these phases have been obtained and used in preparing special EDXD reference patterns. In addition, reference patterns from the Powder Data File have been transformed using a computer program to give the channel number and approximate intensity of each line as it would appear in an EDXD pattern. By comparison of the EDXD patterns of the specimen bar with these reference patterns, reactions such as the hydration of CA to form C_3AH_6, and then $C_4A_3H_3$ can be observed. However, it is the hydration of α-A to form AH (boehmite) and the subsequent dissociation of this compound at higher temperatures that is of particular interest. As was mentioned in the introduction, AH apparently acts as a bonding phase and provides substantial strength to the matrix of the refractory. When it decomposes, the bonding phase loses its structural integrity, causing the flexural strength to decrease by a factor of $\sim$5 and the erosion rate to increase by a factor of $\sim$10.

The ability of the *in situ* x-ray method to follow the dissociation of AH in real time is shown by the series of four x-ray patterns reproduced in Figure 3. The overall change is shown in a difference plot, Figure 4, in which the negative peaks correspond to the AH phase consumed in the reaction, while the positive peaks show the increase in α-A, a reaction product.

Physical properties such as strength and erosion rate, which are affected by the presence of AH, can be expected to follow trends somewhat parallel to the formation or dissociation of this phase. For example, under conditions of this test, the flexural strength of a high purity alumina refractory might be expected to approach a maximum value at or near 354 °C and 5 MPa. At higher temperatures, the strength should gradually decline as AH dissociates. The dissociation of AH can be shown by plotting the intensity of a prominent line such as that in channel 610 for a

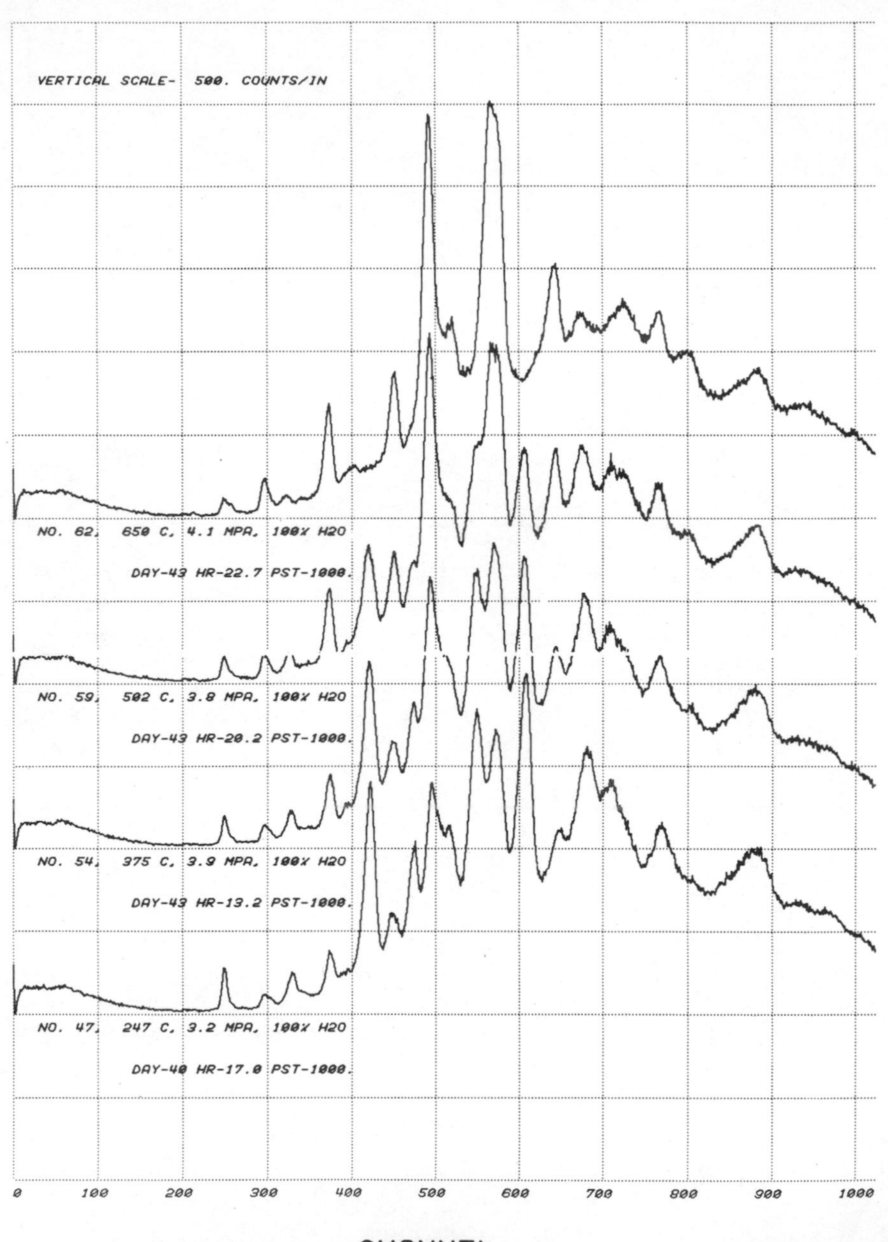

Figure 3. A series of four EDXD patterns showing the dissociation of AH (boehmite) and the formation of additional α-A in the temperature range 247 to 650 °C. Interplanar spacings, d, may be calculated using equations 1 and 2.

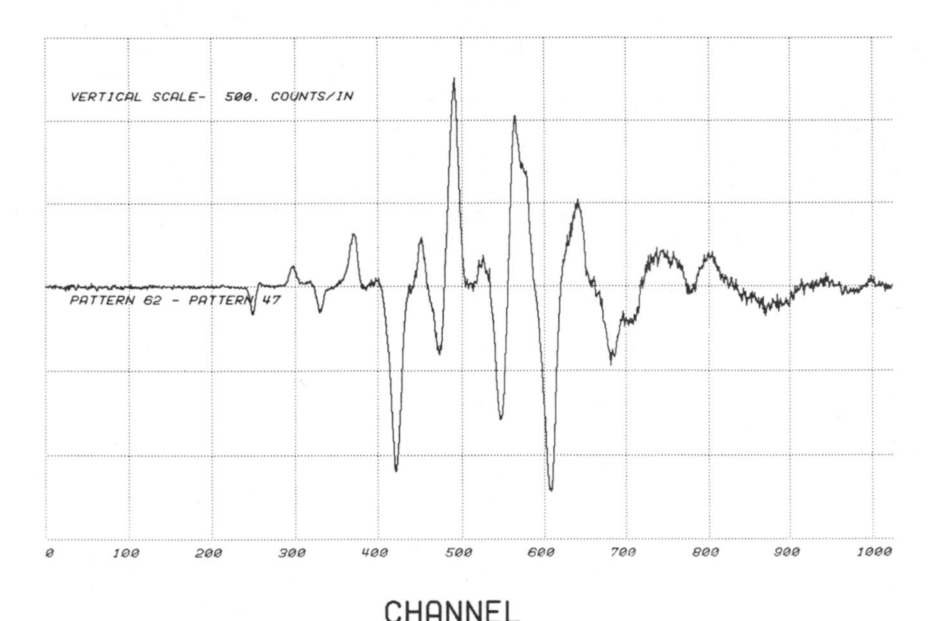

Figure 4. A difference plot obtained by subtracting each ordinate of pattern 47 from the corresponding ordinate of pattern 62. It shows the overall change resulting from hydrothermal treatment in the range from 247 °C @ 3.2 MPa to 650 °C @ 4.1 MPa. The negative peaks correspond to the AH phase consumed in the reaction, while the positive peaks show the increase in the concentration of α-A, a reaction product.

series of patterns covering a range of temperature and pressure. Such plots are easily prepared from the computer file of patterns and are very useful for following changes that occur gradually.

It is interesting to note that AH persists for many hours at temperatures above the reported equilibrium dissociation temperature of $\sim$350 °C at 7 MPa (9). Even at 550 °C it is present as a minor phase. Reaction rates, as well as the location of phase boundaries under equilibrium conditions, obviously influence the phase composition of a specimen under a given set of conditions.

The determination of the changes in phase composition that contribute to changes in the flexural strength is a challenging problem, even when both experiments are performed *in situ*. It will become even more challenging as the scope of the work is expanded to include other refractories and gases which may react with them to degrade their performance.

ACKNOWLEDGEMENT

This work was supported in part by the U. S. Department of Energy under Contract No. EA-77-A-01-6010.

REFERENCES

1. S. M. Wiederhorn, E. R. Fuller, Jr., J. M. Bukowski, and C. R. Robbins, "Effect of Hydrothermal Environments on the Erosion of Castable Refractories," J. Eng. Mater. Technol., 99, 143-146 (1977).

2. B. C. Giessen and G. E. Gordon, "X-ray Diffraction: New High Speed Technique Based on X-ray Spectrography," Science 159, 973-975 (1968).

3. C. J. Sparks, Jr. and D. A. Gedcke, "Rapid Recording of Powder Diffraction Patterns with Si(Li) X-ray Energy Analysis System: W and Cu Targets and Error Analysis," Advances in X-ray Analysis, 15, 240-253 (1972).

4. M. Mantler and W. Parrish, "Energy Dispersive X-ray Diffractometry," Advances in X-ray Analysis, 20, 171-186 (1977).

5. J. Bordas, A. M. Glazer, C. J. Howard, and A. J. Bourdillon, "Energy Dispersive Diffraction from Polycrystalline Materials Using Synchrotron Radiation," Phil. Mag., 35, 311-323 (1977).

6. B. Buras, J. S. Olsen, L. Gerward, G. Will, and E. Hinze, "X-ray Energy Dispersive Diffractometry Using Synchrotron Radiation," J. Appl. Cryst., 10, 431-438 (1977).

7. B. Buras, N. Niimura, and J. S. Olsen, "Optimum Resolution in X-ray Energy-Dispersive Diffractometry," J. Appl. Cryst., 11, 137-140 (1978).

8. E. M. Levin, C. R. Robbins, and H. F. McMurdie, Phase Diagrams for Ceramists - 1969 Supplement, Figure 4033, The American Ceramic Society (1969).

9. G. C. Kennedy, "Phase Relations in the System Al_2O_3-H_2O at High Temperatures and Pressures," Am. J. Sci., 257, 563-573 (1959).

X-RAY DIFFRACTION OF RADIOACTIVE MATERIALS*

D. Schiferl and R. B. Roof

Los Alamos Scientific Laboratory

Los Alamos, New Mexico 87545

ABSTRACT

X-ray diffraction studies on radioactive materials are reviewed. Considerable emphasis is placed on the safe handling and loading of not-too-exotic samples. Special considerations such as the problems of film blackening by the gamma rays and changes induced by the self-irradiation of the sample are covered. Some modifications of common diffraction techniques are presented. Finally, diffraction studies on radioactive samples under extreme conditions are discussed, with primary emphasis on high-pressure studies involving diamond-anvil cells.

INTRODUCTION

X-ray diffraction studies have played and continue to play an important role in increasing our knowledge of materials containing radioactive elements. Diffraction techniques have been used to investigate hundreds of such materials thereby revealing, among other things, the complex chemical crystallography of the actinide elements and their compounds and alloys. This is an area of considerable practical importance and of basic interest as well. The crystal structures and mechanical properties of these materials change with the addition of even trace amounts of other elements. Another problem of practical interest, which is peculiar to radioactive substances, is the damage sustained due to irradiation, especially

---

*Work performed under the auspices of the Department of Energy.

self-irradiation. Some materials are more resistant than others, and the processes involved are not yet completely understood.

This work is intended to provide a brief review of the field and to give an idea of the "state-of-the-art." Only the more severe radioactive materials are treated here; thus the mildly radioactive substances, such as natural uranium and thorium and their compounds, are omitted. Discussed are the problems of safety, modifications of various x-ray diffraction techniques, and some of the special crystallographic properties and problems associated with radioactive materials. Examples are chosen chiefly from work done at the Los Alamos Scientific Laboratory or from work done by associates of the authors at other institutions.

SPECIAL CONSIDERATIONS

Safety

The first problem that should be discussed in the handling of radioactive materials is safety. It should be emphasized at the outset that support from health physics and waste disposal groups is essential for studies on such substances. Exposure to radioactive materials, even in the small quantities required for many types of x-ray experiments, is well known to be dangerous. This point is forcefully brought home when one sees the reaction of a detector to even minute amounts of the heavier transuranic elements. The procedures for dealing with this problem differ according to the type of radiation emitted and the chemical properties of the sample.

Procedures have been developed at Los Alamos for handling radioactive substances, particularly the actinides, in a routine way. Primary emphasis is of course on the prevention of accidents; but equally important is advanced planning of what steps are to be taken when something does go wrong. The procedures described below have been used for over thirty years without serious incident.

Almost all radioactive materials oxidize readily in air, and the oxide spalls off as very fine particles that can become airborne and be inhaled. Materials such as plutonium, which are primarily alpha emitters, are not particularly dangerous unless incorporated in the body. Thus, the first requirement is that the radioactive materials should be handled in an inert-atmosphere glove box 1) to minimize oxidation and 2) to prevent airborne distribution of the finely-divided powder of the oxidation products.

For alpha emitters, it is sufficient to produce sealed samples in an argon-filled glovebox. There is no practical limitation on the time allowed to complete the sample preparation. Loading the sample in a glove box is not nearly as bad a handicap as it might seem. To illustrate these procedures, we give as an example the insertion of plutonium alloy powder into fine fused-quartz (not glass, which is more fragile) capillaries for Debye-Scherrer cameras.

After the plutonium alloy sample is in the argon-filled glove box, the oxide is first carefully filed or brushed off. This is not done for safety considerations, but rather to prevent the appearance of the oxide pattern on the film along with that of the sample. The bare metal is then filed with a separate file so that the filings drop into a glass dish. The filings may then be crushed or ground in a diamond mortar until the particles are fine enough to drop freely down the entire length of the capillary tube.

The capillaries themselves are drawn from 1/8" (inner diameter) fused-quartz tubing, and the 1/8" mouth serves as an adequate funnel for introducing the sample. The capillary is safely held for sample-loading manipulations in a one-hole rubber stopper which is fitted in a large test tube. The test tube has a hole in the side to equalize the inside and outside pressures during the cycling of the glove box airlock.

Once the samples are loaded into the capillary, a fused quartz fiber is dropped down the capillary to hold the sample powder in place. This fiber also keeps the sample from being sucked out during the evacuation of the airlock as the sample is removed from the glove box. The capillary is evacuated and sealed off in a fume hood adjacent to the glove box. The capillary is then loaded into a Debye-Scherrer camera in the open laboratory. The capillaries can almost always be mounted safely with no difficulties; however, precautions must be taken to confine any contamination that might result from accidental breakage during mounting. A pad of cheesecloth with a paper tissue folded into it is placed on the working surface of the bench. The capillary is then inserted into the wax in the brass plug while both are held a small distance directly over the cheesecloth.

If the capillary should break during mounting on the plug it will simply drop onto the cheesecloth pad without shattering. Because the radioactive material is not badly scattered, the pad can be simply folded over the broken capillary and plug and placed in a container for radioactive disposal. The operator examines his hands, clothing and the working area for the presence of any contamination. The rooms in which the samples and films are loaded are monitored with continuous alpha air monitors having audible alarms. If the alarm sounds, operations are terminated and the personnel are evacuated from the laboratory. A health physics group then must decide what procedures are involved to bring the laboratory back to normal operating conditions.

If one is working with strong gamma or neutron emitters, such as americium-243 or curium-244, these same procedures may be followed, but the time allowed is drastically curtailed (sometimes down to a few minutes). The temptation to examine the sample by looking at it at close range should be strictly avoided. Dosimeters must be used to monitor the whole body exposure. It is important to remember that the eyes are much more sensitive to radiation than the body as a whole and that the hands are somewhat less sensitive.

Background Due to the High Gamma Activity of the Samples

The high gamma activity of some materials, particularly the heavier transuranic elements, may blacken films in powder diffraction cameras so much as to make the patterns unreadable. A similar problem is encountered with the detectors on powder diffractometers. Although the single-channel analyzers in the counting electronics are supposed to discriminate against radiation outside a relatively narrow wavelength or energy range, the backgrounds are always considerably higher than one would expect from naive considerations.

Several things can be done to reduce the backgrounds. The samples can be made no larger than absolutely necessary, to minimize the number of gammas produced. It is occasionally possible to choose an isotope which produces fewer gammas than other isotopes available. For many of the transuranium elements, CuKα radiation with a nickel filter over the film provides good diffraction patterns. The high intensity of the CuKα lines reduces the exposure times, and the nickel foil absorbs much of the radiation produced by the specimen. Americium-243, for example, emits copious radiation (including x-ray fluorescence) in the energy range 12-24 KeV as well as radiation at higher energies. The nickel foil absorbs the "band" of lower energy radiation very effectively. Nickel foil can also be used of course to reduce the backgrounds in diffractometers by placing it over the counter window. Still more effective perhaps would be the use of an x-ray monochromator between the sample and a well-shielded counter.

Energy-dispersive x-ray powder diffraction methods are also starting to be explored, a possibility independently conceived by Peterson and co-workers (1), Benedict and co-workers (2) and the authors. In this technique (3), "white" x-radiation is incident upon the sample and the pattern is collected at a fixed diffraction angle with high resolution solid state detectors. The diffraction peaks appear at various energies and the pattern is stored in a multichannel analyzer with the energy calibrated against channel number. The gammas and x-ray fluorescence lines appear largely as sharp spikes at their characteristic energies. It might also be hoped that energy-dispersive methods would be well suited for short-lived isotopes because the entire diffraction pattern is collected simultaneously with high (> 95%) efficiency. Preliminary experiments on americium-243 have revealed some problems (4). There is a high continuous background over the entire energy range up to 50 KeV. In addition, the "band" of radiation between 12-24 KeV makes observation of diffraction peaks in that region nearly impossible, and the AmL_I absorption edge further eliminates the region between 24-30 KeV. The use of smaller samples and more intense "white" x-ray sources should overcome the problem of the high continuous background, and the diffraction angle can be chosen so that the diffraction peaks lie above 30 KeV. Ultimately, it would be desirable to use a synchrotron radiation source because it would be up to 1000 times more intense than conventional x-ray tubes and because it would extend the energy range beyond 70 KeV as well. In that case, the continuum backgrounds would probably be negligible. Most likely, however, all these improvements will not yield substantially better patterns than can already be obtained with film methods. The energy-dispersive technique will probably be most advantageous in high-pressure studies involving diamond-anvil cells where the use of other x-ray techniques is severely restricted, as discussed later on below.

Changes Induced by the Radioactivity of the Material

There are several types of changes that occur as a result of the radioactivity of a sample. These include self-irradiation damage, changes in the sample chemistry due to radiation damage and due to the presence of "daughters" and their compounds, significant raising of the sample temperature, and finally, in some cases, half-lives so short that it is extremely difficult to prepare and study the sample in the short time available. For this last reason, in fact, the structures of a number of elements have not yet been determined. Besides exotic actinides, the list includes radon, astatine and francium.

Self-irradiation produces a variety of changes in the crystal structures. A study of Roof (5-7) on 238(80%)PuO_2 over a three-year period illustrates some of the complexity of self-irradiation damage. A sample disk of this material approximately one centimeter in diameter and 0.25 cm thick was sintered at 1625°C just prior to the experiment. Plutonium-238 is intensely radioactive and the sample remained considerably above room temperature throughout the experiment. (In fact the radioactive self-heating is so great that a sphere of 238PuO_2 about 4 cm in diameter remains red hot due to the alpha activity, a property that allows it to be used for a thermoelectric power source). In these experiments the lattice was damaged by self-irradiation and partly repaired during the course of the experiments. The lattice constant increased by a small amount for about 60 weeks, then decreased to an equilibrium value close to the original value after another 140 weeks. From line-broadening analysis, the strain in the lattice also first increased and then decreased, but to a value not far below the maximum strain; the crystallite size remained unchanged. The damage and recovery processes depend strongly on the previous history of the sample, including initial crystal perfection, the nature of the radiation damage, annealing temperatures and even helium concentration. More detailed discussions of these factors, as well as additional references, are given in recent articles by Turcotte (8) and Kapshukov, *et al*. (9).

The energy liberated by alpha decay vastly exceeds normal chemical bond energies and can break down chemical compounds of radioactive elements. Peterson (1) reports two factors which yielded success with einsteinium halides: 1) a sample size and geometry that maximizes escape of the alphas, and 2) provision for a continual resynthesis system to reverse the chemical effects of self-irradiation and maintain a steady-state concentration of the compound studied.

In fact, it is for the compounds of the intensely radioactive elements, such as radium, polonium or einsteinium, that the art of studying radioactive materials with x-ray diffraction has been developed to its fullest. The high specific alpha activity of these elements limits studies of their compounds to very small amounts of material. For example, 7 mg of polonium produces 32 curies and one watt of heat and converts to lead at the rate of 0.5% per day (10). A further complication is the fact that most of the highly radioactive elements are also highly reactive chemically.

Thus, the x-ray studies on radium, polonium, einsteinium and their compounds must be done on 10-50 microgram quantities in inert atmosphere or vacuum. Moreover, the purification on the radioactive material must be done immediately before use, and both the x-ray studies and identification of the compunds must be done in a short period of time. The common procedure used in these cases is to create the compound of interest from the freshly purified radioactive element directly in the x-ray capillary. A variety of methods have been described for preparing samples, each tailored to the chemical reactions involved and the vapor pressures of the elements and compounds of interest (1,10-17). Procedures have been developed for producing a compound, x-raying it, changing it to another compound, etc. (1).

When chemical synthesis is done on such a small scale, the problem of identification of the compounds produced is not trivial. Witteman and co-workers (10) successfully identified simple polonium compounds with presumably well-defined stoichiometry by comparing the lattice constants with those calculated from ionic radii. Substances may also be recognized from their characteristic powder patterns, and a critical compendium of the wealth of diffraction data on actinides is being prepared by Roof (18) to provide the basis of compound identification. However, it is very difficult to tell the oxides and nitrides of many transuranic elements from the metals themselves from only the lattice constants. For example, americium and curium and their respective oxides and nitrides all have the fcc structure with similar lattice constants (11), to wit:

Am	a = 5.004 ± 0.002 Å	Cm	a = 5.038 ± 0.002 Å
AmN	a = 5.002	CmN	a = 5.041
AmO	a = 5.045	CmO	a = 5.09

Peterson (1) describes a microscope-spectrophotometer to record the absorption and/or luminescence spectrum of the sample as it is being studied with x-rays.

MODIFICATIONS OF COMMON DIFFRACTION TECHNIQUES

Except for monitoring for possible contamination, the operation of Debye-Scherrer cameras containing radioactive materials is not significantly different than for normal materials once the capillaries have been loaded according to the procedures described above.

Radioactive powder samples may also be studied on diffractometers outside of ventilated enclosures. Commonly, a mounted and polished metallographic specimen is used as the sample, but it must be sealed for work done in the open laboratory. This is usually accomplished by wrapping the sample in a thin sheet of a material that contains the contamination, is transparent to x-rays and has an x-ray pattern that will not obscure the pattern of the sample. The large amount of heat generated by plutonium-238 and higher specific activity materials precludes the use of any plastics. Instead the sample is loaded into a metal sample holder (in a safety enclosure) and covered with a thin metal foil window sealed with sticky lead tape. The preparation and subsequent handling must be done very carefully to avoid tearing the foil. When the sealed sample holder has been found free of contamination it may then be placed on the diffractometer in the usual manner.

Many radioactive materials have been studied with single-crystal techniques. A large number of compounds and alloys of plutonium have been studied in this way, and such a study has even been done on $CfBr_3$ (1,19). As with powder samples, the main concern is to prepare the sample so that particles, particularly of the oxide, are not scattered about. It is usually not possible to grow single crystals of plutonium alloys. For this reason, large-grained samples are broken in a mortar in a radiation safety enclosure if the material is sufficiently hard and brittle. The particles are then placed in a small bowl and covered with mineral oil or dilute Duco cement-butyl acetate solution to prevent oxidation. One fragment is then selected and mounted on a standard goniometer head in the usual way while the operation is viewed through a microscope. The bead of Duco cement protects the sample from oxidation, sometimes for years. The radioactivity of the sample can usually be an advantage in this case, because the fragment can almost always be found if it is accidentally knocked off during the loading procedure. Once the sample is mounted on the goniometer head, the usual single-crystal diffractometer or film methods may be used. There is no significant film blackening for patterns of plutonium compounds because plutonium is primarily an alpha emitter and because single-crystal patterns require much less exposure time than do powder patterns.

RADIOACTIVE SAMPLES UNDER EXTREME CONDITIONS

A number of studies at high and low temperatures and at high pressure have been made on radioactive materials. High-temperature studies are complicated by the high chemical reactivity and volatility of some radioactive materials. When small samples are required, even trace amounts of impurities in a nominally inert atmosphere or reactions with the capillary itself can alter the sample drastically. This is particularly vexing when the sample is supposed to be the pure metal. The procedures used to solve these problems depend on the nature of the substance studied, and have been discussed by several authors (1,11-14). Low-temperature studies are more straightforward. The principal problem is in cooling the sample. In practice, this is usually not too difficult for most radioactive substances of interest. To be sure, a material with a sufficiently high specific alpha activity may be impossible to cool very much; but plutonium-239 alloys or americium-243 can be cooled to a few degrees Kelvin without great hardship.

Interest in high-pressure studies on radioactive materials is rapidly expanding, primarily due to recent developments in diamond-anvil cell techniques. Radioactive materials have already been studied to pressures over 190 kbar using diamond-anvil cells. To date only film cassettes have been used in these studies, but several laboratories are setting up energy-dispersive x-ray diffraction systems.

In normal high-pressure studies with diamond-anvil cells, the sample is either 1) placed between the diamonds with no gasket around it, or 2) placed in the hole of an Inconel gasket with a hydrostatic fluid. In the first case very fine x-ray collimators are used so that the x-rays do not see a great pressure distribution. Larger collimators may be used with gasketed samples. Variations of both of these techniques have been successfully applied to the study of radioactive materials.

Roof (18) has used the non-gasketed technique on plutonium up to 190 kbar. Plutonium foil 0.025 mm thick was put between two aluminum foils, each 0.018 mm thick. This sandwich was then placed between two pieces of scotch tape to form the sample coupon, which was then placed in a Bassett (20) diamond cell. The aluminum also provided an internal pressure standard. Exposure times of 300 to 500 hours were required because of the fine collimators employed, but exceptionally clear patterns were obtained. There was no significant film blackening from the plutonium gamma rays.

Burns and Peterson (21) have studied californium up to 140 kbar with a Bassett cell with the gasketed diamond cell technique. The sample of californium metal 0.1 mm in diameter was loaded into a tiny aluminum foil cup which was placed in the hole in the Inconel gasket. The sample was then covered with a small disk of aluminum foil. The operation was performed under a microscope in an inert-atmosphere glove box that was nominally "cold" inside so that the outside of the high-pressure cell was not contaminated. When pressure was applied, the sample was encapsulated in a can of aluminum. This arrangement contained the alpha-emitting californium-249, prevented oxidation by excluding air and provided an internal pressure probe from the aluminum lattice constants. Normal x-ray exposures ran between 18 and 24 hours.

The full advantage of diamond-anvil cells has not yet been used in studies of radioactive materials. We are setting up experiments that will use the diamond-anvil cell design of Mao and Bell (22,23), who have recently achieved 1.72 megabars sustained static pressure. Pressures are measured with the ruby fluorescence technique, which is both rapid and precise (24-26). Energy-dispersive x-ray powder diffraction will be used to explore phase diagrams and determine compressions. Energy-dispersive methods are very well suited for use with diamond cells because strength considerations limit the size of the holes in the tungsten carbide diamond mounts and the optimum energy-dispersive diffraction angles are smaller than the angular apertures of the holes. The energy-dispersive technique also allows extremely rapid exploration of phase diagrams (27,28). Within 20 minutes it is possible to see whether the crystal structure at a new pressure is the same or grossly different from that at a previous pressure setting. Thus it is the ideal tool for the first exploration of the huge area of phase space now available.

CONCLUSIONS

We have endeavored to provide a short review of research involving x-ray diffraction on radioactive materials. We have discussed in detail those matters which we feel we are able to make a useful contribution to and have tried to give adequate references for subjects already treated in depth by other authors. We have emphasized basic safety considerations and have discussed many of the problems peculiar to studies on radioactive substances. We have also pointed out a major developing area; namely, the recently expanded possibilities of studying these materials with diamond cells to pressures possibly in excess of one megabar. We hope that this paper gives useful information on techniques and also gives a feeling for what the state-of-the-art is at the present time.

ACKNOWLEDGMENTS

One of us (D.S.) would like to thank U. Benedict, D. T. Cromer, R. N. R. Mulford, L. Schwalbe, and J. L. Smith for helpful conversations.

REFERENCES

1. J. R. Peterson, "Chemical Identification and Phase Analysis of Transplutonium Elements and Compounds via X-Ray Powder Diffraction," in H. F. McMurdie, C. S. Barrett, J. B. Newkirk, C. O. Ruud, Editors, Advances in X-Ray Analysis, Vol. 20, 75-83 (1976).

2. U. Benedict, private communication.

3. M. Mantler and W. Parrish, "Energy Dispersive X-Ray Diffractometry," in H. F. McMurdie, C. S. Barrett, J. B. Newkirk and C. O. Ruud, Editors, Advances in X-Ray Analysis, Vol. 20, 171-186 (1976).

4. D. Schiferl, L. Schwalbe, J. L. Smith, R. Hagen, unpublished.

5. R. B. Roof, Jr., "The Effects of Self-Irradiation on the Lattice of 238(80%)PuO_2," in K. F. J. Heinrich, C. S. Barrett, J. B. Newkirk and C. O. Ruud, Editors, Advances in X-Ray Analysis, Vol. 15, 307-318, Plenum Press (1971).

6. R. B. Roof, Jr., "The Effects of Self-Irradiation on the Lattice of 238(80%)PuO_2. II," L. S. Birks, C. S. Barrett, J. B. Newkirk, C. O. Ruud, Editors, Advances in X-Ray Analysis, Vol. 16, 396-400, Plenum Press (1972).

7. R. B. Roof, Jr., "The Effects of Self-Irradiation on the Lattice of 238(80%)PuO_2. III," In C. L. Grant, C. S. Barrett, J. B. Newkirk and C. O. Ruud, Editors, Advances in X-Ray Analysis, Vol. 17, 348-353, Plenum Press (1973).

8. R. P. Turcotte, "Alpha Radiation Damage in the Actinide Solids," in H. Blank and R. Lindner, Editors, Plutonium and Other Actinides, pp. 851-859 (1976).

9. I. I. Kapshkov, L. V. Sudakov, E. V. Shimbarev, A. Yu. Baranov and G. N. Yakovlev, "Structural Changes in Actinide Dioxides Under Self- and Reactor-Irradiation," in H. Blank and R. Lidner, Editors, Plutonium and Other Actinides, pp. 861-872 (1976).

10. W. G. Witteman, A. L. Giorgi and D. T. Vier, "The Preparation and Identification of Some Intermetallic Compounds of Polonium," J. Phys. Chem. 64, 434-440 (1960).

11. R. D. Baybarz, J. Bohet, K. Buijs, L. Colson, W. Müller, J. Reul, J. C. Spirlet, J. C. Toussaint, "Preparation and Structure Studies of Less-Common Actinide Metals," in W. Müller and R. Lindner, Editors, Transplutonium Elements, pp. 69-77, North-Holland Publishing Co. (1976).

12. D. B. McWhan, B. B. Cunningham and J. C. Wallmann, "Crystal Structure, Thermal Expansion and Melting Point of Americium Metal," J. Inorg. Nucl. Chem. 24, 1025-1038 (1962).

13. B. B. Cunningham and J. C. Wallmann, "Crystal Structure and Melting Point of Curium Metal," J. Inorg. Nucl. Chem. 26, 271-275 (1964).

14. M. Noé and J. R. Peterson, "Preparation and Study of Elemental Californium-249," in W. Müller and R. Lindner, Editors, Transplutonium Elements, pp. 69-77, North-Holland Publishing Co. (1976).

15. F. Weigel and A. Trinkl, "Zur Kristallchemie des Radiums, I. Die Halogenide des Radiums," Radiochimica Acta 9, 36-41 (1968).

16. F. Weigel and A. Trinkl, "Zur Kristallchemie des Radiums, II. Radiumsalz vom Type $RaXO_4$, X = S, Se, Cr, Mo, W," Radiochimica Acta 9, 140-144 (1968).

17. F. Weigel and A. Trinkl, "Zur Kristallchemie des Radiums, III. Darstellung, Kristallstruktur und Atomradius des Metallischen Radiums," Radiochimica Acta 10, 78-82 (1968).

18. R. B. Roof, Jr., unpublished.

19. J. H. Burns, J. R. Peterson and J. N. Stevenson, "Crystallographic Studies on Some Transuranic Trihalides: $^{239}PuCl_3$, $^{244}CmBr_3$, $^{249}BkBr_3$ and $^{249}CfBr_3$," J. Inorg. Nucl. Chem. 37, 743-749 (1975).

20. W. A. Bassett, T. Takahashi and P. W. Stook, "X-Ray Diffraction and Optical Observations on Crystalline Solids up to 300 kbar," Rev. Sci. Instrum. 38, 37-42 (1967).

21. J. H. Burns and J. R. Peterson, "Studies on Californium Metal at High Pressures by X-Ray Diffraction," for publication in the Proceedings of the International Conference on Rare Earths and Actinides, University of Durham, Durham, England (4-6 July 1977).

22. H. K. Mao and P. M. Bell, "High-Pressure Physics: The 1-Megabar Mark on the Ruby R_1 Static Pressure Scale," Science 191, 851-852 (1976).

23. H. K. Mao and P. M. Bell, "High-Pressure Physics: Sustained Static Generation of 1.36 to 1.72 Megabars," Science 200, 1145-1147 (1978).

24. R. A. Forman, G. J. Piermarini, J. D. Barnett, S. Block, "Pressure Measurement Made by the Utilization of Ruby Sharp-Line Luminescence," Science 176, 284-285 (1972).

25. J. D. Barnett, S. Block and G. J. Piermarini, "An Optical Fluorescence System for Quantitative Pressure Measurement in a Diamond-Anvil Cell," Rev. Sci. Instrum. 44, 1-9 (1973).

26. G. J. Piermarini, S. Block, J. D. Barnett, R. A. Forman, "Calibration of the Pressure Dependence of the R_1 Ruby Fluorescence Line to 195 kbar," J. Appl. Phys. 46, 2774-2780 (1975).

27. K. Syassen and W. B. Holzapfel, "Energy Dispersive X-Ray Diffractometry in High Pressure Research," Europhysics Conference Abstracts, Vol. I: Conference on Electronic Properties of Solids Under Pressure, Leuven, Belgium (1-5 September 1975).

28. E. F. Skelton, I. L. Spain, S. C. Yu, C. Y. Liu and E. R. Carpenter, Jr., "Variable Temperature Pressure Cell for Polycrystalline X-ray Studies Down to 2°K--Application to Bi," Rev. Sci. Instrum. 48, 879-883 (1977).

CHARACTERIZATION OF THIN FILMS BY X-RAY FLUORESCENCE AND DIFFRACTION ANALYSIS

T. C. Huang and W. Parrish

IBM Research Laboratory, 5600 Cottle Road

San Jose, California 95193

ABSTRACT

This paper presents a comprehensive study of various applications of x-ray fluorescence and diffraction techniques for the characterization of thin films. With the proper use of x-ray instruments and techniques, a fairly complete understanding of the chemical and physical structure of thin films was obtained. The x-ray fluorescence (XRF) method was used for the determination of composition, mass-thickness, and density. The x-ray diffraction (XRD) method was used for structural characterization, including: local atomic arrangements of amorphous materials; phase identification, preferred orientation, crystallite size, stacking faults, microstrain, the annealing behaviors of polycrystalline films; and lattice mismatch between the epitaxial film and its single-crystal substrate.

In order to obtain a clear picture of the capabilities of the x-ray method and the properties of thin films, a series of carefully selected specimens representing a wide range of compositions and thicknesses was used. A number of practical x-ray techniques, which are valuable for this type of analysis, are also introduced.

INTRODUCTION

The chemical and physical structure characterizations of thin films are essential requirements for research and development of electronic, magnetic, semiconductive, superconductive and other types of materials. The x-ray method has been used successfully to analyze thin films of various structures, compositions, and thicknesses, and is generally considered to be possibly the most complete and practical analytical method for the characterization of these materials.

This paper presents a comprehensive study of various applications of the XRF and XRD methods for the characterization of thin films. Specimens of various compositions and thicknesses were used for the systematic study of the capability of the x-ray method and of the properties of thin films.

It is demonstrated that the XRF method can be used successfully for the routine analysis of the elemental composition and mass-thickness of a wide variety of films. Methods for determining density and thickness will also be presented. It is followed by applying the XRD method for structural characterization. For amorphous materials, the radial distribution function method is used to determine local atomic arrangements. For polycrystalline films, various x-ray techniques are used to study phase identification, preferred orientation, crystallite size, stacking faults, microstrain, and the annealing behaviors. Finally, for single-crystal films, the double-crystal diffraction method is used to study the lattice mismatches and diffraction profiles of epitaxial garnet films and of the single crystal substrates.

THIN FILM ANALYSIS BY XRF

Chemical composition and mass-thickness are two of the most critical parameters that control film properties. XRF is most suitable for this type of analysis because it is non-destructive, rapid, accurate, and gives information for large areas as compared to point analysis by electron beam microanalysis. However, the application of XRF to routine thin film analysis has been generally limited because of problems of standardization and calibration. The difficulty of preparing and standardizing thin film standards having compositions similar to the unknown, as well as the time consuming process of obtaining calibration curves of intensity versus thickness for the empirical coefficients, make it impractical and sometimes even impossible to use XRF for routine analysis.

The LAMA Program

In response to the need for an XRF technique for routine elemental analysis of thin films, a versatile XRF method designated the LAMA program (1,2) was developed in our laboratory.

The LAMA program was originated by Laguitton and Mantler in 1976 and modified recently by one of the authors (T.C.H.). The fundamental parameters approach (3) is used, and the fluorescence equations for thin specimens are derived from a mathematical model different from that of Sherman (4). Its closest counterpart is the equation used by Pollai, et al. (5). It is assumed only that the specimen is homogeneous, and has a reasonably flat surface. The program uses the measured primary spectral distribution for a given target and operating voltage. The matrix correction and secondary fluorescence enter explicitly for each element, and the composition as well as mass-thickness are determined directly by computer iteration. No intermediate standards and empirical coefficients are needed.

In this program, all necessary fundamental parameters are provided by internal routines, thereby requiring a minimum of input data for an analysis. An important advantage is that the method allows the use of pure element bulk standards and no prior information about the thickness of the unknown sample is needed, thereby avoiding the problem of standardization and calibration. This program is very rapid and takes only seconds of CPU time for the computer to determine both composition and mass-thickness.

Composition

The accuracy of composition determination by the LAMA program was studied by comparing with other independent analytical methods, i.e., "atomic absorption spectroscopy (AAS)," and "electron probe microanalysis (EPM)." Samples of seven Ni-Fe binary alloys of various compositions were used for the comparison. As shown in Table 1, results of all three methods agreed with each other to within 2%. The agreement shown in Table 1, together with results on other kinds of specimens given in Reference 2, demonstrate the accuracy and practicality of the fluorescence method for routine quantitative composition determination.

Mass-Thickness

In addition to composition, the LAMA program also simultaneously determines the mass-thickness (i.e., mass per unit area) of the unknown. Mass-thickness is a basic physical constant of a thin film and is frequently used in calculations. From the mass-thickness, the thickness of a specimen can be obtained if its density is known, either by measurements from other techniques or by calculation based on the compositions and densities of the pure elements. The accuracy of the calculated thickness is generally comparable to the interferometer's value

TABLE I A Comparison of XRF, AAS, and EPM Results on Thin Ni-Fe Films

Desired Composition	Ni (Weight %)		
	XRF	AAS	EPM
$Ni_{95}\,Fe_5$	95.8	95.0	97.5
$Ni_{90}\,Fe_{10}$	90.8	91.0	93.8
$Ni_{80}\,Fe_{20}$	80.6	80.8	80.6
$Ni_{66}\,Fe_{34}$	52.7	51.6	55.5
$Ni_{50}\,Fe_{50}$	40.9	38.7	40.9
$Ni_{34}\,Fe_{66}$	21.1	20.2	21.6
$Ni_{20}\,Fe_{80}$	10.8	10.4	10.8

$\triangle\,(XRF,\ AAS)_{Avg} = 0.8\%$ $\triangle\,(AAS,\ EPM)_{Avg} = 1.9\%$

$\triangle\,(XRF,\ EPM)_{Avg} = 1.2\%$

as discussed in Reference 2. However, in the case of very thin films (about a few hundred Angstroms or less), possible reduction in density may exist and care must be taken in determining the thickness from the calculated density.

Density

Several $Ni_{81}Fe_{19}$ permalloy specimens of various thicknesses (from 50 to 10,000Å) were prepared by evaporation from a single permalloy source on fused quartz discs. The vacuum was in the range of 10^{-7} Torr, and the substrate temperature was set at 300°C. The rate of deposition was 50Å/sec for thickness of 1,000Å or more, and was 2Å/sec for thinner films. They were used for density determination, and later on, for structure analysis by XRD.

By using the mass-thicknesses obtained from the LAMA program and the thicknesses measured by interferometry, their densities were determined. A plot of the calculated density against thickness is given in Fig. 1. It shows that the densities of permalloy films above 1,000Å are close to the bulk value, ρ_o, and drop to 95% at about 500Å and 80% at 50Å. The value obtained is the density averaged over the film thickness and does not reveal the density variation with depth.

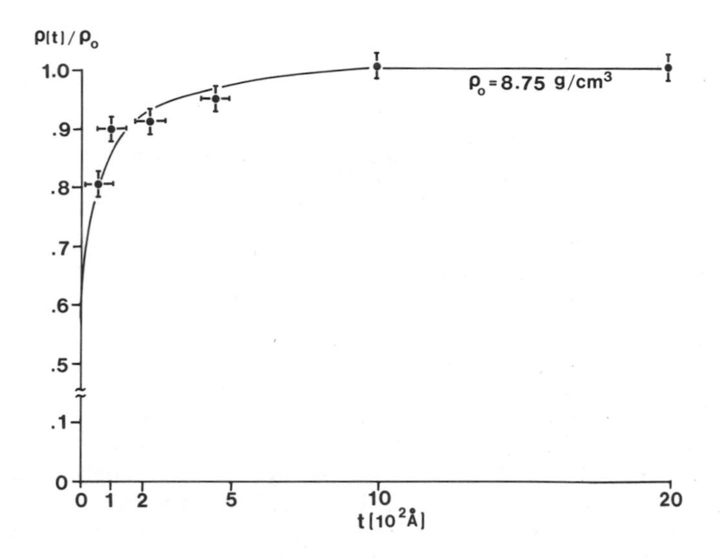

Figure 1. Variation of normalized density for permallowy films of thickness t

STRUCTURE ANALYSIS OF THIN FILMS BY XRD

X-ray diffraction is one of the most widely used methods for the structure characterization of materials. Thin films may be amorphous, polycrystalline, or single-crystal. Therefore, it requires a wide variety of techniques for the characterization.

Amorphous Thin Films

A diffraction pattern of an amorphous material is characterized by diffuse broad peaks with a 2θ half width of a few degrees. These peaks vanish rapidly with increasing diffraction angle. The x-ray method used for analyzing amorphous films is similar to that of bulk materials. This includes: the determination of coherent scattering intensity per atom, $I_{eu}(K)$, expressed in electron units, by correcting, when necessary, the observed intensity for absorption, polarization, inelastic scattering, multiple diffraction and substrate effects. The interference function I(K), is calculated from

$$I(K) = \frac{I_{eu}(K)-(\langle f^2\rangle-\langle f\rangle^2)}{\langle f\rangle^2} \qquad (1)$$

where $K=4\sin\theta/\lambda$ and $\langle f\rangle$ is the average scattering factor of the specimen. The reduced radial distribution function G(r), from which the local atomic arrangement is obtained, is defined as:

$$G(r) = 4\pi r[\rho(r)-\rho_o] = \frac{2}{\pi}\int_0^\infty K[I(K)-1]\sin(Kr)dK \qquad (2)$$

where $\rho(r)$ is the atomic distribution function and ρ_o is the average atomic density of the material.

The detailed procedure for the calculation is rather involved so that computer programs have been found to be very useful. The Fortran program used for this study, with minor modifications, was originated by Wagner (6).

I(K)'s and G(r)'s of the amorphous films with $Co_{77}Gd_{23}$, $Co_{68}Gd_{32}$ (no bias), and $Co_{77}Gd_{17}Ar_6$ (-100V bias on substrate) prepared by sputter deposition on 50μm thick Be substrates (7) are given in Figs. 2 and 3, respectively. Within experimental uncertainties, the I(K) obtained by both reflection and transmission techniques of Mo Kα radiation from a given sample were identical and there is no evidence of structural anisotropy. The diffuse peaks in I(K) (Fig. 2) show surprisingly large line widths at half maximum (ΔK=0.95Å^{-1} for the first peak) in comparison to that of liquid-quenched metal-metalloid alloys (ΔK=0.5Å^{-1}) (8). This may be interpreted as lack of structural order in these Co-Gd films. The position of the first peak decreases with increasing Gd concentration, possibly because the Goldschmidt diameter of Gd (3.6Å) is much larger than Co (2.5Å) (9).

The "dense random packing of hard spheres" model (10) has been found to be the most successful for interpreting the local atomic arrangements from G(r). By using this model, it is possible to associate individual peaks of G(r)'s with distances corresponding to the first and second nearest separations between Co and Gd atoms, shown in the upper part of Fig. 3. Generally, in the amorphous structure determination, only the first two nearest neighbor atomic distributions can be identified and the structure beyond these distances is obscured by overlap of various atomic arrangements.

Polycrystalline Thin Films

There are a number of problems in x-ray structural characterization of polycrystalline thin films which are not encountered in bulk materials. Diffracted intensities are usually low when the specimen is thin, the substrate can cause a high background, and the crystallites may have preferred orientation. It is, therefore, essential to set up the instrumentation properly and to use techniques which give the best possible data.

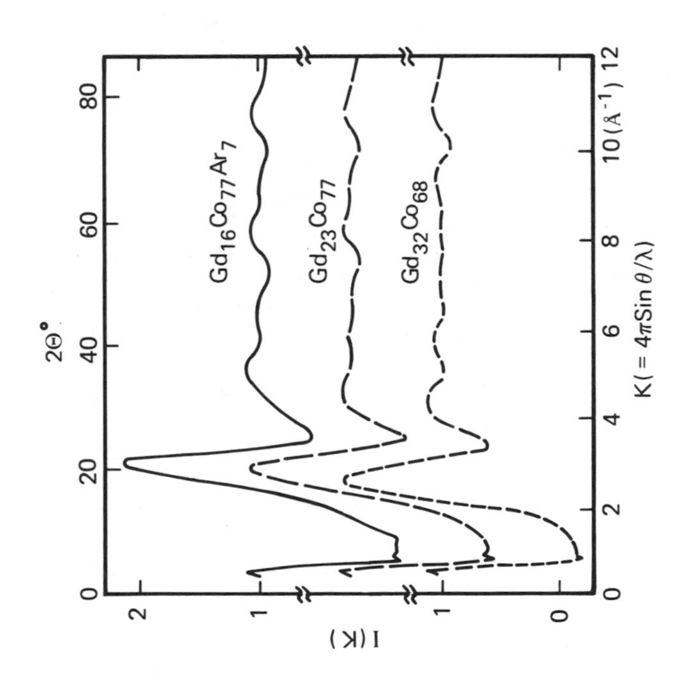

Figure 2. Refined interference function I(K) of Gd-Co alloys as a function of K. Mo Kα radiation was used.

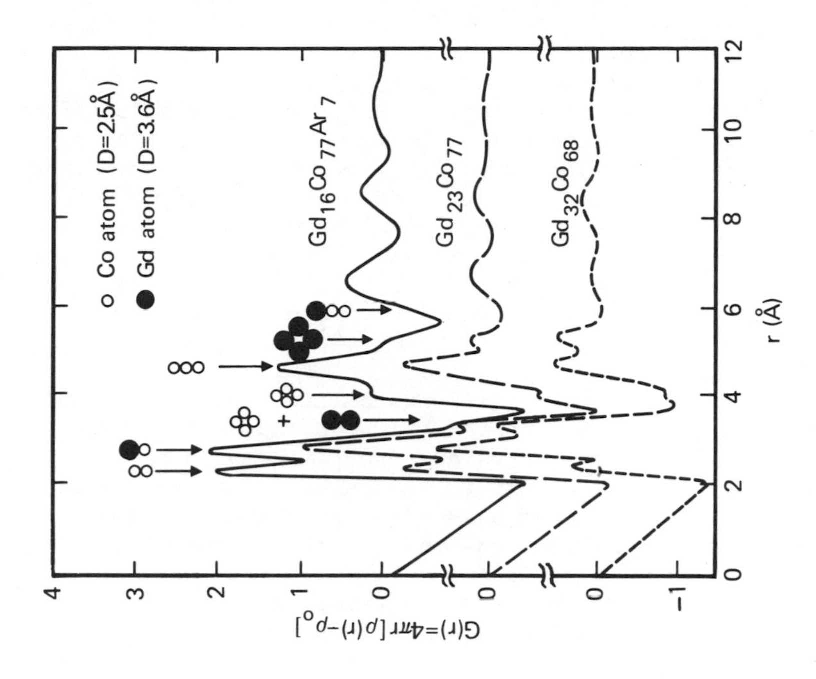

Figure 3. Reduced radial distribution function G(r) of Gd-Co alloys and the proposed local atomic arrangement.

Instrumentation. A conventional vertical scanning Norelco diffractometer was modified (11) to achieve high intensity, large peak to background ratio, and good resolution for quantitative analysis. A long fine focus Cu target x-ray tube is used and the anode is set at a relative large angle of 12° to reduce target self-absorption. The operating voltage is usually chosen to be 50kV. T ogether with the high specific loading and thin Be window of the x-ray tube, this provides high incident intensity. The length (54 mm) and spacing of the incident beam parallel slits are increased such that the total intensity reaching the specimen is high, the distribution is more uniform, and, at the same time, there is no loss of resolution. The divergence slit and the first anti-scatter slit are close to the specimen, which simplifies the elimination of unwanted scattered radiation. The addition of a focusing monochromator in the diffracted beam behind the receiving slit minimizes specimen fluorescence. The use of a bent graphite crystal monochromator results in an increase of intensity by a factor of two over the conventional set-up because the β filter and the diffracted beam parallel slit collimator are eliminated. A further increase in intensity is obtained by using a vacuum along the path of the x-rays to avoid air scattering. This experimental set-up has been routinely used to characterize specimens of thicknesses of 100Å and more.

A θ-2θ scanning diffractometer with transmission geometry, a Seeman-Bohlin diffractometer (S-B) and photographic film technique are also used to complement the conventional diffractometer. The transmission diffractometer allows for the study of the crystal planes normal to the surface of the film. The S-B allows the analysis of those planes inclined to the surface, and is thus capable of providing about ten times more intensity - an attractive factor for thin film analysis (12). It is sometimes advisable to transfer the same specimen to these two diffractometers to obtain sets of complementary data. However, the instrumental aberrations for these two geometries are different so that it is essential to align and use them with extra care if they are to be used for quantitative analysis. Photographic film techniques are sometimes used for qualitative work.

Phase Identification. There are two common problems in thin film analysis that are not often encountered in bulk materials. First, because of preferred orientation, the fixed relative intensities of random powder patterns are no longer applicable. For example, the (111) reflections in face-centered cubic (FCC), (110) in body-centered cubic (BCC) and (001) in hexagonal structures, generally show stronger intensities than in random powders. Use of the transmission and/or the S-B diffractometer(s) to obtain complementary data will be helpful for phase identification. Second, the problem of overlapping peaks from mixed phases could lead to errors in phase determination. Longer wavelength x-ray radiation, e.g., Cr instead of Cu, to disperse the reflections, or the profile fitting method (13) for resolving overlapping peaks are two of the methods that can be used to overcome this problem. Also, a good elemental analysis by XRF will be very helpful for defining the probable phases presented (14).

Examples of the phase identification of various compositions of Ni-Fe binary alloy are given in Fig. 4. When the Fe concentration is close to 100%, the alloy has the α phase with a BCC structure. As the Fe concentration decreases to about 60%, a second γ phase with the FCC structure appears. When the Fe content decreases further, the original α phase disappears, leaving only the γ phase.

Preferred Orientation. Polycrystalline films deposited on substrates generally show preferred orientation (or texture). The degree of preferred orientation depends on the deposition method, film material, chamber pressure, substrate temperature and materials, etc. Figure 5 shows the preferred orientation effect in a 5000Å thick permalloy film before and after annealing at 450°C for 21 hours in vacuum. For purposes of comparison, the pattern of Ni powders is also given. The deposited film shows a strong (111) texture with its orientation factor, R=I(111)/I(200), increasing to about four times higher than the Ni powders. After annealing, the texture changes to an even stronger (111) preferred orientation with R increasing by another 50%. The texture of these permalloys was found to be independent of their thicknesses.

The textures of metallic films deposited on substrates tend to have rotational symmetry around the normal of the film surface. A transmission pinhole photographic method is best for this purpose, and it shows uniform and continuous diffraction rings for a permalloy film with substrate removed.

For texture with rotational symmetry, a one-dimensional pole distribution curve of the normal crystallographic axis, the (111) pole in this case, proves to be a simple and effective method for quantitative preferred orientation analysis. There is no need of recording a complete pole figure. As illustrated in the upper right hand side of Fig. 6, the pole distribution curve is obtained with a conventional diffractometer by tilting the specimen to various angles of ϕ with respect to its normal position, and the integrated intensities, e.g., I(111) in this case, is recorded. For quantitative study, the observed intensity must be corrected for defocusing, volume of diffracting material,

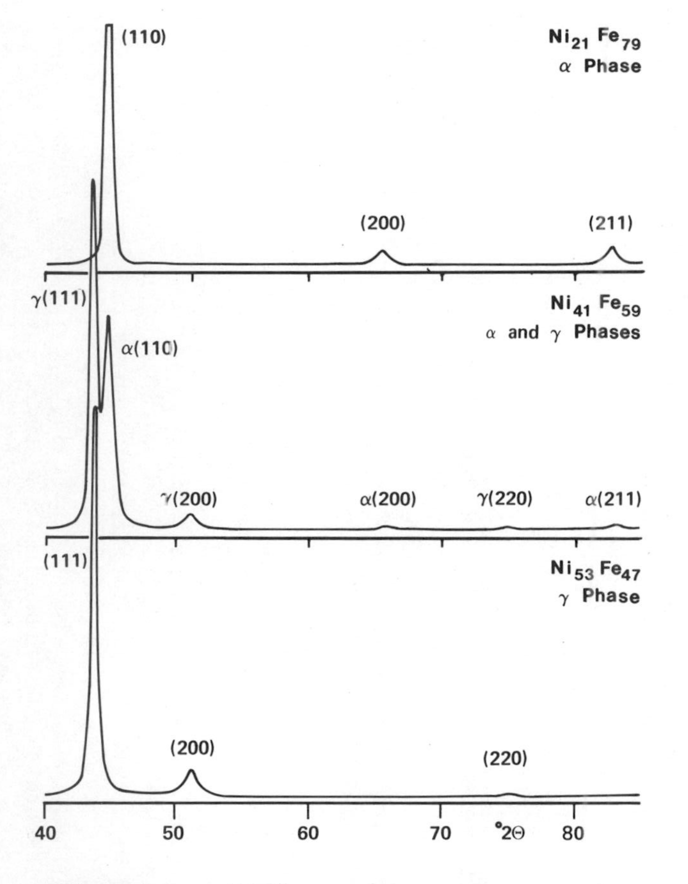

Figure 4. Diffractometer patterns of Ni-Fe alloys using Cu Kα radiation.

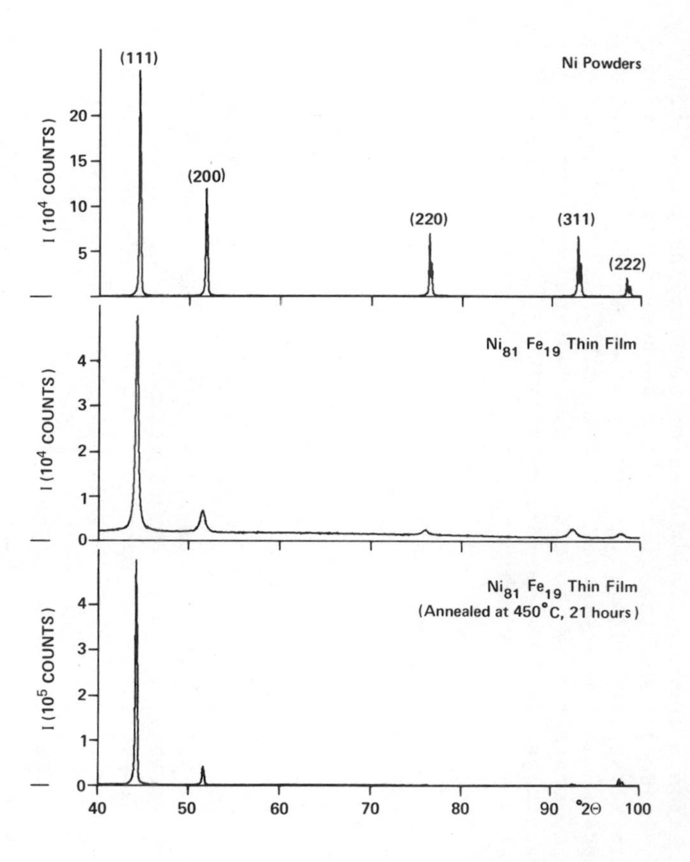

Figure 5. Diffractometer patterns of Ni powders and a 5,000Å thick permalloy film before and after annealing. Cu Kα radiation was used.

and absorption. The defocusing aberration is minimized by using a narrow divergence slit to confine the incident beam to the focusing circle, by removing the receiving slit, and by using the integrated intensity instead of peak intensity. Volume and absorption corrections are necessary because the variation in ϕ causes variations in both the volume of diffracting material and the path length of the x-ray within the film. Two one-dimensional (111) pole distribution curves obtained by the method described above are given in Fig. 6. It shows that more than 50% of the (111) poles in the as-deposited film are within 17° of the substrate normal. Annealing causes more (111) poles to align with the normal and the spread of the distribution curve reduces to only 6°.

X-Ray Line Broadening Analysis. Broadening of an x-ray diffraction profile is normally caused by small particle size, by faulting on certain crystal planes, and by non-uniform strains within the crystallites. A number of methods have been proposed for analyzing these profiles and they have had various degrees of success. They are the Scherrer equation for crystallite size determination (15), the method of integral breadth (16), and the Warren-Averbach method (17) for crystallite size and strain separation. Since the Fourier method of Warren-Averbach has been most successful and commonly used for quantitative analysis, it will be used in this study.

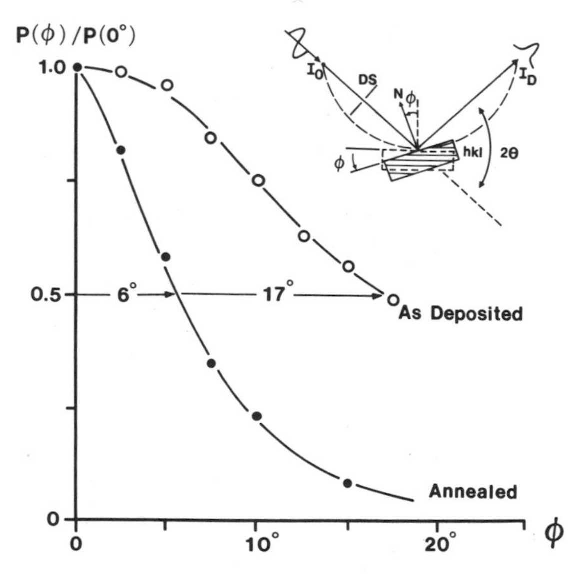

Figure 6. One-dimensional (111) pole density distribution curve of a permalloy film before and after annealing. The experimental method is described schematically on the upper right hand corner.

The true diffraction profile from the specimen, P(S), corrected for instrumental aberration and wavelength dispersion, can be expressed in the form of a Fourier series as:

$$P(S) = \sum_{L=-\infty}^{+\infty} [A_L \cos 2\pi L(S-S_o) + B_L \sin 2\pi L(S-S_o)] \tag{3}$$

where $S=2\sin\theta/\lambda$, L is the distance normal to the reflection plane (hkl) and equal to the product of the order n and the lattice constant, a. The cosine coefficient A_L is the product of a size coefficient, A_L^S, and a strain coefficient, A_L^D:

$$A_L = A_L^S \cdot A_L^D \tag{4}$$

For small values of L and strain,

$$A_L^S = 1 - \frac{L}{D(eff)} = 1 - L\left[\frac{1}{D} + \frac{1.5\alpha+\beta}{a} V_{hkl}\right] \tag{5}$$

$$= 1 - L\left[\frac{1}{D} + \frac{1}{D(F)}\right] \tag{5'}$$

$$A_L^D = 1 - \frac{2\pi L^2 \langle\varepsilon_L^2\rangle (h^2+k^2+l^2)}{a^2} \tag{6}$$

where D(eff) and D(F) are, respectively, the effective crystallite size and the fictitious particle size due to faulting in the direction normal to the (hkl) plane, D is the true crystallite size, α and β are, respectively, the deformation and twin fault probabilities, and $\langle\varepsilon_L^2\rangle$ is the mean square strain component normal to the (hkl) plane and averaged over L. V_{hkl} is a constant depending on the (hkl) plane and is equal to $\sqrt{3/4}$ for (111) plane and 1.0 for (200) plane in the FCC lattice.

Since A_L^S is independent of (hkl), it can be separated from A_L^D if multiple orders of reflections are available. In the case when only one order of reflection is measurable, a common case in thin films, A_L^S's are determined by approximating from A_L of the first order reflection. A number of methods have been proposed for this purpose (18). The true crystallite size D can be determined from the D(eff)'s of two different kinds of reflections by Eq. (5).

The mathematical procedure for these calculations is quite involved so that computer programs, again, are found to be essential. The Fortran program used for this study, with modifications, was originated by Wagner (19).

Seven evaporated Ni-Fe permalloy samples on glass substrates with film thicknesses from 100Å to 1μm were used for the determination of crystallite size and strain by the Warren-Averbach method. These were then annealed at 450°C for 21 hours in vacuum and the annealing behaviors were studied.

Crystallite Size. Crystallite sizes and non-uniform strains were determined along both the [111] and [200] directions. For as-deposited films with thicknesses of 1000Å and higher, D(eff, 111)'s were obtained from the initial slope of A_L^S against L curves as shown in Fig. 7. A_L^S's were determined from the size coefficients of two orders of reflections, the (111) and (222). It was found that all the D(eff, 111)'s obtained from A_L^S vs. L curves were approximately 30% larger than D(eff. 111)'s obtained from the uncorrected A_L vs. L curves of the first order (111) reflection. Because the (222) reflections were too weak to be measured, this approximation was used for the determination of D(eff, 111)'s for specimens with thicknesses of 500Å and less. All D(eff, 111)'s are given in Column 2 of Table 2.

Similarly, D(eff, 200)'s for 1.0 and 0.5μm films were obtained from A_L^S vs. L curves using the (200) and (400) reflections, and others were approximated from A_L of the (220) reflection with a 20% correction. The (200) reflection from films of 200Å or less were too weak to be measured and no D(eff, 200)'s were obtained. All calculated D(eff, 200)'s are listed in Column 5 of Table 2. Notice that all the D(eff, 200)'s are smaller than the corresponding D(eff, 111)'s, which is an indication of the presence of faulting. This will be discussed later.

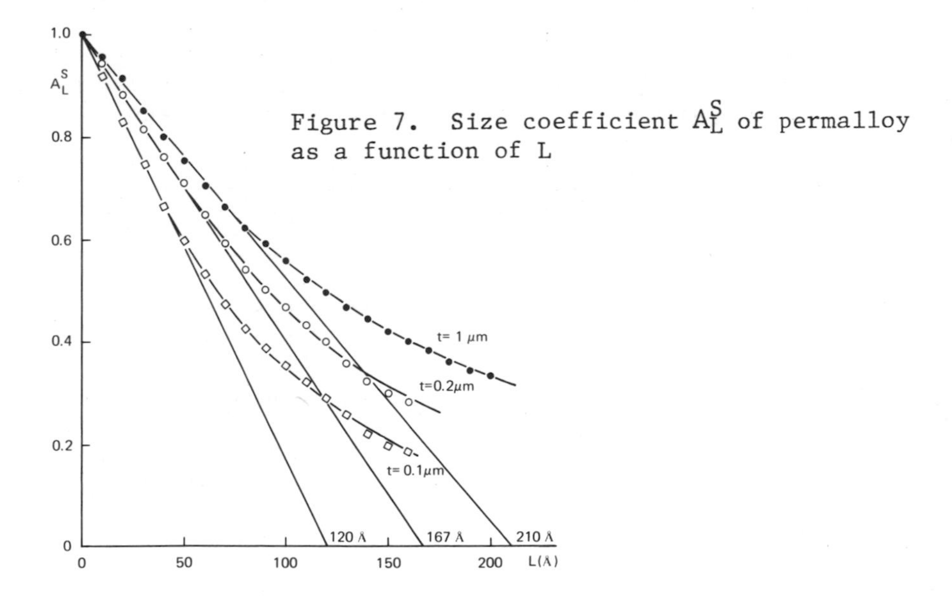

Figure 7. Size coefficient A_L^S of permalloy as a function of L

The true crystallite size was calculated from D(eff, 111) and D(eff, 200) using Eq. (5). The values are listed in Column 8 of Table 2. All D(eff, 111)'s, D(eff, 200)'s and D's are plotted as a function of thickness in Fig. 8. It shows that these crystallite sizes increase rapidly at small thickness, and then gradually level off with increasing thickness. When the thickness reaches about 5000Å and higher, the crystallite sizes remain almost unchanged.

As pointed out by Langford and Wilson (15), care must be taken when the crystallite sizes obtained from diffraction broadening are compared with the "particle size" given by other methods. The crystallite size determined by the diffraction method is the length, as measured in the direction of the reciprocal-lattice vector, along which the diffraction is coherent. The individual particle observed by other methods, e.g., S EM or light optics, may contain several crystallites or domains having different orientations. The diffraction technique is the only method which can investigate the true crystallite size within which atoms are crystallographically related.

Grain Growth. When polycrystalline films are annealed, grain growth accompanied by a change in texture frequently occurs. After annealing, the observed intensities increased by almost an order of magnitude as a result of grain growth. Since the intensities were so much higher than before, all (222) reflections from specimens of thicknesses of 500Å and more could be measured. D(eff, 111)'s were then obtained from the A_L^S vs. L curves. Similarly, D(eff, 200) of the 500Å thick specimen was approximated from the (200) reflection with a correction factor of 25%.

TABLE II Calculated Crystallite Sizes for Deposited and Annealed Permalloy Thin Films of Various Thicknesses

Thickness (A)	D(eff,111)			D(eff,200)			D			D(F,200)		
	Dep.	Ann.	R=Ann/Dep	Dep.	Ann.	R	Dep.	Ann.	R	Dep.	Ann.	R
10,070	210			120			490			159		
4,900	190	1,850	10	107	1,040	10	466	4,565	10	139	1,347	10
2,200	165	1,040	6	104	620	6	299	2,155	7	160	872	6
1,050	120	545	5	83	345	4	182	975	5	152	533	4
450	90	270	3	70	180	3	115	437	4	178	305	2
200	57	115	2		75			194			122	
100	40											

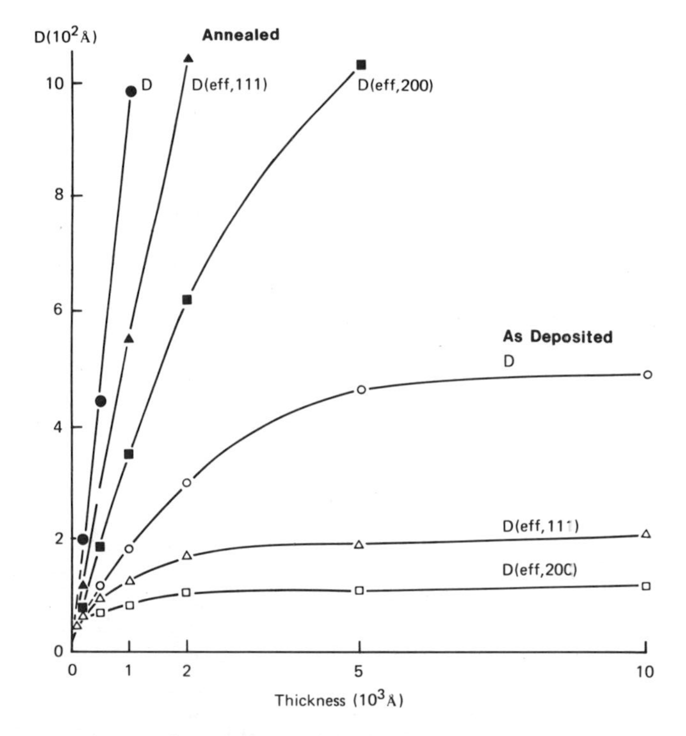

Figure 8. Crystallite size of permalloy as a function of film thickness. Notice the tremendous grain growth due to annealing.

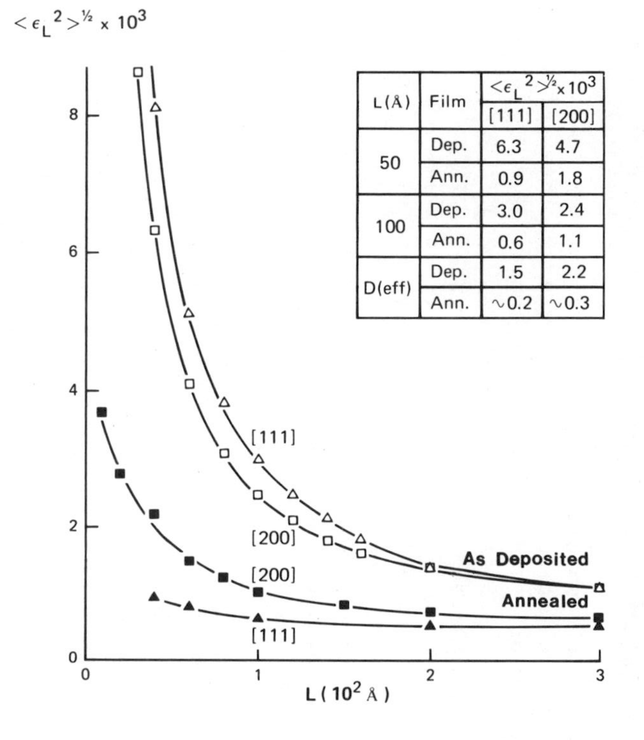

L (Å)	Film	$<\epsilon_L^2>^{1/2} \times 10^3$ [111]	[200]
50	Dep.	6.3	4.7
	Ann.	0.9	1.8
100	Dep.	3.0	2.4
	Ann.	0.6	1.1
D(eff)	Dep.	1.5	2.2
	Ann.	∿0.2	∿0.3

Figure 9. RMS strain as a function of L for permalloy before and after annealing.

The annealed D(eff,111)'s, D(eff, 200)'s and D's are listed in Columns 3, 6 and 9 of Table 2, and are plotted in Fig. 9. Notice that a tremendous growth of the crystallites is observed and the true crystallite size approaches the thickness of the film. The ratios of grain growth due to annealing are listed in Columns 4, 7 and 10. D(eff, 111), D(eff, 200) and D grew at the same rate, depending only on the thickness. This rate varies from about ten times at a thickness of 5000Å to three times at a thickness of 500Å. The smaller ratios for thinner films seem to be a result of the physical limit of the thickness. The crystallite size of the 1μm thick specimen was not obtained because the crystallites become too large, exceeding the limit for which the approximation of the Warren-Averbach method is still valid.

Non-Uniform Strain in the Warren-Averbach method is expressed in the form of a root mean square (RMS) strain, i.e., $[\langle\varepsilon_L^2\rangle - \langle\varepsilon_L\rangle^2]^{1/2}$ or $\langle\varepsilon_L^2\rangle^{1/2}$, averaged over the length L. It can be calculated from Eqs. (4), (5) and (6). For the as-deposited films, RMS strains have been determined for specimens 1000Å and thicker along the [111] direction, and for 5000Å and 1μm thick films along the [200] direction. The curves of $\langle\varepsilon_L^2\rangle^{1/2}$ vs. L for the 5000Å thick specimen along both directions are given in Fig. 9. The RMS strains are relatively higher at small L, decrease with increasing L, and reach an asymptotic value for large L. This kind of distribution of RMS strain seems to indicate the influence of faulting and dislocations on strain as well as the existence of large strains in small domains. Along the [111] or [200] direction, the distribution of RMS strain as a function of L is found to be approximately the same for all thicknesses, except when L is very small at about 50Å or less.

The RMS strains along the [111] and [200] directions averaged over 50Å, 100Å and the effective crystallite sizes of the 5000Å thick film are listed in the upper right hand side of Fig. 9. The unequal values of $\langle\varepsilon_L^2\rangle^{1/2}$ along the [111] and [200] directions indicate that they are elastically anisotropic. As shown in Fig. 9 , when L becomes greater than the effective crystallite size at about 200Å or more, both RMS strains along the [111] and [200] directions gradually approach and coincide with each other. It indicates that when $\langle\varepsilon_L^2\rangle^{1/2}$ is calculated beyond the dimension of each individual grain in a polycrystalline material, its directional dependence gradually disappears.

Strain Relaxation. The effect of annealing on strain relaxation is shown in the lower two curves of Fig. 9. The RMS strains along the [111] and [200] directions do not coincide with each other as early as for the as-deposited films because the dimensions of the crystallites are so much larger after annealing.

Stacking Faults. The contribution of faulting to the line broadening is given by the second term on the right hand side of Eq. (5'). Column 11 of Table 2 lists the calculated D(F, 200)'s. Notice that all D(F, 200)'s of the as-deposted films are approximately the same dimension, indicating that the density of faulting is independent of the thickness.

After annealing, the D(F, 200)'s grew to different sizes for different thicknesses. As shown in Column 12 of Table 2, the thicker the film, the larger the D(F, 200), and the less the probability of finding a stacking fault.

Single Crystal Films

There has been increasing interest in measuring the lattice mismatch of single crystal films grown epitaxially on single crystal substrates. In the magnetic garnet bubble technology, the lattice mismatch $\Delta a/a=(a_s-a_f)/a_s$, determined by x-ray diffraction is an important parameter for the determination of stress-induced anisotropy, one of two sources of magnetic uniaxial anisotropy in garnet films (20). $\Delta a/a$ is about 10^{-4}, and a double crystal diffraction method with high resolution is needed to separate the diffraction peaks.

A conventional horizontal powder diffractometer was modified and converted into a single- and double-crystal diffractometer. The modifications include: a narrow incident-beam collimator with a pinhole (6 mil in diameter); specimen holders using the Bond adjustable barrel holder (21); adjustable detector arm for the second crystal; and automatic computer control with $\Delta\theta\geq1'$ per step. This set up was designed by Dr. M. Hart and it allows precise and direct measurements of lattice parameter and mismatch (22). Figure 10 shows the double-crystal diffraction profile with Cu $K\alpha_1$ radiation of a high density magnetic bubble garnet film epitaxially grown on a $Gd_3Ga_5O_{12}$ substrate, using GGG(888)-GGG(888) reflections. Notice the difference in line widths and profile shapes between the film and substrate. The full width at half maximum (FWHM) of the substrate's profile is about 10". The profile of the film is relatively broad, indicating the existence of non-uniform strains. By measuring the separation between these two peaks, the lattice mismatch and stress-induced anisotropy can be determined. Another example is the specimen with a second magnetic film sandwiched between the original magnetic bubble film and its substrate. Double-crystal diffraction profiles of this multilayer film is shown in Fig. 11. All three peaks are well separated, and as a result, precise measurement of their lattice mismatches can be obtained with high confidence. Again, the mean strain on each layer can be calculated from the peak separation, and the non-uniform strain is obtained from the profile shapes. The upper layer has the

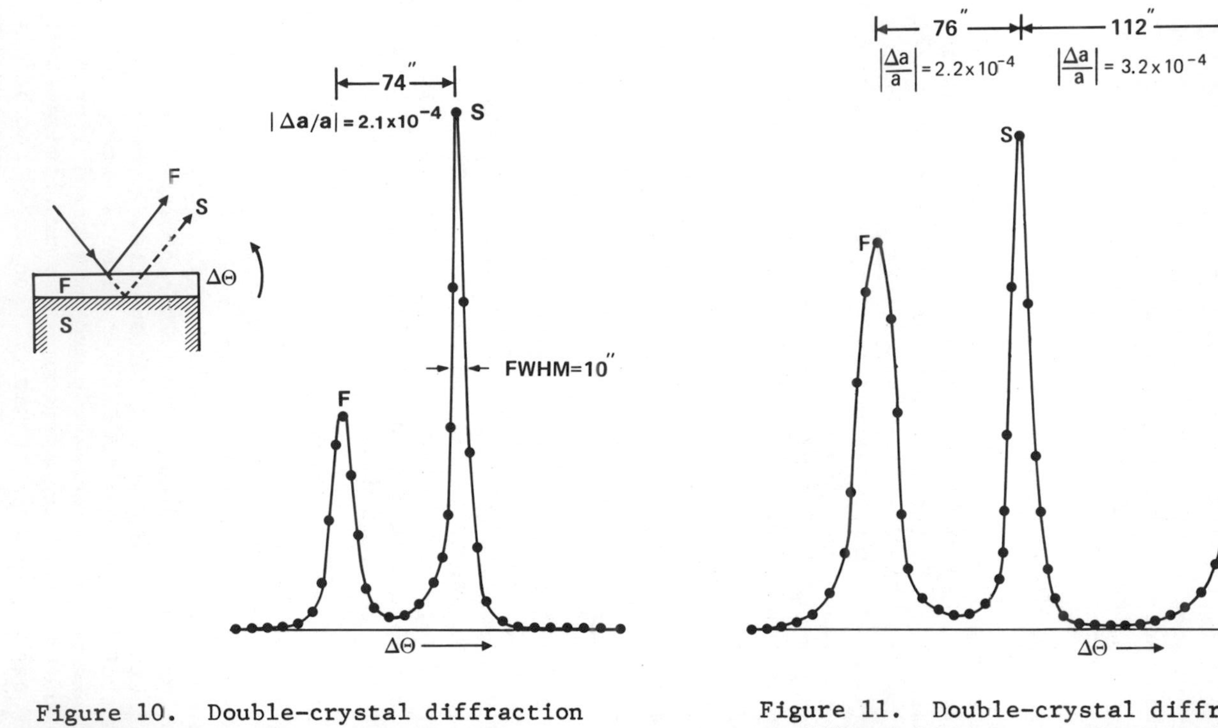

Figure 10. Double-crystal diffraction profile with Cu $K\alpha_1$ radiation of an epitaxial garnet film F, and its substrate S.

Figure 11. Double-crystal diffraction profile with Cu $K\alpha_1$ radiation of a double-layer epitaxial film F and C, and its substrate S.

largest line width indicating that it is under the strongest non-uniform strain.

SUMMARY AND CONCLUSION

A comprehensive study of a wide variety of applications of XDF and XRD techniques for the characterization of thin films has been presented. By the proper use of x-ray instrumentation and techniques, a reasonably complete analysis was obtained.

XRF was shown to be a rapid, accurate and non-destructive method for routine composition and mass-thickness analysis. XRF has proved to be very accurate in comparison to AAS and EPM for elemental analysis. The density of very thin films has also been determined from values of mass-thickness by XRF and thickness by interferometry.

XRD was used successfully for structural characterization of thin films. For amorphous materials, the local atomic arrangement was obtained from the diffuse broad diffraction pattern, the interference function, and the reduced radial distribution function. For polycrystalline materials, phase identification and transition studies were made with a modified conventional 2:1 vertical scanning powder diffractometer. Qualitative and quantitative analysis of preferred orientations were obtained with the diffractometer and a one-dimensional pole distribution curve method was used. Crystallite size, stacking and microstrain were analyzed with the Warren-Averbach method. Their annealing behaviors were also studied. For single-crystal films, studies of lattice mismatch and diffraction profiles of epitaxial films grown on single-crystal substrates were made using the double-crystal diffraction method.

ACKNOWLEDGMENT

We would like to express our thanks to Dr. A. H. Sporer, Head of the Applied Science Department, whose continued support made this study possible. We also thank Mr. J. Carothers and Dr. M. Lee for preparing and annealing the Ni-Fe films; Dr. H. Wieder, Mr. B. Olson, and Mr. R. Tom for EPM, AAS, and interferometer measurements, respectively; Mr. E. Moore for preparing the magnetic garnet films. Special thanks are due to Dr. K. Lee whose constructive criticism in reviewing this manuscript has been most valuable. Finally, we want to express our appreciation to Drs. C. Ruud and G. McCarthy for inviting us to present this work.

REFERENCES

1. D. Laguitton and M. Mantler, "A General Fortran Program for Quantitative X-Ray Fluorescence Analysis," in H. F. McMurdie, C. S. Barrett, J. B. Newkirk and C. O. Ruud, Editors, Advances in X-Ray Analysis, Vol. 20, p. 515-528, Plenum Press (1977).

2. D. Laguitton and W. Parrish, "Simultaneous Determination of Composition and Mass-Thickness of Thin Films by Quantitative X-Ray Fluorescence Analysis," Anal. Chem. 49, 1152-1156 (1977).

3. J. W. Criss and L. S. Birks, "Calculation Methods for Fluorescent X-Ray Spectrometry. Empirical Coefficients vs. Fundamental Parameters," Anal. Chem. 40, 1080-1086 (1968).

4. J. Sherman, "The Theoretical Derivation of X-Ray Intensities from Mixtures," Spectrom. Acta 7, 283-306 (1955).

5. G. Pollai, M. Mantler and H. Ebel, "Die Sekundaranregung bei der Rontgenfluoreszenzanalyse ebener dunner Schichten," Spectrom. Acta 26B, 747-759 (1971).

6. C. N. J. Wagner, "A Fortran IV Program for the Calculation of the Radial Distribution Function of Binary Alloys," Technical Report No. 2, NSF GP 3213.

7. C. N. J. Wagner, N. Heiman, T. C. Huang, A. Onton and W. Parrish, "The Structure of Amorphous Gd-Co Alloy Films," in J. J. Becker, G. H. Lander and J. J. Rhyne, Editors, Magnetism and Magnetic Materials, AIP Conf. Proc. No. 29, p. 188-189, Am. Inst. of Physics (1975).

8. B. C. Giessen and C. N. J. Wagner, "Structure and Properties of Non-Crystalline Metallic Alloys Produced by Rapid Quenching," in S. Z. Beer, Editor, Liquid Metals, Chemistry and Physics, p. 633-695, Marcel Dekker (1972).

9. R. P. Elliot, "Constitution of Binary Alloys," First Suppl., p. 870, McGraw-Hill (1965).

10. G. S. Cargill III, "Structure of Metallic Alloy Glasses," in F. Seitz, D. Turnbull and H. Ehrenreich, Editors, Solid State Physics, Vol. 30, p. 295-318, Academic Press (1975).

11. W. Parrish, "X-Ray Diffractometry Methods for Complex Powder Patterns," in H. van Olphen and W. Parrish, Editors, X-Ray and Electron Methods of Analysis, p. 1-35, Plenum Press (1968).

12. W. Parrish, "Role of Diffractometer Geometry in the Standardization of Polycrystalline Data," in C. L. Grant, C. S. Barret, J. B. Newkirk and C. O. Rudd, Editors, Advances in X-Ray Analysis, Vol. 17, p. 97-105, Plenum Press (1974).

13. T. C. Huang and W. Parrish, "Qualitative Analysis of Complicated Mixtures by Profile Fitting X-Ray Diffractometer Patterns," in C. S. Barrett, D. E. Leyden, J. B. Newkirk and C. O. Rudd, Editors, Advances in X-Ray Analysis, Vol. 21, p. 275-288 (1978).

14. R. Jenkins, "Interdependence of X-Ray Diffraction and X-Ray Fluorescence Data," in C. S. Barrett, D. E. Leyden, J. B. Newkirk and C. O. Rudd, Editors, Advances in X-Ray Analysis, Vol. 21, p. 7-22 (1978).

15. J. I. Langford and A. J. C. Wilson, "Scherrer After Sixty Years: A Survey and Some New Results in the Determination of Crystallite Size," J. Appl. Cryst. 11, 102-113 (1978).

16. C. N. J. Wagner, "Analysis of the Broadening and Changes in Position of Peaks in an X-Ray Powder Pattern," in J. B. Cohen and J. E. Hillard, Editors, Local Atomic Arrangements Studied by X-Ray Diffraction, p. 218-268, Gordon and Breach Science (1966).

17. B. E. Warren, "X-Ray Diffraction," p. 251-314, Addison-Welsley (1969).

18. A. Gangulee, "Separation of the Partical Size and Microstrain Components in the Fourier Coefficients of a Single Diffraction Profile," J. Appl. Cryst. 7, 434-439 (1974).

19. C. N. J. Wagner, "A Fortran IV Program for the Analysis of Position and Profiles of X-Ray Powder Pattern Peaks," Technical Report No. 15, Office of Naval Research, NONR 609 (43) (1966).

20. J. E. Davies and E. A. Giess, "The Design of Single Crystal Materials for Magnetic Bubble Domain Applications," J. Mat. Sci. 10, 2156-2170 (1975).

21. W. L. Bond, "Device for Preparing Accurately X-Ray Oriented Crystals," J. Sci. Instr. 38, 63-64 (1961).

22. M. Hart and K. H. Lloyd, "Measurement of Strain and Lattice Parameter in Expitaxial Layers," J. Appl. Cryst. 8, 42-44 (1975).

CHARACTERIZATION OF LATERITES BY X-RAY TECHNIQUES

D. Chandra, C. O. Ruud and C. S. Barrett
University of Denver Research Institute
Denver, Colorado 80208

R. E. Siemens
Albany Metallurgy Research Center
Albany, Oregon 97321

ABSTRACT

To assure an adequate supply of such critical metals as nickel and chromium, extraction procedures must be developed to process low grade domestic sources. In optimizing these procedures it is essential to use suitable analytical procedures to characterize the materials, identify phase transformations, and determine metal and mineral association of the critical metallic elements through all stages of the process. Evaluation of complex sources such as laterites requires special material handling techniques coupled with X-ray diffraction and with optical and SEM analyses of many individual particles. A joint study by the Bureau of Mines and the Denver Research Institute using these procedures has resulted in optimizing a new modification of an extraction process.

INTRODUCTION

The demand for nickel in the United States is increasing rapidly, while nickel production is increasing very slowly. It is anticipated that U.S. demand will be 500 million pounds while the U.S. supply will be only 50 million pounds by 1980 (1). To bridge the gap between supply and demand, maximum utilization of all feasible nickel-bearing ores will be necessary. One such source is laterite.

Laterites are low grade nickel-bearing ores formed by continuous weathering of serpentinized ultramafic rock (2). The weathering is primarily a hydrolysis reaction with rain water contributing hydrogen cations to form silicic acid which leaches out the more easily soluble cations such as magnesium and silicon. These cations are carried away in solution while all the relatively insoluble nickel, cobalt, chromium and ferric ions are residually concentrated to form laterites. The ores with high iron content are called nickeliferous laterites.

Laterites are primarily used to extract nickel and cobalt values, although valuable elements such as chromium, manganese and titanium are always present in these ores. The chromium, manganese and titanium values are usually left behind in the tailings or in the leached residues of the various nickel extraction processes. The extraction of chromium from laterites has recently become very important for economic and strategic reasons. New methods of chromium extraction are being investigated at the University of Denver.

The technology of nickel extraction from laterites is still in the developmental stages. The mineralogical constitution of laterites dictates the type of extractive metallurgical process. Therefore, a thorough understanding of their chemical and mineralogical character as to how the elements of interest are associated in the minerals constituting the ore body is vital to the determination of the process parameters.

The mineralogy of laterites is rather complex and the weathered products are, by and large, fine grains in a cryptocrystalline state. The identification of the contained minerals, in particular nickel and chromium, by straightforward bulk X-ray diffraction techniques alone is not possible because of their low concentration in the minerals and because of X-ray diffraction peak overlaps. To make matters worse, there are peak broadening effects due to the small crystallite size of the ore and due to lattice distortions in the minerals. To establish how the valuable elements such as nickel, cobalt and chromium are associated in the minerals, a logical combined application of X-ray diffractometry, electron microscopy and electron microbeam analysis, in conjunction with conventional petrography, is used.

The general characterization techniques for the raw ore and for treated products from a Bureau-patented ammoniacal leaching process (3) with prior carbon monoxide reduction are discussed in this paper.

EXPERIMENTAL

Materials Investigated

1. Lateritic ore from Northern California, Southern Oregon, New Caledonia and the Philippine Islands.

2. Lateritic ores from Northern California and Southern Oregon; laterites treated with carbon monoxide gas in reduced state.

Scheme of Experiments

The characterization scheme is outlined in Figure 1. X-ray diffraction and chemical analysis of the ores in the as-received condition were performed prior to other experimentation to get a general idea of the elemental distribution and the minerals associated with each element.

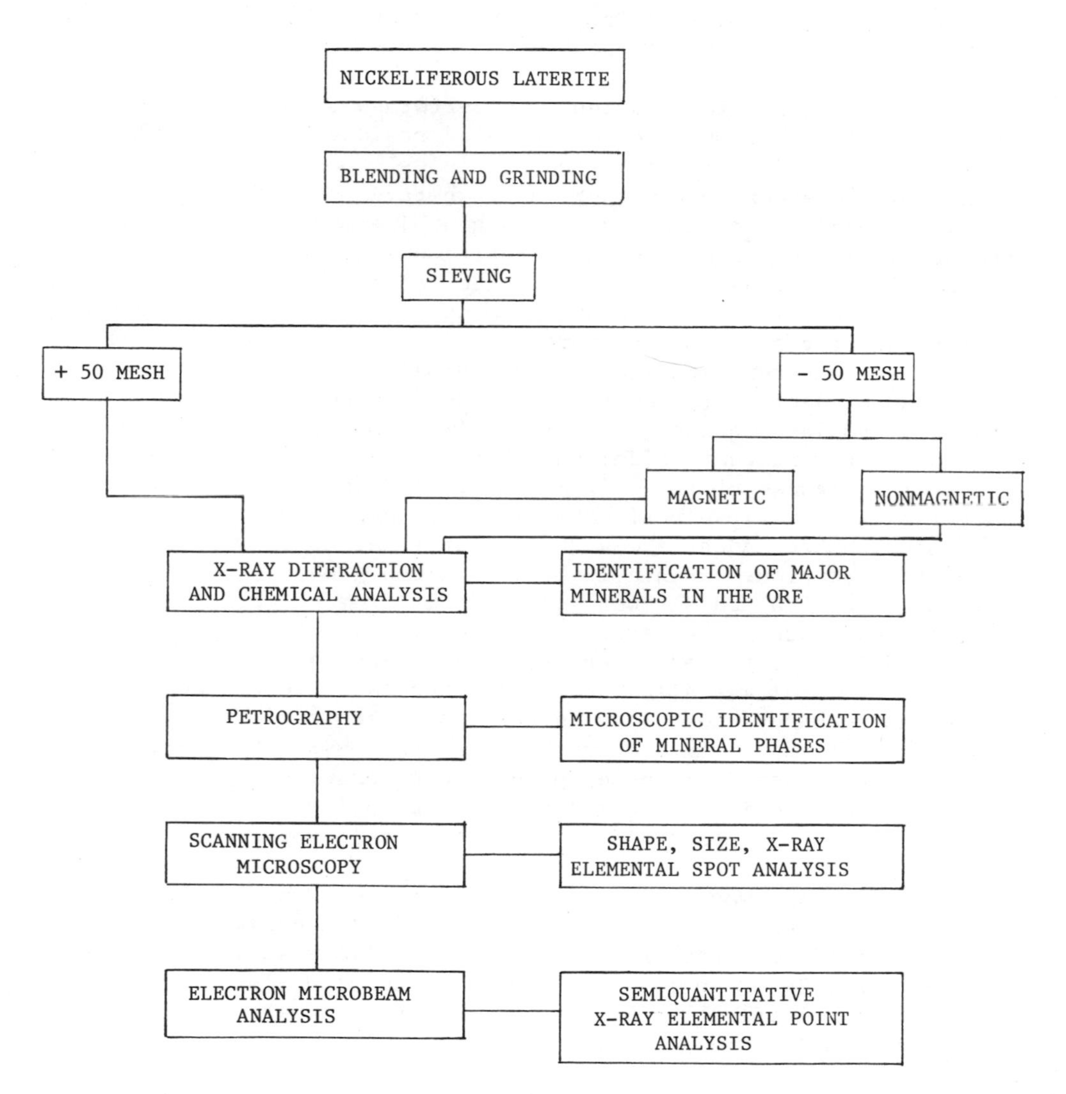

Figure 1. Schematic of the Characterization Procedure

The laterites were thoroughly blended in a mixer and several representative samples were obtained by cone and quarter method. These were ground for 15 minutes with the grinding time optimized to obtain a wide range of particle sizes. The ground samples were sieved through a 50 mesh Tyler screen. The -50 mesh samples were used for magnetic separation. Although physical concentration methods such as magnetic and density separation are not commercially applicable to laterites, they are important for characterization purposes. A modified Davis tube (4) was found to be most effective and the products were characterized. Hydrochloric acid was used to dissolve soluble iron ions in order to identify some minor mineral phases which are inert to hydrochloric acid, such as chromite, α-quartz, and serpentine.

X-ray diffraction studies were performed on the magnetic, non-magnetic, hydrochloric acid treated residues, sieved fractions, and also on hand-picked isolated particles. Both diffractometer and Debye-Scherrer methods were used wherever applicable. A Philips X-ray diffractometer with a graphite monochromator was used to obtain the patterns. Copper radiation was preferred because higher X-ray intensities were obtained at higher power rating than with chromium radiation. Standard compaction techniques for the diffractometer samples and quartz capillaries (0.1 mm) were used for Debye-Scherrer techniques. The samples in reduced condition posed an oxidation problem as they were pyrophoric. Thus all the reduced samples required special handling to prevent oxidation in ambient air. Standard aluminum diffractometer sample holders were preferred because of the difficulty of handling glass capillary material in the glove box. The aluminum sample holders were polished on both sides and a thin mylar sheet cover was used to prevent oxidation. A thin layer of silicon grease was applied to the holder. Trapped air was squeezed out from between the holder and the mylar sheet and excess grease removed. A parlodion coating was used to seal the edges of the mylar sheet, as shown in Figure 2. A parlodion seal was also used on the reverse side of the mylar sheet as shown by the arrows in the cross section. The mylar sheets were temporarily backed by glass slides and the sample holders were then introduced in the glove box with argon atmosphere. The reduced materials were inserted in the holders and compacted; the holders were then sealed with a warm sealing wax.

Petrography was performed on polished sections by embedding particles of the raw or reduced laterites in an epoxy. Several hundred particles were examined and color slides were obtained by using a Carl Zeiss optical microscope. The minerals and the metallics were identified by their characteristic reflected color, texture, and their transparency or opacity.

A scanning electron microscope (AMR-900) equipped with a Kevex

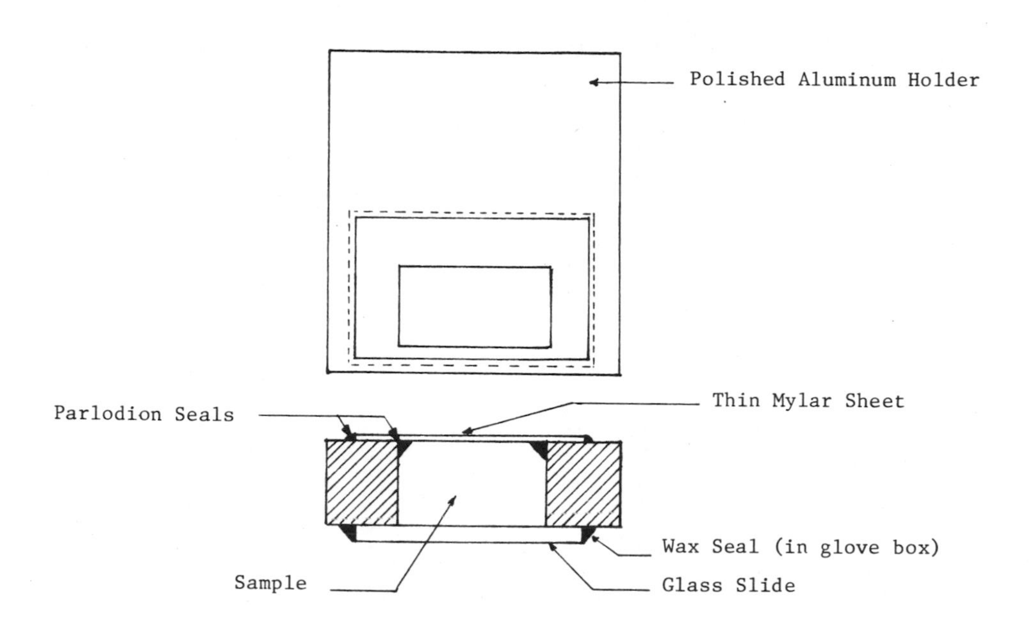

Figure 2. Sample Holder for Reduced Material

X-ray energy dispersive analyzer was used for this study. An electron microprobe (Norelco AMR-3) equipped with a wavelength dispersive vacuum spectrometer was used to detect low concentrations of elements and to yield semiquantitative elemental analyses.

STUDIES ON NICKELIFEROUS LATERITES

These laterites were generally yellowish brown in color, generally high in iron (30 to 45%), low in nickel (0.4% to 1.2%), cobalt (0.02% to 0.2%) and chromium (0.5% to 3.0%). X-ray diffraction analyses of some of the laterites examined are listed in Table 1. The analyses, however, were not direct but were arrived at by using concentration techniques described above. The domestic laterites from Oregon and California were generally higher in magnesium and silicon content than those of the New Caledonian or Philippine laterites, indicating more silicate can be expected to be present in the domestic ores.

X-ray diffraction data obtained from the bulk samples were not very encouraging because only goethite and quartz could be identified, while the other minerals were not. This was because of their lower concentration and because of diffraction peak overlaps. For example, the strongest hematite reflection occurring at

Table 1. X-Ray Diffraction Analyses of the Lateritic Ores

	Oregon Laterite	California Laterite	New Caledonian Laterite	Philippine Laterite	Structure	Lattice Parameters
Goethite [α-FeO(OH)]	High	High	High	High	Orthorhombic	a= 4.596Å b= 9.957Å c= 3.012Å
Hematite [α-Fe_2O_3]	Low	Low	Low	Medium	Rhombohedral	a= 5.4228Å α=55.17Å
Quartz [α-SiO_2]	High	Medium	--	Low	Hexagonal	a= 4.913Å c= 5.405Å
Magnetite [Fe_3O_4]	Low	Trace	Trace	Trace	Face-Centered Cubic	a= 8.396Å
Chromite [FeO($CrAl_2O_3$)]	Trace	Trace	Trace	Trace	Face-Centered Cubic	a= 8.30Å
Maghemite [γ-Fe_2O_3]	--	--	--	Low	Face-Centered Cubic	a= 8.350Å
Serpentine (Antigorite) [$Mg_3Si_2O_5(OH)_4$]	--	Low	--	Trace	Monoclinic	a=43.5Å b= 9.25Å c= 7.26Å β=91°23'
Chlorite [$Mg_{2.6}Fe_2AlSi_{2.8}Al_{1.2}O_{10}(OH)_8$]	Low	--	--	--	Monoclinic	a= 5.30Å b= 9.20Å c=14.30Å β=97°

Table 1 (Continued)

	Oregon Laterite	California Laterite	New Caledonian Laterite	Philippine Laterite	Structure	Lattice Parameters
Tremolite $[Ca_2Mg_5Si_8O_{22}]$	Trace	--	--	--	Monoclinic	a= 9.84Å b=18.02Å c= 5.27Å β=104°57'
Talc $[Mg_3Si_4O_{10}(OH)_2]$	--	--	Low	--	Monoclinic	a= 5.287Å b= 9.171Å c=18.964Å β=99.61

Q = Quantity based on intensity. 'High' indicates the intensity of the X-ray diffraction peak was over 80, 'Medium' indicates between 55 and 45, 'Low' indicates below 40, and 'Trace' indicates below 5.

2.69Å (I/I_o = 100) overlaps with a goethite reflection (I/I_o = 30). The diffraction peaks for chromite, magnetite and maghemite (all ferromagnetic minerals) are also overlapped because of their similarity of structure and lattice parameters (Figure 3 and Table 1).

In order to identify the small amounts of silicate minerals present, the sieved (+50 mesh) product was washed and reground; X-ray diffraction analyses were then carried out. Coarse, greenish, fibrous serpentines were observed and later identified as antigorite serpentine. With the -50 mesh products subjected to magnetic separation, the X-ray diffraction charts of the magnetic concentrate matched chromite, but also matched magnetite and maghemite patterns, thus there was uncertainty in identification by this means.

In order to positively identify the minerals, especially the nickel- and chromium-bearing minerals, microscopic techniques were used to supplement the X-ray diffraction data. Optical microscopy on the polished sections revealed that the mineral goethite (identified by X-ray diffraction techniques (Table 1) was always found to be cryptocrystalline and identifiable by its characteristic yellowish internal reflections. Hematite was found to be present in both coarse-grained and cryptocrystalline varieties. The coarse-grained polished sections exhibit dull white reflections and the cryptocrystalline variety exhibits blood-red internal reflections. The mineral chromite exhibits dark gray to purple reflections and magnetite exhibits brownish reflections. The above characteristics were used for identification, but are not presented in detail in this report.

A typical example of a cluster of particles found in a laterite is shown in the scanning electron micrograph in Figure 3. The particles or clusters of cryptocrystalline particles labeled A, B, C and D have been optically identified as magnetite, chromite, cryptocrystalline goethite and hematite minerals. X-ray diffraction charts of some of these synthetic materials are reproduced on the left hand side of Figure 3 to show how peak overlaps can occur. The X-ray energy dispersive elemental spot analysis on each particle supports the mineral identification. It has been established (4,5) that there are no discrete nickel oxides in the nickeliferous laterites; the majority of the nickel was found to be associated with the mineral goethite. The X-ray energy dispersive spectra of goethite particles show that a small amount of nickel, aluminum and silicon are present as impurities. A typical elemental electron microprobe analysis of goethite minerals showed Fe = 56.3 %, Ni = 0.9%, Al = 2.0 %, Si = 2.8% and oxygen = 38.0%. The iron content in the mineral is much lower than the theoretical value of 62.8%. If one assumes that nickel, aluminum and silicon are present in the lattice of goethite mineral, then the sum of the weight fractions of Fe, Ni,

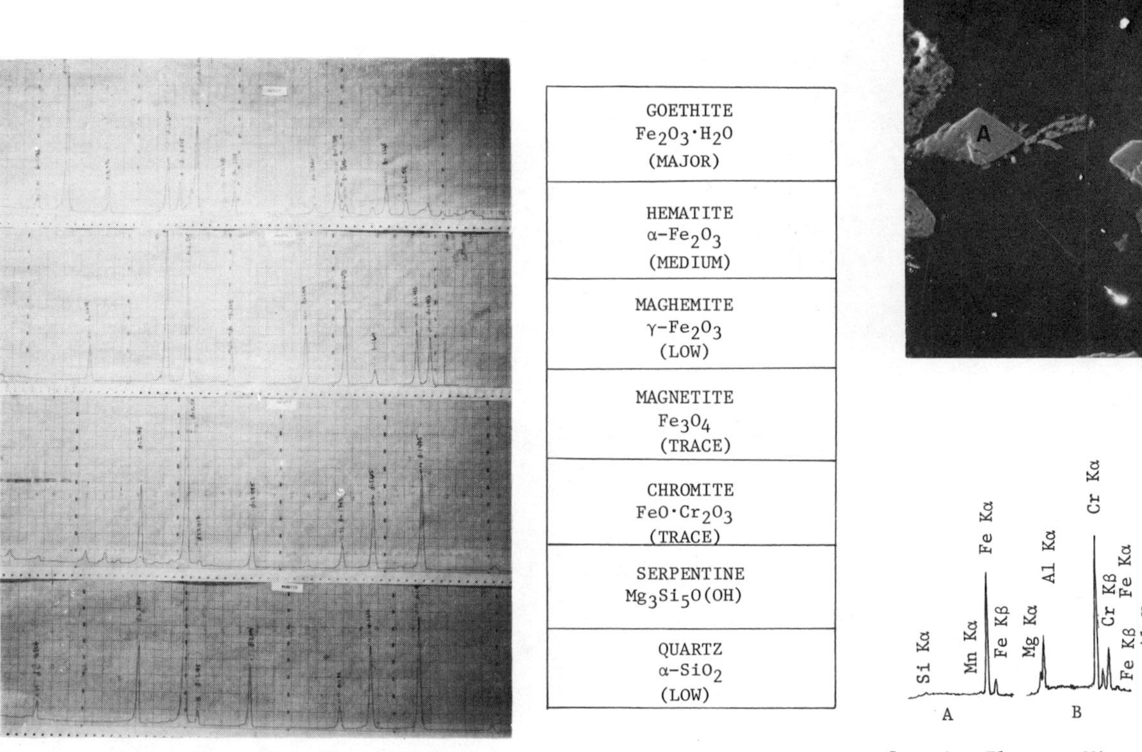

X-Ray Diffraction Charts of Synthetic Compounds which have Diffraction Peak Overlaps, Commonly Found in Laterites

GOETHITE $Fe_2O_3 \cdot H_2O$ (MAJOR)
HEMATITE α-Fe_2O_3 (MEDIUM)
MAGHEMITE γ-Fe_2O_3 (LOW)
MAGNETITE Fe_3O_4 (TRACE)
CHROMITE $FeO \cdot Cr_2O_3$ (TRACE)
SERPENTINE $Mg_3Si_5O(OH)$
QUARTZ α-SiO_2 (LOW)

Composition of a Typical Nickeliferous Laterite

Scanning Electron Micrograph of Optically Identified Magnetite (A), Chromite (B), Cryptocrystalline Goethite (C) and Hematite (D) Particles Found in a Typical Laterite (Top) X-Ray Energy Dispersive Spot Analyses on the Particles Labeled A, B, C and D are Shown in the X-Ray Spectra a, b, c and d.

Figure 3. X-Ray Identification of Minerals in the Scanning Electron Micrograph

Al and Si should agree with the theoretical cation content of goethite. Since in a typical particle Fe, Ni, Al and Si summed amount to 62.0%, this assumption seems to be reasonable.

By combining the X-ray diffraction, optical and scanning electron microscopic and electron microbeam analysis results, it was possible to identify all minerals and determine the elemental associations. Based on studies of hundreds of particles, it was found that the nickel was mainly incorporated in the lattice of the mineral goethite. Chromium was found to be mainly associated with the mineral chromite, but was found occasionally also to be associated with the mineral goethite.

STUDIES ON REDUCED NICKELIFEROUS LATERITES

The characterization of the reduced laterites was performed in a similar manner to that of the raw laterites; however, precautions were taken to prevent oxidation of the particles. The minerals goethite, hematite and maghemite are reduced to magnetite, metallic iron and ferronickel alloys, with magnetite being the major transformation product. These identifications were confirmed by X-ray diffraction analysis. In the reduced laterites, nickel is associated with either metallic iron (4,5,6) or complex iron silicate compounds. Several particles were examined and showed that the ferronickel alloys formed during the reduction are either found as free discrete particles or sometimes encapsulated in a sheet of silicate matrix. It was observed from prior optical examination that ferronickel alloys exhibit bright white reflections, while oxides of iron such as magnetite show a dull brown color (7). However, this test was not conclusive for identification of metallic or alloy phases. A scanning electron micrograph of a typical reduced laterite particle is shown in Figure 4. By relating the X-ray energies to the X-ray diffraction results, the bright particles labeled A and B were suspected of being ferronickel alloy particles with encapsulating sheet C being a complex iron-silicate compound. To determine whether these particles were ferronickel alloys or oxides, an EMP quantitative spot analysis was performed on the particle labeled B, which showed 61.09% Ni and 38.1% Fe. Although the weight percents do not add up to 100%, it is within the expected variance and established the identification as ferronickel alloy. A typical composition of a complex silicate is as follows: Fe = 41.9%, Ni = 0.9%, Si = 1.1%, Cr = 0.3%, Al = 2.8%, Co = 1.4%, Mn = 2.4% and O = 49.2%.

Based on a study of several particles under different conditions (5) it can be concluded that the nickel was in the form of ferronickel containing up to 70% nickel and in a matrix of a refractory complex iron silicate of slightly varying compositions.

PHASE CHANGES IN THE REDUCED LATERITES

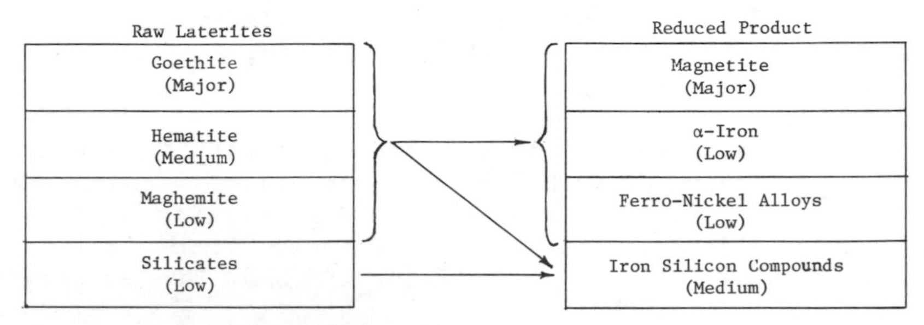

Iron-Silicate (C)
Ferro-Nickel Alloys (A and B)

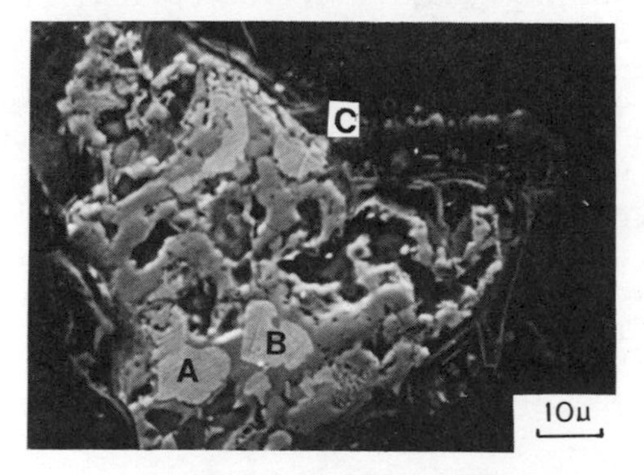

Scanning electron micrograph showing metallic, oxides, and silicate phases

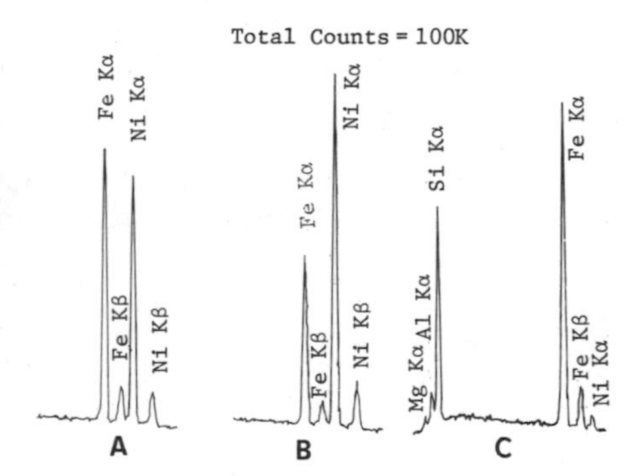

X-ray energy dispersive spot analysis on the particles A, B and C

Figure 4

In conclusion, the authors wish to emphasize that a combination of X-ray and optical techniques is necessary to characterize laterites.

REFERENCES

1. *Minerals Facts and Problems*, U.S. Dept. of Interior, Bureau of Mines, 1970.

2. H. E. Zessink, "The Mineralogy and Geochemistry of a Nickeliferous Laterite Profile (Greenvale, Queensland, Australia)," Mineral Deposita (Berl) 4, 132-152 (1969).

3. R. E. Siemens, "Process for Recovery of Nickel from Domestic Laterites," paper presented at the 1976 Mining Convention (1976).

4. D. Chandra, "Characterization of Lateritic Nickel Ore by Electron-Optical and X-Ray Techniques," Ph.D. Thesis, University of Denver, Denver, Colorado.

5. D. Chandra, R. E. Siemens, and C. O. Ruud, "Electron-Optical Characterization of Laterites Treated with a Reduction-Roast Ammoniacal Leach System," paper presented at the AIME Annual Meeting in Denver, Colorado, 1977.

6. A. Gosh and G. R. St. Pierre, "Isobaric Ternary Phase Diagrams in the Iron-Carbon Oxygen System," Indian Journal of Technology, Vol. 18, 1970.

7. B. Uyten and E. A. J. Burke, "Tables for Microscopic Identification of Ore Minerals," Elsevier Publ., Amsterdam, 1971.

MINIMIZATION OF PREFERRED ORIENTATION IN POWDERS BY SPRAY DRYING

Steven T. Smith, Robert L. Snyder and W. E. Brownell

NYS College of Ceramics at Alfred University

Alfred, NY 14802

ABSTRACT

The spray drying of powders is a method of forming spherical or torroidal shaped agglomerates. A relatively simple method is given for preparing spray dried samples of the quantities used in x-ray diffraction analysis. This technique is shown to minimize preferred orientation effects on diffraction intensities from materials of widely differing symmetry and crystallite habit.

INTRODUCTION

The use of x-ray powder diffraction for quantitative and, to a lesser extent, qualitative analysis has been severely limited by the quality of the diffraction intensities. If experimental intensities were reliable to 10% or better, ambiguities in qualitative multiphase analyses would be greatly reduced. This would allow the development of a second generation of computer search algorithms which could routinely produce unique phase identifications in complex mixtures. Quantitative analysis of samples with reliable intensities could become as easy as a simple comparison of the $I/I_{corundum}$ ratios (1-3). Use of the more conventional methods for quantitative analysis could be extended routinely into such complex systems as the clay minerals.

Diffraction intensities are affected by both instrumental and sample difficulties (4). The principal effects arise from particle orientation, microabsorption, crystallinity and solid solution, with preferred orientation usually being by far the most important effect. Orientation effects have been discussed in nearly every paper on quantitative analysis. Orientation errors of at least a few percent have been observed in every pattern tested (including a number of the carefully prepared NBS patterns) with the orientation analysis program PREF (5).

Spray drying (6) is a technique which has received some attention as a means of reducing preferred orientation. Spray drying is an industrial technique used for the drying of, among other things, milk, soaps and detergents, chemicals and clay slips. It has several advantages. Particle size may be varied. Generally, hollow spheres are obtained which have superior flow properties during subsequent materials-handling procedures. There are several other industrial advantages along with some disadvantages. One of the disadvantages is that 100% product recovery may be a problem.

Fundamentally, spray drying is the atomization of a slurry containing the solid of interest into a high temperature gas zone which allows for rapid evaporation of the liquid portion, leaving spherically shaped agglomerates (i.e., shaped like the droplets before evaporation). Atomization is essential to spray drying; it creates a high surface to mass ratio in the liquid phase allowing drying to take place in air. For our purpose, which was to create tiny spheres, an organic binder is dissolved in the suspension medium. This binder holds the particles together in a spherical shape after drying which allows packing in an x-ray sample holder.

Three previous studies have used spray drying in x-ray sample preparations. Florke and Saalfeld (7) spray dried materials using an organic binder and benzene as a volatile solvent. The volatile solvent obviated the need for a heat source. They obtained spherically shaped agglomerates of about 50μm diameter for various platy materials which they said showed no preferred orientation. Jonas and Kuykendall (8) reported producing randomly oriented montmorillonite powders by spray drying and collecting the dust in an electrostatic precipitator. Hughes and Bohor (9) spray dried clay minerals including illite, kaolinite and attapulgite. Within the accuracy of their x-ray techniques and test methods they concluded that their minerals had been randomized.

These studies each showed that preferred orientation had been greatly reduced by spray drying, but did not quantitatively evaluate the extent of randomization. Further, these studies were applied only to platy materials leaving possible doubts about the generality of the technique. This study was undertaken to quantitatively assess the effect of spray drying on minimizing preferred orientation in materials of different crystallite shapes.

EXPERIMENTAL PROCEDURE

The Spray Dryer

A diagram of the spray drying apparatus is shown in Fig. 1. The drying cylinder is constructed from thin (1 mm) aluminum sheets. A Kanthol heating element, supported by transite standoffs, is placed in the lower half of the cylinder and directly connected to a 120 VAC outlet. It draws about 1500 watts. The spraying apparatus is supported 2.5 cm above the drying chamber so that the spray enters through a 2 cm hole in the top cover. The atomizer consists of a funnel for the slurry, connected to a series of telescoping brass tubes, soldered together, ending in a 0.8 mm orifice. Above this nozzle is another brass tube conducting compressed air at 34.5 k Pa. As the suspension flows by gravity through the small orifice, the air stream from above breaks the slurry into a fine spray. The size of the droplets can be controlled by varying the air pressure or the viscosity of the slurry.

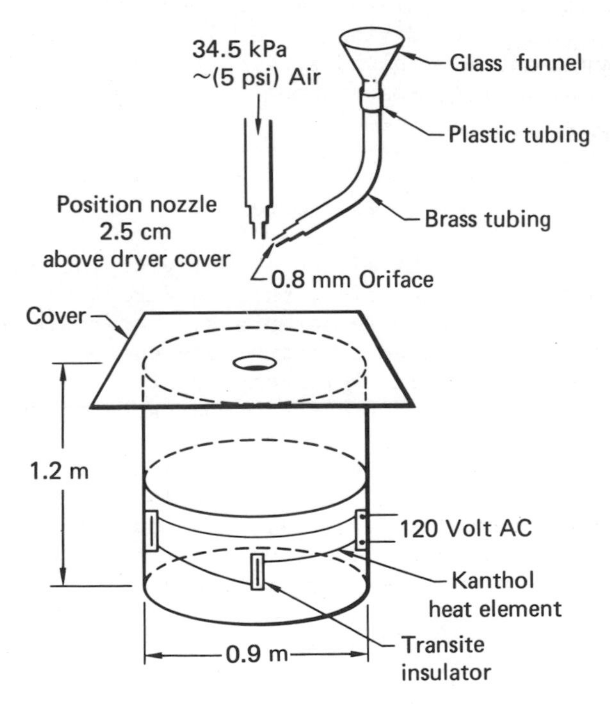

Figure 1. Schematic Diagram of Spray Dryer

The spray is injected into the drying chamber which is held at about 130°C when water is used as the liquid vehicle. The suspension medium evaporates before the droplets strike the bottom of the dryer. The organic binder will also dry during this interval leaving hard spherical particles, typically in the 40-50μm range, on the bottom plate of the dryer. The sample is collected by lifting the dryer off the bottom aluminum sheet, slightly curling this sheet and pouring the dry particles into a beaker. Only those particles which roll off unaided are used. Any particles which have not fully dried before hitting the bottom will be splattered flat and will not roll off with the rest of the sample.

Materials are prepared for spray drying by grinding and then either Stokes Law settling or sieving. A particle size between one and ten μm is desired. Three to four grams of this fine powder are suspended in 6 to 7 ml of a 1% aqueous solution of Polyvinal Alcohol (PVA). In this study water was chosen as the suspension medium and the water soluble PVA was chosen as the binder. In general the suspension medium must not react with or dissolve the material to be spray dried, must volatilize rapidly at the dryer temperature and must be x-ray amorphous. The volume of 1% PVA solution used was adjusted to control the viscosity of the slurry. Thick suspensions will clog the nozzle and too dilute ones will not dry before striking the bottom. The lack of drying in the larger droplets is the principal source of material loss which typically ran about 50%. The spray drying process takes about one minute for two to four gram samples.

Procedure

In order to assess the effect of spray drying on the removal of preferred orientation, three diffraction patterns were obtained for each of five materials. Materials of different crystallite habit and symmetry were chosen. The experimental diffraction patterns were compared to the "ideal" intensities of the calculated diffraction patterns obtained using a local version of D. K. Smith's program (10). In addition the Powder Diffraction File (PDF) (11) pattern was compared with the calculated pattern. Experimental patterns were obtained using a computer controlled vertical Phillips diffractometer with a diffracted beam graphite monochromator using CuKα radiation. Scanning electron micrographs were obtained using an ETEC Autoscan.

Materials

I. Muscovite ($KA\ell_2(Si_3A\ell O_{10})(OH)_2$): A sample from Effingham, Ontario was ground first with a Spex Mill impact grinder, then with a Polytron shearing device. These techniques were used in an attempt to avoid the well known problems of grinding mica (12). The final powder was separated using a 20 μm electrodeposited sieve. PDF pattern 6-0623 was used for a non-spray dried comparison. The ideal pattern was calculated using the parameters of Gatineau (13). This material (monoclinic: $C_{2/c}$) is extremely platy and shows {001} cleavage.

II. Wollastonite ($CaSiO_3$): A grade C-1 sample from Interpace Corp. (P.O. 4293, N.Y., NY, 10249) was ground and the fines separated by aqueous settling. PDF pattern 27-1064 was used for a non-spray dried comparison. The parameters used for the calculated pattern were obtained from Buerger and Prewitt (14). This material (triclinic: $P\bar{1}$) has a rod like acicular habit and shows {001} and {100} cleavages.

III. Iron Pyrite (FeS_2): A sample from Rico, Colorado was ground to pass a 325 mesh sieve. PDF pattern 6-710 was used for a non-spray dried comparison. The parameters for the calculated pattern were taken from Parker and Whitehouse (15). This material (cubic: P_{a3}) displays a pronounced cubic habit.

IV. Siderite ($FeCO_3$): A sample from Roxbury, Connecticut was ground to less than 325 mesh. Since a diffractometer pattern has not been published, a conventional pattern was determined to compare with that of the spray dried. The calculated pattern was determined from the parameters of Graf (16). This material($R\bar{3}c$) shows a rhombohedral habit and cleavage.

V. Corrundum ($A\ell_2O_3$): A sample was obtained from Dr. C. Hubbard, of the National Bureau of Standards, who is investigating its use as an intensity standard (4). This finely divided material was spray dried without treatment. The parameters used for the calculated pattern are from Newnham and DeHann (17). The non-spray dried pattern was taken from the PDF (10-173). This material (hexagonal: $R\bar{3}c$) is well known not to show preferred orientation.

Results and Discussion

The spray drying process yielded rounded agglomerates for each of the materials tested. Figures 2 and 3 show Wollastonite before and after spray drying. These scanning electron

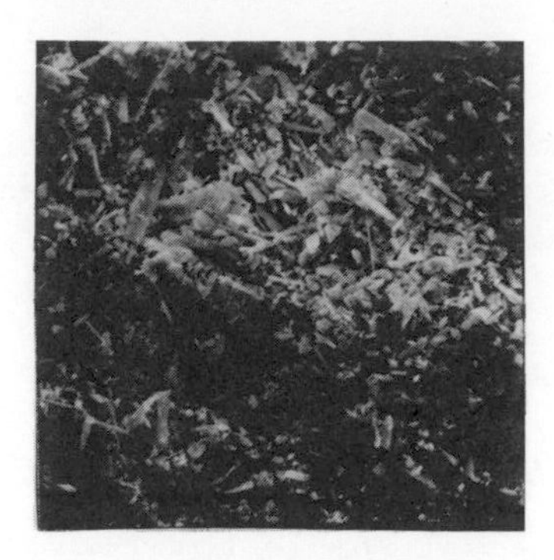

Fig. 2: Wallostonite before spray drying (800x)

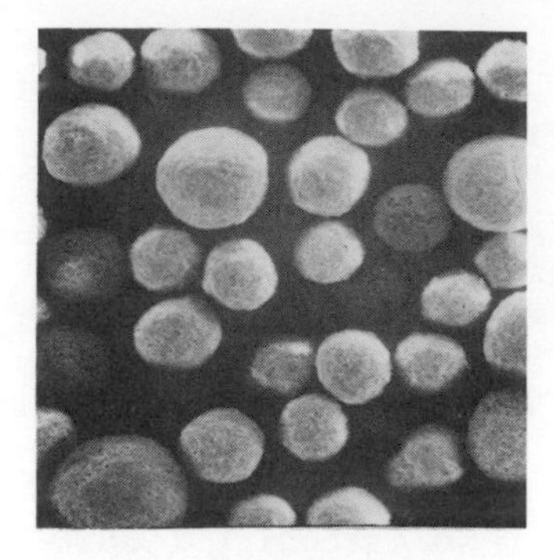

Fig. 3: Spray dried wallostonite (200x)

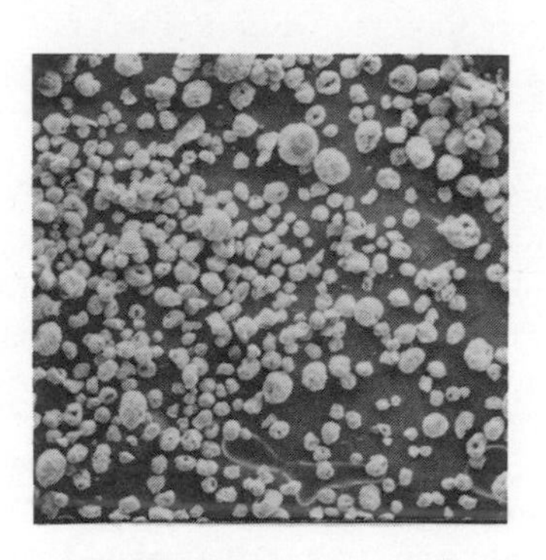

Fig. 4: Spray dried pyrite (80x)

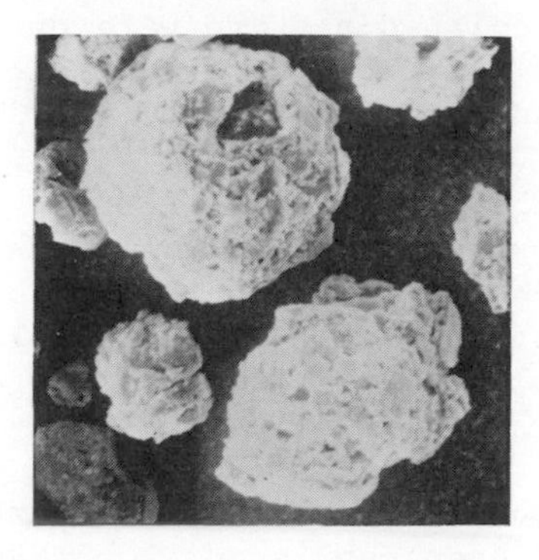

Fig. 5: Spray dried siderite (600x)

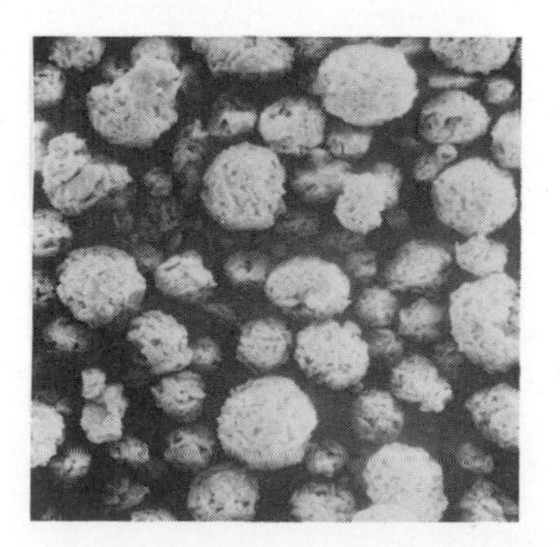

Fig. 6: Spray dried Alumina (80x)

Fig. 7: Fractured agglomerate of spray dried muscovite (800x)

micrographs dramatically show the sphere creating effect of spray drying even on rod shaped crystallites. Figure 4 shows spray dried pyrite. A number of torroidal agglomerates may be seen. The presence of torroids in addition to spheres is a common observation in spray dried powders. Figure 5 shows the spray dried siderite sample. The crystallite size in this powder was larger than in the other materials, and the large crystallites can be seen distorting the shape of the agglomerates. Figure 6 shows the spray dried alumina sample. Here as in figure 4 a rather wide variation in agglomerate size can be seen. Figure 7 shows a close up of a fractured spray dried sphere of muscovite. This shows the microstructure of the agglomerates to be hollow spheres with a high degree of particle orientation randomization.

Tables 1 through 5 list the powder patterns of non-spray dried powders, calculated "ideal" values and the spray dried samples. In each case the agreement between the spray dried and calculated intensities is remarkable.

The well known basal reflection distortion of muscovite has been eliminated. Even the intensity distortions due to the rod shaped crystallites in non-spray dried Wollastonite have been eliminated. The less prominent preferred orientation effects in cubic pyrite and rhombohedral siderite have been reduced. Alumina which shows concoidal fracture yields no significant improvement in the intensities on spray drying (Table 5). This supports its selection as an intensity standard and shows that spray drying does not adversely affect the intensities.

At the bottom of each Table a quantitative agreement index is given. This index called R is defined as

$$R = \frac{\Sigma \; |I - I_c| \; I_c}{\Sigma I_c^2}$$

where I is the intensity from the non-spray dried or spray dried patterns and I_c is the calculated intensity. This function is a weighted relative error. The weighing of the relative error with I_c reduces the large effect of variations in weak reflections. The R values are under 4% for the essentially random alumina patterns. For each of the other materials, with the exception of muscovite, the R values of the spray dried samples are about 8%. As these samples are natural minerals with solid solution effects and since the calculation of the "ideal" patterns involved assumptions for both the atomic scattering

Table 1. MUSCOVITE
$K\ Al_2(Si_3Al)O_{10}(OH)_2$

d(Å)	H	K	ℓ	NSD	I/I. CAL	SD
9.974	0	0	2	95	85	64
4.987	0	0	4	30	19	20
4.458	-1	1	1	20	90	91
4.404	0	2	1	0	8	23
4.294	1	1	1	4	19	19
4.113	0	2	2	4	15	16
3.961	1	1	2	6	8	10
3.877	-1	1	3	14	47	43
3.735	0	2	3	18	48	34
3.575	1	1	3	0	3	1
3.489	-1	1	4	20	58	57
3.347	0	2	4	25	43	59
3.325	0	0	6	100	50	73
3.198	1	1	4	30	55	49
3.118	-1	1	5	2	6	7
2.990	0	2	5	35	60	46
2.859	1	1	5	25	40	36
2.789	-1	1	6	20	32	24
2.677	0	2	6	16	41	44
2.565	1	1	6	55	100	100
2.506	-1	1	7	8	6	7
2.494	0	0	8	14	5	9
2.486	1	3	2	0	4	0
2.465	-1	3	3	8	23	19
2.441	2	0	2	8	11	10
2.410	0	2	7	0	3	0
2.393	-2	0	4	10	9	23
2.381	1	3	3	25	22	31
2.265	-1	1	8	0	1	0
2.257	0	4	0	10	2	0
2.249	-2	2	1	0	9	14
2.243	0	4	1	4	10	11
2.231	-1	3	5	0	5	7
2.206	2	2	1	8	9	11

R For Non-Spray Dried Pattern
= .5177
R For Spray Dried Pattern
= .1385

Table 2. WOLLASTONITE
$CaSiO_3$

d(Å)	H	K	ℓ	NSD	I/I. CAL	SD
7.687	1	0	0	25	13	9
7.037	0	0	1	0	1	0
5.459	-1	0	1	8	1	0
4.958	1	0	1	0	4	3
4.704	1	1	0	0	2	0
4.679	-1	1	1	0	3	0
4.429	1	-1	1	0	2	0
4.052	-1	-1	1	0	8	5
3.843	2	0	0	85	22	19
3.770	-2	1	0	0	5	0
3.519	-2	0	1	75	29	29
3.443	-2	1	1	0	7	5
3.323	-1	0	2	100	35	27
3.244	2	0	1	13	2	0
3.206	-1	2	1	0	3	0
3.183	0	-1	2	0	6	3
3.107	-1	1	2	0	11	0
2.981	-2	2	0	30	100	100
2.928	-2	-1	1	0	8	5
2.801	-2	2	1	5	4	5
2.745	2	1	1	0	1	0
2.730	-2	0	2	0	6	3
2.722	1	1	2	30	4	5
2.687	1	2	1	0	2	0
2.613	-3	1	0	0	1	0
2.531	0	-2	2	0	3	5
2.477	1	-2	2	25	14	16
2.381	3	-1	1	0	1	0
2.356	-3	2	0	0	7	0
2.352	2	2	0	0	9	0
2.346	0	0	3	0	9	13
2.339	-2	2	2	30	7	0
2.306	-1	0	3	50	16	12

R For Non-Spray Dried Pattern
= 1.0017
R For Spray Dried Pattern
= .0737

NSD = Intensities from this normal non spray dried pattern
CAL = Intensities from the calculated pattern
SD = Intensities from the spray dried pattern

Table 3. IRON PYRITE FeS_2

d(Å)	H	K	ℓ	I/I. NSD	I/I. CAL	I/I. SD
3.1280	1	1	1	35	37	43
2.7080	2	0	0	85	100	100
2.4230	2	1	0	65	55	59
2.2110	2	1	1	50	42	49
1.9152	2	2	0	40	42	53
1.8057	2	2	1	0	1	0
1.6333	3	1	1	100	76	67
1.5638	2	2	2	14	12	10
1.5024	0	2	3	20	11	15
1.4478	3	2	1	25	14	21
1.2427	3	3	1	12	6	8

R For Non-Spray Dried Pattern = .1935
R For Spray Dried Pattern = .0852

Table 4. SIDERITE $FeCO_3$

d(Å)	H	K	L	I/I. NSD	I/I. CAL	I/I. SD
3.5900	0	1	2	3	52	34
2.7910	1	0	4	100	100	100
2.3440	1	1	0	3	17	24
2.1320	1	1	3	4	16	17
1.9627	2	0	2	2	21	19
1.7950	0	2	4	2	9	8
1.7371	0	1	8	6	17	19
1.7296	1	1	6	1	27	27
1.5269	2	1	1	1	2	0
1.5048	1	2	2	6	11	9
1.4378	1	0	10	1	2	2
1.4251	2	1	4	4	7	8
1.3956	2	0	8	0	3	4
1.3806	1	1	9	0	2	2
1.3730	1	2	5	0	1	2
1.3533	3	0	0	0	7	10

R For Non-Spray Dried Pattern = .2975
R For Spray Dried Pattern = .0808

Table 5. ALUMINA Al_2O_3

d(Å)	H	K	ℓ	Relative Intensities: Non-Spray Dried	Calc.	Spray Dried
3.4800	0	1	2	75	74	69
2.5510	1	0	4	90	94	96
2.3790	1	1	0	40	42	38
2.1650	0	0	6	0	1	0
2.0850	1	1	3	100	100	100
1.9642	2	0	2	2	1	0
1.7400	0	2	4	45	42	44
1.6014	1	1	6	80	84	80
1.5466	2	1	1	4	2	2
1.5148	1	2	2	6	3	6
1.5108	0	1	8	8	7	7
1.4045	2	1	4	30	29	31
1.3738	3	0	0	50	42	45
1.3360	1	2	5	2	1	0
1.2754	2	0	8	4	1	0
1.2390	1	0	10	16	11	12

R For Non-Spray Dried Pattern = .0383
R For Spray Dried Pattern = .0359

factors and thermal parameters, these values for R indicate excellent agreement. This becomes more apparent when the R values for the non-spray dried samples are examined. They range from 20 to over 100%.

The spray dried muscovite pattern shown in Table 1 has the highest R value (14%). While this is considerably better than the 52% of the non-spray dried pattern, it appears to indicate some degree of orientation. However, an examination of the Miller indices shows no consistent orientation effect. It is more probable that the increased R value is due to structure distortion during the grinding and shearing operations (12).

The largest differences between the spray dried and calculated intensities for muscovite and wollastonite occur at low diffraction angles. The spray dried patterns were obtained using a 1° divergence slit over the entire 2θ range. Because a narrower divergent slit at low angles was not used some intensity distortions will occur (18). This effect appears to be present in our patterns.

On considering all the sources of error in both the experimental and calculated patterns we conclude that the R values indicate that preferred orientation effects have been minimized and possibly eliminated.

Conclusions

The spray drying of powders, of varying symmetry and crystallite habit is an effective means of reducing preferred orientation. Orientation effects for materials with rod, plate, cube and rhombohedral habits were reduced to the levels of error in the calculated patterns used for comparison. Spray drying of gram size samples can be carried out in a relatively simple apparatus in a few minutes.

Acknowledgments

This work was supported in part by U.S. Energy Research and Development Administration Contract Nos. EY-76-C-05-5207 and W-7405-ENG-48.

REFERENCES

1. J. W. Visser and P. M. De Wolff, "Absolute Intensities," Report No. 641.109 Technisch Physische Dienst. Delft, Netherlands (1964).
2. F. H. Chung, "Quantitative Interpretation of X-ray Diffraction Patterns. III. Simultaneous Determination of a Set of Reference Intensities," J. Appl. Cryst. 8, 17 (1975).

3. C. R. Hubbard, E. H. Evans and D. K. Smith, "The Reference Intensity Ratio, I/I_c, for Computer Simulated Powder Patterns," J. Appl. Cryst. 9, 169-174 (1976).
4. C. R. Hubbard and D. K. Smith, "Experimental and Calculated Standards for Quantitative Analysis by Powder Diffraction", Adv. in X-ray Anal. 20, 27-39 (1976).
5. R. L. Snyder and W. Carr, "A Method for Determining the Preferred Orientation of Crystallites Normal to a Surface," in Surfaces and Interfaces of Glass and Ceramics, V. D. Frechette, et. al., ed., Plenum Press, New York, 1974, p. 85.
6. W. R. Marshall Jr., "Atomization and Spray Drying," Chemical Engineering Progress Monograph Series #2, American Institute of Chemical Engineers, Pub., Vol 50 (1954).
7. O. W. Florke and H. Saalfeld, "Ein Verfahren Zur Herstellung Texturfreier Rontgen-Pulverpraparate," Z. Krist. 106, 460 (1955).
8. E. J. Jonas and J. R. Kuykendall, "Preparation of Montmorillonites for Random Powder Diffraction," Clay Miner. Bull. 6, 232-5 (1966).
9. R. Hughes and B. Bohor, "Random Clay Powders Prepared by Spray Drying," Amer. Miner. 55, 1780-1786 (1970).
10. C. N. Clark, D. K. Smith and G. G. Johnson Jr., "A Fortran IV Program for Calculating X-ray Powder Diffraction Patterns-Version 5," Department of Geosciences, Pennsylvania State Univ., University Parks, PA, 16802.
11. JCPDS, International Centre for Diffraction Data, 1601 Park Lake, Swarthmore, PA.
12. R. C. Mackenzie and A. A. Milne, "The Effect of Grinding on Micas: I. Muscovite," Min. Mag. 30 222 , 178-185 (1953).
13. L. Gatineau, "Localisation des Remplacements Isomophiques dans la Muscovite," Compt. Rend. 256 22 , 4648-9 (1963).
14. M. J. Buerger and C. T. Prewitt, "The Crystal Structures of Wollastonite and Pectolite," Proc. Natl. Acad. Sci., U.S. 47, 1884-8 (1961).
15. H. M. Parker and W. J. Whitehouse, "X-ray Analysis of Iron Pyrites by the Method of Fourier Series," Phil. Mag. 14, 939-961 (1932).
16. D. L. Graf, "Crystallographic Tables for the Rhombohedral Carbonates," Am. Miner. 46, 1283-1316 (1961).
17. R. E. Newnham and Y. M. DeHaan, "Refinement of the αAl_2O_3, Ti_2O_3, V_2O_3, Cr_2O_3 Structures," Z. Krist. 117, 235 (1962).
18. W. Parrish, Advances in X-ray Diffractometry of Clay Minerals, Proc. 7th Natl. Conf. on Clays and Clay Minerals, A. Swineford ed., Pergamon Press, New York, 1960.

LOW TEMPERATURE X-RAY DIFFRACTOMETER WITH CLOSED CYCLE REFRIGERATION SYSTEM

U. Benedict, Y. Cornay, C. Dufour

Commission of the European Communities, Joint Research Centre, European Institute for Transuranium Elements, P.O. Box 22 66, D-7500 Karlsruhe 1

ABSTRACT

An assembly consisting of an X-ray tube, a quartz crystal monochromator, and a vertical X-ray goniometer with proportional counter was mounted in a glove box for work with actinide metals and compounds. An evacuated camera containing the expander tube of a closed-cycle cryogenic system was fitted onto the goniometer. A thin layer of the powdered sample was fixed on the end-plate of the expander tube. Beryllium windows were provided in the camera wall for the incident and for the diffracted beam.

The cooling camera is suitable for use with thin powder layers or metallic foils. Thicker layers can be studied, but a strong thermal gradient perpendicular to the specimen surface is expected. The camera has been applied to the study of lattice contraction and phase transformation at low temperature in actinide metals and compounds.

INTRODUCTION

Low temperature X-ray diffraction work is required in the Actinide Research Project of the European Institute for Transuranium Elements (EITU) for several purposes.

First, low temperature measurements of properties such as magnetic susceptibility, electrical resistivity, thermal expansion, and specific heat often show anomalies at temperatures between absolute zero and 200 K. It is important to know whether these are related to changes in crystallographic structure. Our first aim is thus to look for low temperature phase transformations.

Second, we want to know the coefficients of thermal expansion which are useful information for other work in actinide research.

A third application will be the study of self-irradiation damage at low temperatures which requires a constant cryogenic temperature for a period of at least several weeks.

DESCRIPTION OF THE LOW TEMPERATURE CAMERA

As an initial step we set up a low temperature camera allowing studies in the range 50 to 350 K. Useful information can be obtained in this temperature range, especially in the case of compounds, for which anomalies and transformations generally occur at higher temperatures than with the actinide metals. Extension of the experiments to lower temperatures is planned for the future.

We decided to mount a closed-cycle helium cooling device on an X-ray diffractometer with a quartz crystal monochromator positioned in a glove box. A glove box was required to protect workers from health hazards of even minute amounts of the actinides which are strong alpha emitters. Fig. 1 is a drawing of the diffractometer box. An X-ray tube, a monochromator stand, and a vertical Philips goniometer are mounted on a table in the glove box. The vacuum equipment is also mounted inside the glove box, with the exception of the rotary pump which is connected by means of tubing containing two absolute paper filters to avoid outside contamination.

A section through the low temperature camera is shown in Fig. 2. A cylindrical stainless steel camera is fixed on the axle of the goniometer. The X-ray beam enters and leaves the camera through windows of 0.3 mm thick beryllium plate glued with an epoxy resin. The expander tube of a closed cycle refrigeration system is mounted on the camera wall. Helium is compressed to about 250 psi in a compressor. Its flow to and from the expander tube is controlled by an automatic valve assembly. The maximum cooling is obtained at the copper endplate of the expander tube, where the sample is spread as a powder or fixed as a thin foil. Evacuation of the camera is required to prevent the sample from oxidizing, to thermally insulate the expander tube, and to prevent frost formation on the expander tube. The camera is closed by a stainless steel cover with an O-ring seal which can be fixed in seconds by four coarse-threaded nuts. A rapid closing procedure is important for samples sensitive to the glove box atmosphere.

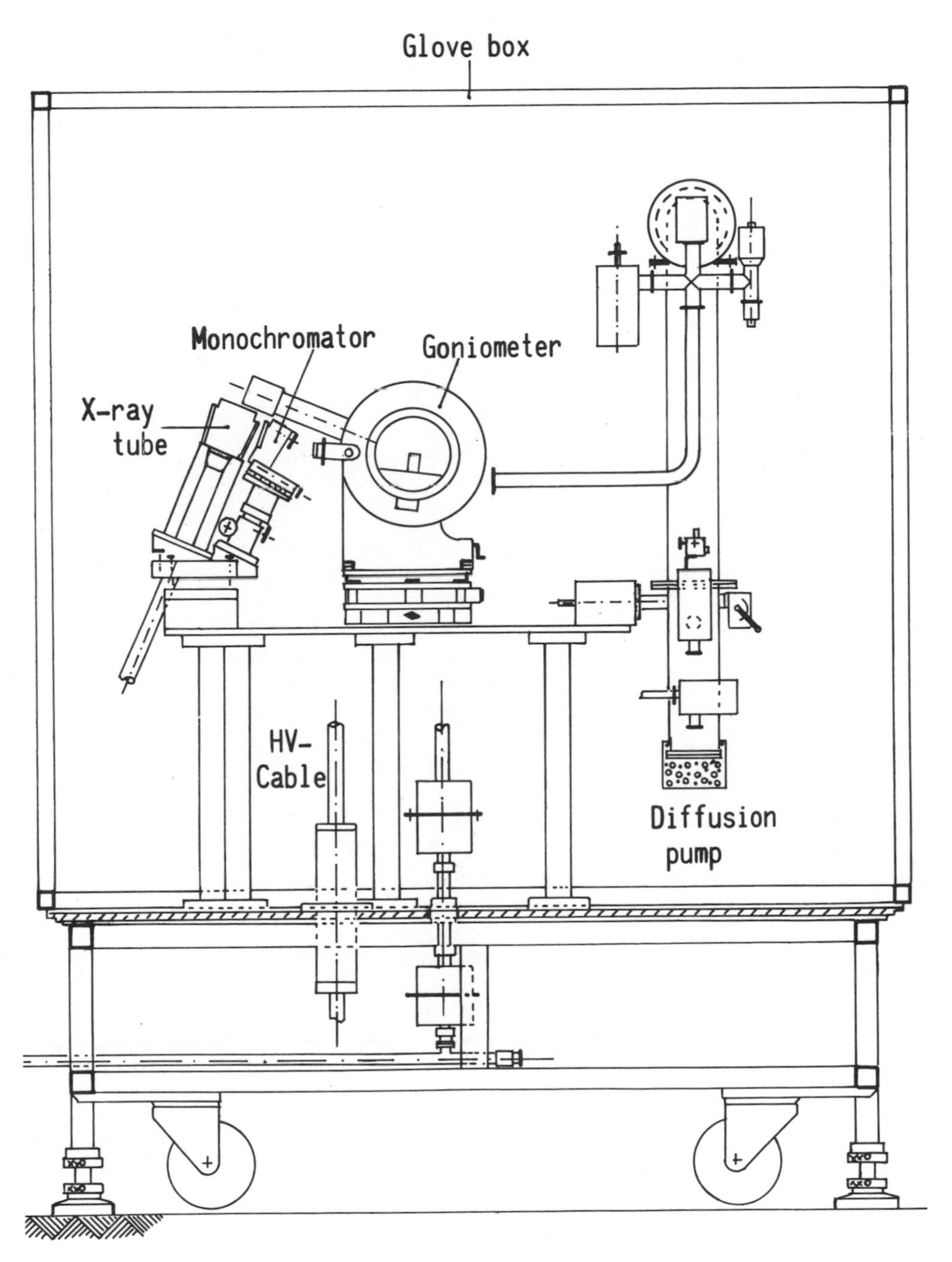

Fig. 1. Drawing of the low temperature diffractometer glove box

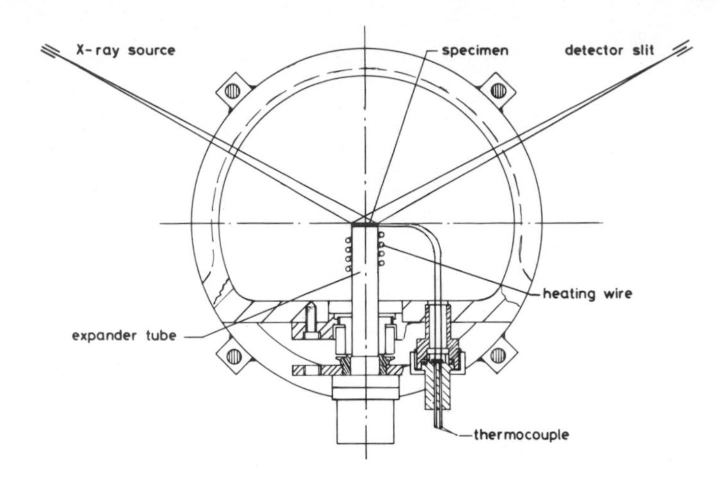

Fig. 2. Section through low temperature camera

Adjustment of the expander tube is achieved by four screws in the plate where the tube is fixed. A flexible but vacuum-tight connection between the tube and the plate is obtained by use of a metal bellows. The specimen-supporting surface of the tube is positioned in the camera centre by means of a gauge which represents the camera diameter and which can be clamped to two opposite places in the periphery. The surface is made parallel to the direct beam by placing on it a hood containing a 0.1 mm wide horizontal slit which transmits maximum X-ray intensity when parallel to the beam.

The code number of the instrument in the classification proposed by Rudman (1) is 353271 when used for powders, and 453271 when used for compact metal surfaces.

The Quartz Crystal Monochromator

The CuKα_2 component of the X-radiation was completely eliminated by a quartz crystal monochromator inserted between the X-ray tube and the sample. This simplifies the interpretation of patterns with closely neighbouring lines. The reflection angle for the CuKα_1 radiation at the $10\bar{1}1$ planes of the quartz monochromator crystal is 26°40'; hence, the X-ray tube had to be inclined by this angle. For exact adjustment, the tube, the monochromator, and the goniometer were placed on devices which allowed them to be adjusted vertically and horizontally. The inclination of the tube is also variable.

The monochromator housing was mechanically not stable enough for the precision adjustment which is necessary to eliminate the

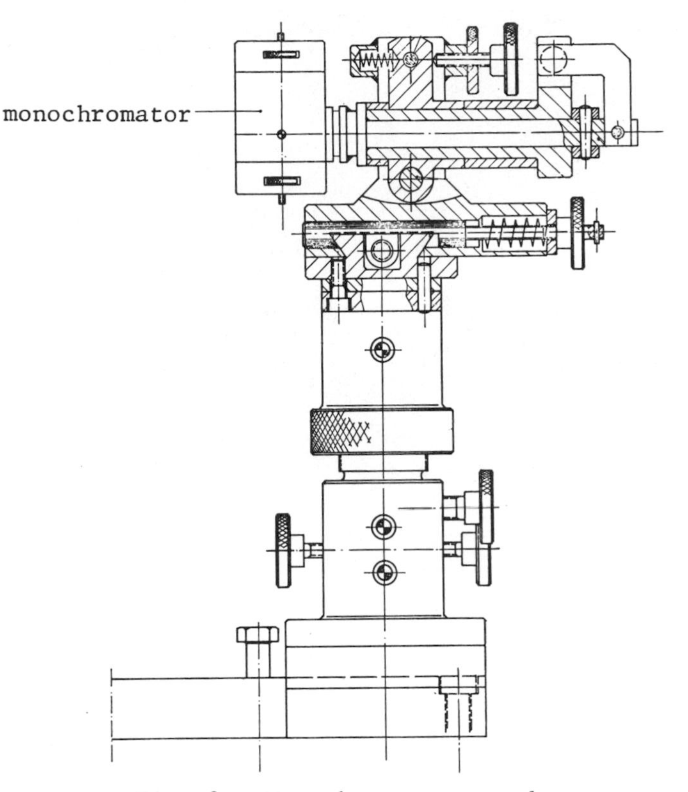

Fig. 3. Monochromator stand

$K\alpha_2$ component. A support was constructed which is mechanically very stable and allows adjustment and fixation in all directions (Fig. 3). Much more reliable adjustment was obtained than when the monochromator was fixed on the tube shield.

The entire assembly is shown in the photograph of Fig. 4. The expander tube is visible in the open camera. The valve assembly which controls the helium flow between compressor and expander is also mounted inside the glove box because the high pressure plastic hoses connecting the valve to the expander were not suitable for leading through the box wall. The compressor was placed outside the box and linked to the valve by means of a copper tube lead-through. Porous metal filters are inserted into this tubing to prevent outside contamination.

Since the high voltage generator was outside, and the X-ray tube inside the glove box, a special leadthrough had to be made for the high voltage cable (Fig. 5). This consists of a back-to-back mounting of two electrically-linked standard high voltage plug sockets in a steel tube filled with an epoxy resin. The resin has

Fig. 4. Photograph of X-ray tube, monochromator, and goniometer, with the low temperature camera open

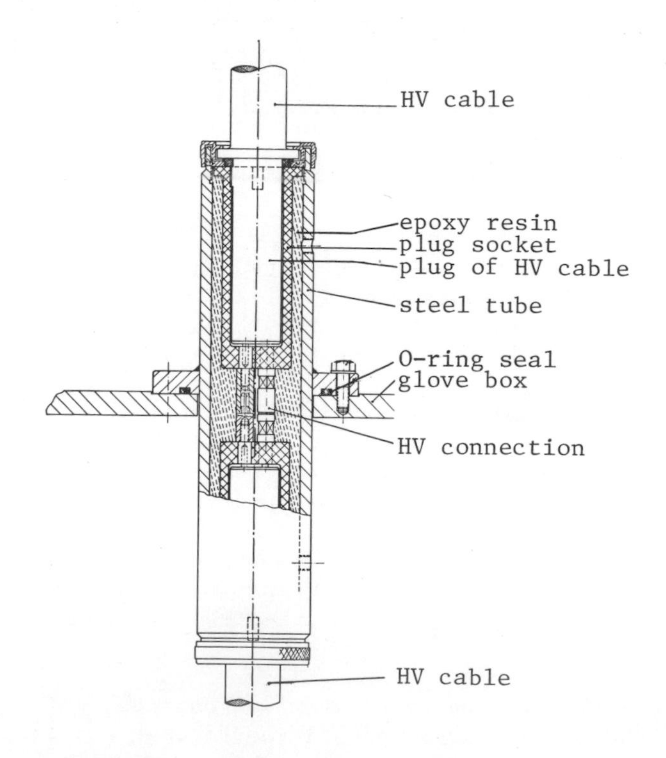

Fig. 5. High voltage lead

to be poured into the tube gradually in separate portions to avoid formation of cavities and thermal strain.

TEMPERATURE MEASUREMENT AND CONTROL

The temperature is measured by a gold-0.03% iron/chromel thermocouple inserted into a radial bore in a 2 mm thick copper plate soldered onto the top copper plate of the expander tube. The additional copper plate was then used as the specimen-supporting surface. To achieve good thermal contact, the thermocouple tip was embedded in high-conductivity silver paste. The upper part of the expander tube was surrounded by a heating wire which, together with a temperature indicator and controller, allowed temperature control between 50 and 345 K. Higher temperatures cannot be obtained because a low-melting solder is used to fix the end-plate to the expander tube. The long- and short-term temperature fluctuations were ± 1K. Ice water and liquid nitrogen were used for temperature calibration. Because only a thin layer of sample was used, the thermal gradient normal to the sample surface is assumed to be negligible. No systematic line broadening was observed at low temperature, with respect to the room temperature lines. Hence it is concluded that the "radial" temperature gradient in the plane of the sample is also negligible.

SAMPLE PREPARATION

High-conductivity silver paste was also used for embedding the sample powder. To do this, a thin layer of silver paste was spread on the copper endplate, and the sample powder was pressed into the paste before it had dried completely. Thorium metal powder was directly fixed onto the copper plate with dilute collodion solution. For protactinium metal powder, which is sensitive to oxidation, a special technique was used for transferring the sample from the argon-filled glove box where it had been powdered: silver paste was spread on a 0.03 mm thick copper foil, the sample powder pressed into the paste, and covered by a 13 μm thick Hostaphan foil which was fixed to the copper foil by a ring of double sided adhesive tape. Thus oxygen was excluded while the sample was transferred to the low-temperature camera. The whole assembly was then mounted on the expander tube end-plate.

LENGTH CHANGE OF THE EXPANDER TUBE

The effect of a length change of the expander tube on the position of the diffraction lines was studied by simulation. A micrometer device allowing small specimen displacements perpendicular to the specimen surface was mounted on the specimen support of a room temperature diffractometer. The line positions of a gold standard were measured for displacements between + 0.3 mm and -0.3 mm with respect to the correct zero position of the specimen. The formula

$$\Delta\theta = \frac{S}{R} \cos \theta \qquad (1)$$

(S: specimen displacement, R: goniometer radius) given by Neff (2) is confirmed by our measurements. The slope of the straight line $\underline{a}$ versus $\cos^2\theta$ used for extrapolation to $\theta = 90°$ of the lattice parameter of gold varies regularly with the specimen displacement (Fig. 6).

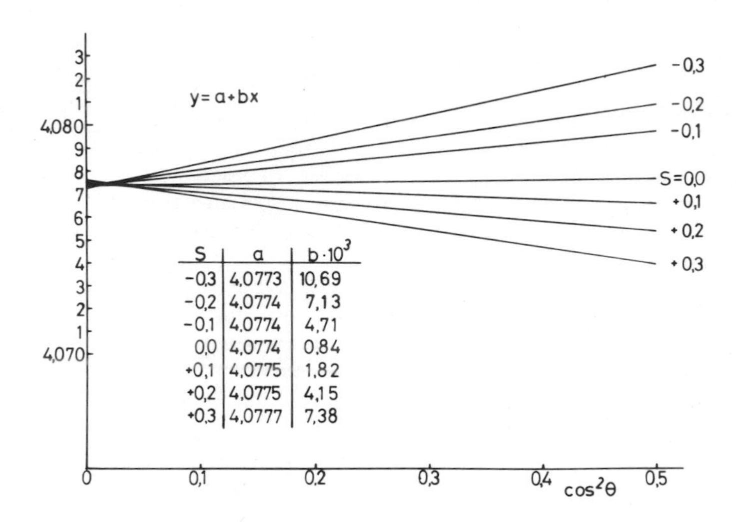

S	a	b·10³
-0,3	4,0773	10,69
-0,2	4,0774	7,13
-0,1	4,0774	4,71
0,0	4,0774	0,84
+0,1	4,0775	1,82
+0,2	4,0775	4,15
+0,3	4,0777	7,38

Fig. 6. Extrapolation slopes of gold at room temperature

From this variation, and from the extrapolation slopes of cubic phases determined in the low temperature camera (Fig. 7), the specimen displacement by contraction of the expander tube can be estimated not to exceed 0.4 mm. Moreover, for cubic substances, this specimen displacement error is eliminated when extrapolation procedures are used to determine the lattice parameter.

ADVANTAGES AND DRAWBACKS

Advantages

The cooling device chosen has the following advantages: it was easier to install on the goniometer which was already in the

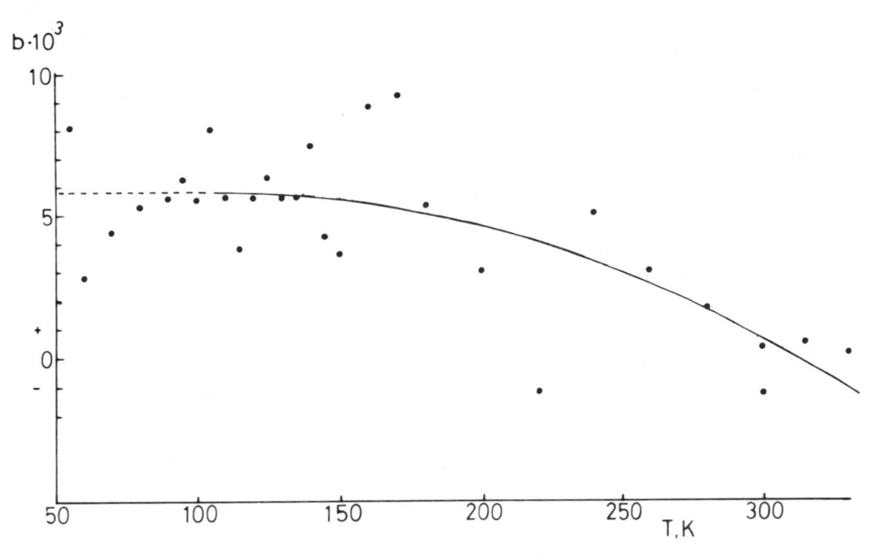

Fig. 7. Extrapolation slopes b of PuN vs. temperature

alpha-contaminated glove box. The need for periodic replenishing of cryogenic liquids is avoided which simplifies operation in the alpha box. Generally, the closed-cycle cooling system needs less maintenance and is more suitable for long-time service than systems with liquid cryogens.

Drawbacks

A drawback is the fact that only the smallest closed-cycle device available could be fitted to the existing diffractometer-glove box assembly. A temperature of 40 K is specified by the supplier, but we were never able to achieve less than 50 K with the device mounted in the glove box, though outside the glove box 41 K was reached with the same assembly. The difference is probably due to the additional resistance to helium flow caused by the spiral-shaped copper tubing used between the compressor and the valve in the glove box to increase the flexibility of the tubing, and by the porous metal filters ($\leq$ 8 μm pore diameter) which had to be inserted in the helium circuit to avoid contamination of the external parts of the circuit.

APPLICATIONS

Some results obtained with the instrument will be briefly described.

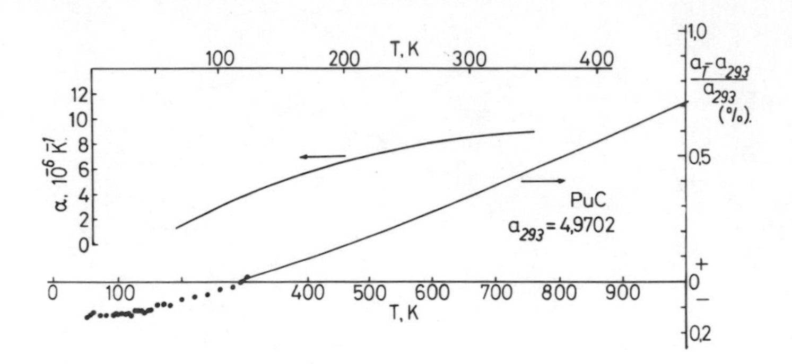

Fig. 8. Thermal linear expansion, $\Delta a/a$, and coefficient of thermal linear expansion, α, vs. temperature, for PuC

The lattice parameter variation between 50 and 350 K was measured for PuC, PuN, and Pu(C,N) (3) as well as for UC, PaC, Th, and Pa (4). It was found that the thermal contraction levels off below 100 K for the fcc compounds PuC, UC, and PuN, but not for the fcc metal Th. As an example, Fig. 8 shows that for PuC, the thermal linear expansion curve becomes nearly horizontal, and the coefficient of thermal linear expansion approaches zero, below 100 K.

Tetragonal protactinium metal has a negative coefficient of thermal linear expansion parallel to the a axis for $T < 250$ K (Fig. 9). The volume expansion coefficient derived from the least squares fit to the data points for the a and c parameters is also negative at $T < 200$ K.

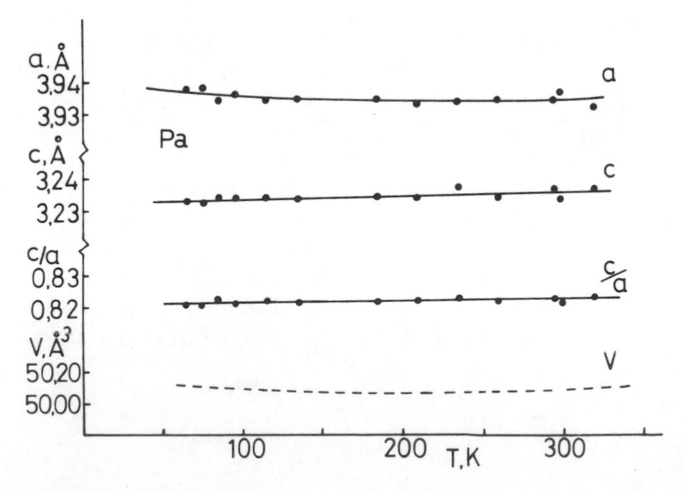

Fig. 9. Lattice parameters and unit cell volume of protactinium vs. temperature

REFERENCES

1. R. Rudman, "The Proper Description of Low Temperature X-Ray Diffraction Apparatus," J. Appl. Cryst. 10, 209-210 (1977).

2. H. Neff, "Grundlagen und Anwendung der Röntgen-Feinstruktur-analyse," Oldenbourg München 1959.

3. U. Benedict, C. Dufour, O. Scholten, "Lattice Parameter Variation of PuC, PuN, and Pu(C,N) between 50 and 350 K," J. Nucl. Mat. 73, 208-212 (1978).

4. U. Benedict, C. Dufour, K. Mayne, "Low Temperature XRD Study of Actinide Metals and Compounds," Proc. 3rd Int. Conf. Electronic Structure of the Actinides, Grenoble, France, Aug. 30 - Sept. 1, 1978, to be published in J. Physique.

X-RAY POWDER DIFFRACTION OF EINSTEINIUM COMPOUNDS*

R. G. Haire

Transuranium Research Laboratory, Oak Ridge National Laboratory, Oak Ridge, TN 37830

and

J. R. Peterson

Department of Chemistry, University of Tennessee, Knoxville, TN 37916 and Transuranium Research Laboratory, Oak Ridge National Laboratory, Oak Ridge, TN 37830.

ABSTRACT

X-ray powder diffraction (XRPD) analysis has proven to be extremely valuable in the study of the transplutonium elements. The high specific activity and the thermal energy associated with the radioactive decay of einsteinium isotopes make XRPD analysis of einsteinium samples very difficult. The major problems are destruction of the samples' crystallinity and alteration of their chemical composition. Blackening of the X-ray film also compromises the analyses. By applying certain guidelines given here, a limited amount of success has been achieved in obtaining diffraction data on einsteinium and some of its compounds.

INTRODUCTION

XRPD has proven to be a valuable tool for studying the transplutonium elements; a considerable amount of data on these elements

---

*Research sponsored by the Division of Nuclear Sciences, Office of Basic Energy Sciences, U. S. Department of Energy under contracts W-7405-eng-26 with Union Carbide Corporation and EY-76-S-05-4447 with the University of Tennessee (Knoxville).

was obtained by XRPD when the total amount of material that existed was less than one milligram. The classic work of B. B. Cunningham using a single ion exchange resin bead to concentrate initially the actinide element is a prime example [1]. Very often XRPD is the only method available to characterize these scarce actinide elements and their compounds. Even with the availability of greater quantities of these elements, XRPD still remains one of the most used means of establishing that a particular compound or phase has been prepared.

One of the major reasons for interest in einsteinium and its compounds is that it is the last of the actinide elements expected to be available in greater than picogram quantities. Knowledge of its chemistry is therefore useful for extrapolation to higher members of the series, especially since there is a tendency for greater stability of lower valence states (< III) in going across the second half of the actinide series.

The scarcity of einsteinium isotopes and the high specific activity of the most available isotope, ^{253}Es (5.6×10^{10} $\alpha \cdot min^{-1}$ μg^{-1}, 6.6 MeV α) require that microtechniques be used for preparing and studying einsteinium. The thermal energy associated with the radioactive decay (1 watt$\cdot mg^{-1}$) is also detrimental to a sample's composition. The blackening of X-ray film poses an additional problem. Thus, XRPD analysis of einsteinium materials creates a situation more difficult than previously encountered with the other actinides. The particular preparation and handling procedure used for an einsteinium sample will determine whether or not useful X-ray data can be obtained from it.

EXPERIMENTAL

Many of the microtechniques used for handling einsteinium have been described elsewhere [2-4] as applied to other transplutonium elements. As with any research problem, the investigator desires to work with the purest samples that can be obtained. With einsteinium the task is most difficult since in one day after purification the Es-253 already contains ∼ 3% of its daughter, Bk-249.

The oxides and halides of a new element are usually the first compounds that are prepared and studied. Further, many of the synthetic routes to other compounds start with the oxide. Our early attempts to prepare einsteinium oxide by the resin-bead technique [1] failed for one of two reasons: (1) the einsteinium decomposed the resin before it could be calcined; or (2) the exchange sites in the resin were occupied to only ∼ 20%, and upon calcination, very small, fragile pieces of oxide were formed. In recent years, we have used a microprecipitation method, that provides upon calcination in air individual Es_2O_3 particles 2-10 μg in size [4]. In principal, these Es_2O_3 particles could be studied by XRPD, but in practice X-ray patterns have not been obtained from these samples, the analyses always indicating the samples were not crystalline.

These oxide particles have also been used as starting materials for the synthesis of other einsteinium materials, notably the anhydrous halides, oxyhalides and the metallic state. The anhydrous trihalides of einsteinium (chloride, bromide, iodide) are prepared in quartz capillaries by reacting Es_2O_3 with the respective hydrogen halide [4]. Reduction of these trihalides with hydrogen yields the dihalides [5,6]. Recently we have prepared EsF_3 by treating the Es_2O_3 with ClF_3 or F_2 in a monel system. The trihalides, excluding the trifluoride, can be melted or sublimed in the quartz capillaries to provide well-crystallized samples for X-ray and/or absorption spectrophotometric analyses [4]. Einsteinium metal can be prepared by reducing Es_2O_3 with lanthanum metal, distilling the volatile einsteinium metal and condensing it on a quartz fiber or a tungsten wire. The fiber or wire is then placed in a quartz capillary and examined by XRPD. Because it is necessary to make the above transfers in an inert-atmosphere gloved box, the time elapsed between preparation and X-ray analysis is much greater for the metal than for the halides. Further, the reactivity of the metallic state precludes extensive heating of it in a quartz capillary. Thus, re-crystallization of a sample in an X-ray capillary is best achieved with the halide preparations.

DISCUSSION

Five guidelines are given in Table I for XRPD of einsteinium metal and einsteinium compounds. One of the most important is the

Table I

Guidelines for XRPD of Einsteinium and Its Compounds

1. XRPD analysis must begin immediately after preparation or annealing of the sample.

2. Exposure times should be limited to 1/2-1 hour.

3. Sample sizes must be kept between 1-3 μg of Es.

4. Shielding of the film with nickel foil is important (Cu radiation).

5. Thin deposits of the sample are preferred over bulk forms.

examination of the sample as soon as possible after preparation, to minimize destruction of the material and/or its crystallinity due to self-irradiation and/or the associated thermal energy. In absorption spectrophotometric studies of einsteinium halides, we have shown that the loss of crystallinity begins in 5-15 minutes after annealing. Since exposure times for XRPD usually take greater than 30 minutes, even without considering handling time before the analysis, the sample's crystallinity can be destroyed before the end of the X-ray examination. If the material is resistant to decomposition, or can be re-synthesized in the X-ray capillary, then the crystallinity can be restored by annealing immediately prior to or during the XRPD analysis. The annealing can take the form of a single heat treatment, be pulsed allowing for cooling time between pulses (for ambient temperature measurements), or the sample can be held at elevated temperatures during XRPD. The first X-ray diffraction data on einsteinium compounds were obtained for $EsCl_3$ and EsOCl by continuously resynthesizing and annealing the materials at 430°C during the X-ray analysis [7]. Annealing the sample (at 400°C for 3 min) immediately prior to X-ray analysis provided sufficient crystallinity to collect limited diffraction data on $EsBr_3$ [8].

The use of a 57.3 mm Debye-Scherrer camera modified with a small heating coil [3] permitted annealing the $EsBr_3$ sample only seconds before collecting X-ray data. Annealing of Es_2O_3 or einsteinium metal samples in the quartz capillaries has not been as successful. The reactivity of the metal prohibits annealing it in the X-ray capillary, and the higher temperature (~900°C) required to recrystallize the oxides poses experimental problems (capability of the miniature heating coils plus possible reaction of the sample and the quartz). To date, XRPD data have not been obtained for einsteinium oxide or metal.

Protecting the X-ray film from the einsteinium's radioactivity is also necessary. Whereas the radioactivity of other transplutonium elements used in research work can be tolerated in 57.3 mm cameras for 2-8 hours, 1-3 μg samples of ^{253}Es require limiting exposures to 1/2-1 hour. The ideal quantity of sample used for X-ray work is a balance between the sample's radiation and the amount of material necessary to give suitable XRPD data, utilizing a fine focus or microfocus X-ray source. Although a longer-lived Es isotope (Es-254) is known, sufficient quantities of it are not available, and in addition, the short-lived ^{250}Bk daughter in equilibrium with it would also irradiate the film significantly. The use of a beryllium cup around the sample [9] protects the capillary during the film loading and unloading and also affords protection from any alpha particles escaping the thin glass walls of the capillary. The use of 0.008 mm thick nickel foil (for CuKα radiation), either around the beryllium cup or immediately in front of the film, offers the best protection from gamma irradiation. In this case, the usual nickel filter (to remove CuKβ radiation) for the X-ray source is not needed, and thus a greater intensity of X rays reach the sample. Unfortunately, the Es-253 isotope emits a number of gamma rays, ranging up to ∿ 1 MeV in energy, which are not significantly attenuated by the nickel foil.

The last guideline in Table I was determined through experience, where it was found that samples in the form of their sublimates or thin films retained their crystallinity longer than those in bulk. From the standpoint of making micro-transfers, it is necessary to have samples of 1-10 μg in mass; if additional transfers of the material being synthesized are not required, then thin films or deposits are acceptable for analysis by X rays. XRPD data on EsI_3 [10] were recently obtained by analyzing sublimates, made by heating one portion of sealed capillaries, each containing ∿4 μg of EsI_3, and allowing the sublimed EsI_3 to condense in the cooler part of the capillaries. The crystallinity of these samples was retained for longer periods of time than that observed for bulk forms of EsI_3. This aspect of thin samples was also demonstrated in the case of Es_2O_3, where 2-3 μg pieces yielded no XRPD data, but 50-100 ng samples, consisting of individual 400-800Å Es_2O_3 particles, were readily analyzed by electron diffraction [11] and remained crystalline for many hours. Similarly, 100-200 ng samples of einsteinium metal in the form of 200-400 Å films remained crystalline for a few hours [12], whereas thicker films appeared amorphous by XRPD.

An explanation for the observed difference in behavior of thin deposits and bulk forms of samples is that a significant fraction of the einsteinium's alpha particles escape the small crystallites in the thin deposits, releasing the major part of their energy outside of the crystallites themselves. Damage due to recoiling nuclei would still be encountered, although this process may produce some self-annealing by localized thermal spikes. Thus, by using thin deposits made up of very small crystallites, a large portion of the radiation damage to the sample may be avoided.

It is beyond the scope of this paper to discuss in detail electron diffraction or to compare the advantages/disadvantages of electron and X-ray diffraction analyses. It is worthwhile to note, however, that with regard to einsteinium samples, electron diffraction analysis has three advantages: (1) very short exposure times (10 sec) for diffraction patterns; (2) the possibility of analyzing samples in ∿ 10 minutes after preparation; and (3) blackening of the film is avoided due to the greater distance between the film and sample (∿ 1 meter). There are also some general disadvantages to electron diffraction analysis: these are the need for repeated calibration of the electron's wavelength; the need to examine many different portions of a sample to assure the areas examined are representative of the entire material; and the need for very thin samples, since the penetrating power of the electron is low. For einsteinium samples, the wavelength calibration creates an interesting problem. Normally when precise electron diffraction measurements are desired, an internal standard is used. However, with einsteinium samples, one must not only contend with radiation damage to the einsteinium itself, but also that damage which may be done to the internal standard used for calibration!

In Table II are summarized the einsteinium materials for which X-ray or electron diffraction data have been collected. Hopefully, this list will be expanded in the future. We are continuing to use XRPD for analyzing einsteinium samples, especially for Es(II) halide compounds, where their extreme hygroscopic nature and reactivity with oxygen present major difficulties for electron diffraction analysis.

It is also planned to attempt the rapid analysis of einsteinium compounds by energy-dispersive XRPD. This technique, amenable to our present preparative and handling procedures, may afford the capability of attaining diffraction data on a sufficiently short time scale to allow monitoring structural changes in an einsteinium sample brought about by its intense self-irradiation.

Table II

Diffraction Data on Einsteinium and Its Compounds

Material	Method	Date	Reference
$EsCl_3$ $EsOCl$	XRPD	1969	7
Es_2O_3	Electron Diffraction	1973	11
$EsBr_3$	XRPD	1975	8
EsI_3	XRPD	Unpublished	10
Es Metal	Electron Diffraction	1978	12

REFERENCES

1. B. B. Cunningham, "Submicrogram Methods Used in Studies of the Synthetic Elements," Microchem. J. Symp. Ser., 69-93 (1961).
2. J. R. Peterson, "Chemical Identification and Phase Analysis of Transplutonium Elements and Compounds via X-ray Powder Diffraction," Advances in X-Ray Analysis, Vol. 20, H. F. McMurdie, C. S. Barrett, J. B. Newkirk and C. O. Ruud, Editors, Plenum Pub. Corp., New York (1977), p. 75.
3. J. N. Stevenson and J. R. Peterson, "Some New Microchemical Techniques Used in the Preparation and Study of Transplutonium Elements and Compounds," Microchem. J., 20, 213 (1975).
4. J. P. Young, R. G. Haire, R. L. Fellows, and J. R. Peterson, "Spectrophotometric Studies of Transcurium Element Halides and Oxyhalides in the Solid State," J. Radioanal. Chem., 43, 477 (1978).
5. J. R. Peterson, D. D. Ensor, R. L. Fellows, R. G. Haire and J. P. Young, "Preparation, Characterization and Decay of Einsteinium(II) in the Solid State," Proc. 3rd International Conf. on Electronic Structure of the Actinides, Grenoble, France, Aug. 30-Sept. 1, 1978; to be publ. in J. de Physique.
6. R. L. Fellows, J. R. Peterson, J. P. Young and R. G. Haire, "The First Preparation of an Einsteinium(II) Compound, $EsCl_2$," The Rare Earths in Modern Science and Technology, G. J. McCarthy and J. J. Rhyne, Editors, Plenum Press, New York (1978), p. 493.
7. D.K. Fujita, B. B. Cunningham and T. C. Parsons, "Crystal Structures and Lattice Parameters of Einsteinium Trichloride and Einsteinium Oxychloride," Inorg. Nucl. Chem. Lett., 5, 307 (1969).
8. R. L. Fellows, J. R. Peterson, M. Noé, J. P. Young and R. G. Haire, "X-ray Diffraction and Spectroscopic Studies of Crystalline Einsteinium(III) Bromide, $^{253}EsBr_3$," Inorg. Nucl. Chem. Lett., 11, 737-742 (1975).
9. R. L. Sherman and O. L. Keller, "Modified Debye-Scherrer X-Ray Diffraction Camera for Radioactive Compounds," Rev. Sci. Instrum., 37, 240 (1966).
10. R. G. Haire, J. P. Young, J. R. Peterson and R. L. Fellows, "Spectrophotometric and X-ray Data of $^{253}EsI_3$," to be published, Inorg. Nucl. Chem. Lett.
11. R. G. Haire and R. D. Baybarz, "Identification and Analysis of Einsteinium Sesquioxide by Electron Diffraction," J. Inorg. Nucl. Chem., 35, 489-496 (1973).
12. R. G. Haire and R. D. Baybarz, "Studies on Einsteinium Metal," J. de Physique (in press; see ref. 5).

IDENTIFICATION OF MULTIPHASE UNKNOWNS BY COMPUTER METHODS: ROLE OF CHEMICAL INFORMATION, THE QUALITY OF X-RAY POWDER DATA AND SUBFILES

Gregory J. McCarthy and Gerald G. Johnson, Jr.

Materials Research Laboratory, The Pennsylvania State University, University Park, PA 16802

ABSTRACT

Three variables in the identification of crystalline multiphase unknowns by computer search/match methods were examined: (1) presence or absence of chemical information; (2) varying quality of d-I diffraction data; (3) use of a large data base *vs* a data base having about 10% as many entries--both contain the unknown phases. The results showed that the search/match using average quality d-I data with chemical information was quite successful, while that obtained using higher quality d-I data alone missed one of four phases. It took only half the time to obtain the average d-I data plus chemistry as it did to obtain the higher quality data. Searching the smaller data base with the higher quality data alone resulted in an identification of all phases with high reliability.

INTRODUCTION

It is widely appreciated in x-ray powder diffraction analyses that complex multiphase unknowns can be identified with greater speed and confidence when high quality powder data and/or chemical (elemental, functional groups, etc.) information are utilized. This generalization is especially true in the case of computer search/match (s/m) routines where the analyst often must choose from among a long list of potential "answers." It is the nature of computer s/m routines to select or "sieve" from the large data base, the *Powder Diffraction File* (*PDF*) (1), those phases that have interplanar spacing-relative intensity (d-I) coincidences with the powder data of the unknown. Obviously, the more precise the match of d-I data in the unknown and in the data base, the greater will be the reliability reported in the output of the computer routine because smaller error tolerances can be established. In addition, knowing some or all of the chemical make-up of the unknown specimen can eliminate extraneous phases with a large number of accidental d-I matches and further increase confidence in "answers" with good d-I matches.

For those laboratories that do not possess automated/computer-interfaced diffractometers, obtaining high quality powder data is a time-consuming procedure requiring careful specimen preparation and separate scans for maximizing I-sensitivity and d-precision (made accurate by the use of an internal standard). Most of these same laboratories do have x-ray fluorescence (XRF) or scanning electron microscopy with energy dispersive x-ray spectrometry (SEM/EDX) elemental analysis capability and either computational center facili-suitable for implementing search/match computer routines in house or access to an on-line commercial version (2) of this routine via time sharing. Because maximizing through-put is often the chief consideration, the question of the best balance of time spent in obtaining powder data and elemental information for input into a computer s/m routine needs to be addressed.

In the identification of multiphase unknowns, time would also be saved by searching a smaller set, or "subfile," of the complete PDF if it were known that most or all of the unknown phases were present in that subfile. As indicated in Table 1, the PDF currently contains 34,534 powder patterns and is growing by about 2000 patterns per year. To date, six subfiles of the PDF have been made available and the powder patterns from these subfiles have been encoded on the complete computer data base tape. Of particular interest in identifying general unknowns is the Frequently Encountered Phases (FEP) subfile. The concept underlying the FEP subfile is that a large percentage (perhaps 70-90%) of the phases in a general multiphase unknown would occur in this 2,400 powder pattern subfile. Thus, in most cases, less time would be required for manual or computer s/m and results would be more reliable (i.e. fewer accidental d-I matches would occur).

The second "Search/Match Round Robin" (3,4) provided an opportunity to examine the balance of time spent in collecting powder data and obtaining elemental data and to compare computer s/m with the full inorganic subfile and the FEP subfile. A multiphase unknown specimen, provided as part of this exercise, was used to collect data of two types:

- "Chemistry-oriented data" (C-data): obtained with minimum time spent on diffraction data gathering;

"Diffraction-only data" (D-data): additional care taken in obtaining more precise and accurate d- and I-data.

These two types of data were used as input to the Johnson Version-18 s/m program (5-7). Both the full inorganic subfile and the FEP subfile for sets 1-28 of the PDF were searched. The s/m runs were made both with and without chemical elemental information on the specimen. The results illustrate several important aspects of the identification of multiphase unknowns by computer methods.

EXPERIMENTAL PROCEDURES

C-Data

The chemistry oriented data were obtained with routine qualitative phase identification methods on a Siemens diffractometer equipped with a scintillation detector, graphite diffracted beam monochromator and solid state electronics using CuKα (λ = 1.54056Å for $CuK\alpha_1$) radiation. The diffractometer had been calibrated with a silicon standard (NBS SRM-640, a_o = 5.43088Å) several weeks earlier and the data were corrected according to the current calibration curves. Specimen 7B was prepared using a collodion-amylacetate mixture to form a slurry on a glass slide* and two separate 2°2θ per minute scans from 18 to

*For a detailed discussion of specimen preparation in x-ray diffraction see the paper by Smith and Barrett in this volume.

TABLE 1

PDF Statistics--Sets 1-28

Component	Powder Patterns
Full PDF	34,534
Inorganic Subfile[a]	24,971
Organic Subfile[a]	10,721
Metals and Alloys Subfile[a,b]	5,700
Minerals Subfile[a,b]	3,039
Frequently Encountered Phases Subfile[a]	2,378
National Bureau of Standards Subfile[a,b]	947

a. Available as a search manual from the JCPDS (1).
b. Available in book form with accompanying search manual from the JCPDS (1).

70°2θ (a typical "routine" scan range) were made. The 2θ and I data were averaged from the strip chart output of the two scans. Before converting to d, the 2θ values were corrected with the external Si standard calibration curve for the instrument. The precision of the 2θ values obtained by this method would be expected to be no better than ±0.04°2θ. Although the I-values could be read from the strip charts with good precision, their accuracy could vary widely because no care was taken in specimen preparation to avoid the effects of preferred orientation, differential sedimentation, microabsorption, and so on. The time required to obtain the data and read the charts was about two hours.

D-Data

A quite different procedure was used to obtain these higher quality data. Separate specimen preparation and diffractometer procedures were used to obtain I-data and 2θ-data on the same Siemens instrument. For the I-data, specimen 7B was ground thoroughly and dusted through a -400 mesh sieve onto a glass slide made sticky with a thin film of petroleum jelly. This method is especially useful for minimizing preferred orientation effects. Relative intensities were averaged from strip charts made by scanning this specimen twice over a 5-70°2θ range at 1°2θ/min. The density of this "dusted" specimen was relatively low, resulting in a reduced diffraction signal. Thus, a slurry-mounted specimen of ground 7B was used to collect d-data. To increase the 2θ precision compared to the C-data, the scanning rate was reduced from 2° to 1°2θ/min. It was known that specimen 7B contained NBS SRM-640 silicon, so this phase was used as an internal standard to correct the 2θ data before conversion to d values. With this scan rate and the internal standard, it is estimated that the 2θ-data were accurate to about ±0.02°2θ. Approximately 5 hours were required for specimen preparation, data collection and reading of the strip charts.

Chemical Elemental Data

The mounted diffractometer specimen used to collect the C-data was given a carbon coating, placed in a scanning electron microscope (SEM) and analyzed for elements heavier than neon by energy dispersive x-ray spectrometry (EDX). This procedure required about one-half hour, giving a total time for obtaining C-data and chemical elemental information of 2.5 hours, which is about half that for obtaining the D-data set.

RESULTS AND DISCUSSION

Experimental Data

Specimen 7B contained four phases:

$BaCl_2 \cdot 2H_2O$ ZnO KI Si.

The container indicated that one of the phases was NBS SRM-640 silicon and that a "poisonous" Ba salt was included. A portion of the C- and D-diffraction data is given in Table 2, along with the corresponding data for the four phases. The SEM/EDX indicated that, for elements heavier than neon, only Ba, Cl, Zn, K, I, and Si were present as major or minor constituents.

Examination of Table 2 indicates that the level of agreement between the d values of the PDF standard and those of the C-data is generally only 0.001Å worse than those of the D-data. Apparently, use of the Si external standard to correct the data from the faster 2°2θ scan brought the data close in accuracy to that of the 1°2θ scan corrected with the Si internal standard. As expected, the D-data had somewhat better resolution. The greatest benefit of the extra care taken in obtaining the D-data came in the relative intensities. It can be seen that the I values of the D-data scale much better with the I's of the PDF standards than do those of the C-data. Since the reliability of the match in computer s/m is a function of both d and I, the D-data should give more reliable "answers" in the absence of chemical data. Another interesting feature of the data in Table 2 is that several of the $BaCl_2 \cdot 2H_2O$ reflections that were present on the strip chart were apparently overlooked during the more rapid collection of the C-data.

Computer s/m

The two sets of d-I data were inputed into the s/m program with key parameters set as follows:

DHI: 4.938 for C-data; 20.0 for C-data
DLO: 1.352 for C- and D-data
BACKG: 2 for C-data; 1 for D-data
WINDOW: 4 for C-data; 2 for D-data
POSITIVE ELEMENTS: Si, Ba, K, I, Zn, Cl
MAJOR ELEMENTS: Si, Ba, K, I, Zn, Cl
UNDETERMINED ELEMENTS: H, Li, Be, B, C, N, O, F
NEGATIVE ELEMENTS: all other elements.
Subfiles: Inorganic (MAXI) or Frequently Encountered Phases (MINI).

---

*See references (5), (6) and (7) for an explanation of these parameters.

TABLE 2

Portion of the Experimental and PDF Standard Diffraction Data

C-Data		D-Data		$BaCl_2 \cdot 2H_2O$ (25-1135)*		ZnO (5-664)*		KI (4-471)*		Si (27-1402)*	
d	I	d	I	d	I	d	I	d	I	d	I
		5.98	2	5.96	7						
		5.74	4	5.72	12						
		5.46	42	5.45	85						
		4.95	40	4.94	60						
4.84	10	4.84	13	4.84	20						
4.50	20	4.50	42	4.50	55						
4.43	24	4.43	71	4.42	100						
4.34	12	4.34	10	4.33	17						
				4.23	2						
4.08	16	4.09	26					4.08	42		
3.66	16	3.67	36	3.661	45						
3.63	8	3.62	11	3.622	17						
				3.569	15						
3.54	28	3.54	59					3.53	100		
3.399	16	3.391	37	3.390	55						
		3.360	13	3.359	25						
3.239	12	3.242	11	3.240	12						
				3.213	30						
3.210	12	3.209	27	3.200	25						
3.138	96	3.136	49							3.1355	100
				3.048	3						
3.004	4	2.999	6	3.003	10						
		2.945	43	2.948	50						
2.927	40	2.930	52	2.928	65						
2.909	40	2.910	61	2.908	90						
2.863	16	2.861	38	2.861	45						
2.819	48	2.817	67			2.816	71				
2.711	20	2.711	36	2.712	50						
2.672	8	2.670	8	2.671	15						
2.605	32	2.603	51			2.502	56				
2.548	36	2.545	56	2.547	75						
		2.525	8	2.527	18						
2.480	100	2.497	42					2.498	70		
		2.476	100			2.476	100				

*PDF card number.

Two of these parameters have important effects on the computer results. The d window was set rather wide at ±4 parts per thousand for the C-data because of lower confidence in its accuracy. This means that the computer will have to consider a larger number of possible answers, and it allows more phases to be listed as possible answers because of larger numbers of coincidences. The smaller window used for the D-data permits fewer of these coincidences, and allows the computer to rate as more reliable the answers it does find. The chemical certainty implied in setting the "positive," "major" and "negative" elements will be important in calculating the reliability of answers when chemistry is used.

The relative value of time spent on obtaining chemical plus average quality diffraction data vs higher quality diffraction data is illustrated with the computer output in Figure 1. This is one of several "reports" given in the s/m program. It is based on computer modification of the d-I data as follows: the d-I data of the most reliable answer from among the "best 50" matches is subtracted from the total set of d-I data and the remaining 49 possibilities are searched for the best match, and this becomes the second answer. The process is repeated until the computer has accounted for all of the inputed d-I data or until ten answers are obtained. The order of the "best 50" is strongly dependent on chemistry when this is specified. The parameter "scale" is a relative measure of the amount of the phase present and is a function especially of the I-match. The closer the value to 1.0, the more likely is this phase to be present in the unknown, i.e. the greater is the reliability of the answer.*

The C-data plus chemistry gave five answers with large values for "scale" (Figure 1a). Four of these are the phases in specimen 7B. The fifth, $SiCl_4$, occurs first because all of its reflections have coincidences in the 2θ range covered:

$SiCl_4$ (10-220)		C-data	
2.821	100	2.819	48
2.228	25	2.232	12
1.996	45	2.001	8
1.823	15	1.824	8
1.410	10	1.409	8

*Care must be taken in accepting the results of this subtraction routine exclusively. It is most valid when high quality d-I data are used. If the computer chooses a coincidentally better, but wrong, phase for an early subtraction, subsequent answers will probably also be wrong. The analyst should consult each of the reports in evaluating computer s/m answers.

PDF	SCALE	FORMULAE
50565	1.000	(SI)
100220	0.507	SI CL4
40471	0.604	K I
251135	0.465	BA CL2 . 2 H2 O
50664	0.515	(ZN O)
20326	0.122	K CL O4
240835	0.111	K C24
10776	0.128	K I O3
110252	0.090	SI O2
191129	0.105	(SI B6)

a. C-Data; Chemistry

PDF	SCALE	FORMULAE
271402	0.707	(SI)
50664	1.000	(ZN O)
251135	0.697	BA CL2 . 2 H2 O
181190	0.383	AG3 S I
110521	0.408	LI2 S O4
30707	0.115	AG2 F
251222	0.092	(CU4 IN MG)
230215	0.080	CU3 FE (C N)6 . 4 H2 O
80346	0.057	(TE TH)
170567	0.095	(RH NB)

b. D-Data; No Chemistry

PDF	SCALE	FORMULAE
271402	1.000	(SI)
120513	0.692	(RH SI)
100220	0.507	SI CL4
230213	0.615	CU2 FE (C N)6 . 2 H2 O
90340	0.133	(LI H)
31215	0.190	(AL LI)
50664	0.552	(ZN O)
80341	0.104	NB RU
90237	0.186	(CD3 CE)
191046	0.155	(RH SM)

c. C-Data; No Chemistry

PDF	SCALE	FORMULAE
50565	0.712	(SI)
251135	0.706	BA CL2 . 2 H2 O
50664	0.976	(ZN O)
40471	0.557	K I
200957	0.127	K2 ZN2 (S O4)3
200852	0.084	K 2 (N H4)2 (N O3)2 S O4
230815	0.048	BA2 B5 O9 CL
221140	0.117	LI2 B4 O7

d. D-Data; Chemistry

Figure 1. Selected Computer Output After Searching the Full Inorganic Subfile.

and secondly, because both of its elements were entered as "positive elements." The analyst would probably eliminate this answer immediately because it is well known that $SiCl_4$ is a liquid between -70 and +58°C. Thus, the C-data plus chemistry yielded the correct four answers with a high reliability.

The D-data, without chemistry, yielded three of the answers with high reliability, but missed KI (Figure 1b). The overlaps in some of the d-I data between $BaCl_2 \cdot 2H_2O$ and KI were such that if $BaCl_2 \cdot 2H_2O$ is found first, the computer routine rates KI lower in the "best 50" and subtracts too many of its key reflections in preparing the "report" shown in Figure 1. The KI phase appears in position 26 of the "best 50."

When the D-data are used with chemistry (Figure 1d), the four phases are found and listed with the highest reliability. The extraneous "answer," $SiCl_4$, does not appear because of the higher accuracy, and consequently the better match, of the d-I data to the PDF standard data. Here the KI appears in the first four with high reliability because its two elements were entered as positive chemistry.

If the average quality C-data were to be used without chemistry, there would be too many d-I coincidences in the full Inorganic Subfile and the results would have little value. Figure 1c illustrates this case. Only two of the phases (Si and ZnO) occurred in the "best 50" portion of the s/m output. When the C-data (without chemistry) were used with the FEP subfile (Figure 2a) where fewer d-I coincidences were possible, three of the phases were listed among the ten answers in the "report." The fourth, ZnO, appeared high in the "best 50" listing. But, when the computer routine subtracted the d-I data for Si and $SiCl_4$, several important ZnO reflections were eliminated from further consideration. This occurrence illustrates the importance of the analyst's consulting all of the computer output. When the D-data are used with the FEP subfile, the fourth phase, KI, is now included among the first four, high reliability, answers.

Comparison with "Hand" Searching

The D-data were also employed in a "hand" search using the Hanawalt Search Manual to the full Inorganic Subfile. At the time the search was made, no chemical information (other than the presence of Si and Ba indicated on the bottle) had been obtained. The phases were identified in the order ZnO, Si, $BaCl_2 \cdot 2H_2O$, KI. The complete search plus match procedure took about 2.5 hours.

For comparison, it took about 0.5 hour to enter either the C- or D-data into the computer and about one hour of turn-around on

PDF	SCALE	FORMULAE
271402	1.000	(SI)
100220	0.507	SI CL4
221189	0.589	(NI O)
40471	0.604	K I
60261	0.323	(HG S)
50682	0.439	(TI)
90104	0.393	(CS2 O)
251135	0.291	BA CL2 . 2 H2 O
150765	0.169	CD (C N)2
200470	0.243	NA CA MG FE AL SI H O

a. C-Data; No Chemistry

PDF	SCALE	FORMULAE
271402	0.707	(SI)
50664	1.000	(ZN O)
251135	0.697	BA CL2 . 2 H2 O
40471	0.560	K I
250177	0.145	CA TI SI O5
10206	0.082	CU S C N
40593	0.063	(CE O2)

b. D-Data; No Chemistry

Figure 2. Selected Computer Output After Searching the FEP Subfile.

the University's IBM 370/168 computer. Assuming reasonably easy and rapid access to the computer, perhaps only a third to a half of the analyst's time would be required for identifying the components of the multiphase unknown. Use of C- or D-data plus chemistry in the computer s/m gave the same result as the hand s/m.

Questions of cost and computer time are so site and machine specific that they will not be detailed here. One might expect that the most complex run described above would cost less than $50 and involve only a few minutes of CPU time.

Several additional comments concerning computer *vs* hand s/m procedures can be made:

- The computer, even though extremely fast, works using a methodology different than manual (or book) search, and the small mental variations in delta d and I used in the manual search to obtain an "answer" will not be tolerated by a computer. For the computer, all parameters are mathematically fixed for an entire identification.

- The relationship of phases which normally occur together (such as from rock specimen or a corrosion product) are not known by the computer program. Thus the manual searcher is using this information in a subtle manner while the computer cannot.

- The results from the computer need to be interpreted by the analyst in a scientific manner utilizing additional information on the sample (e.g. non-diffraction data such as range of temperature or pressure stability, magnetic or electronic properties, color, density) which has not been told to the computer.

- The "answers" from the computer cannot consist of phases that are not present in the current PDF. However, because isostructural compounds are printed by the search portion of the output, but are eliminated by the match portion (due to being marked chemically absent by the input of the user), certain useful information regarding structures can be obtained from even an incomplete data base.

CONCLUSIONS

The combination of average quality diffraction data (C-data), plus chemistry took less of the analyst's time to obtain (compared to the D-data) and resulted in identification, with greater confidence, of the four components of the multiphase unknown. It can be concluded that in using this popular computer s/m program, the analyst's time would be better spent on obtaining chemical information than on making improvements in diffraction data obtained with a well-aligned diffractometer. The results also point out that without chemical information, high quality of both d and I data are absolutely necessary in order to obtain a successful identification of multiphase unknowns. It was also demonstrated that searching subfiles that contain the unknown phases results in more successful, rapid and reliable identifications.

ACKNOWLEDGEMENT

This research was supported in part by the JCPDS, International Center for Diffraction Data.

REFERENCES

1. W. F. McClune, Managing Editor, Powder Diffraction File, JCPDS, International Center for Diffraction Data, Swarthmore, PA.

2. The "Diffraction Data Tele Search (2 DTS)," JCPDS, International Center for Diffraction Data, Swarthmore, PA.

3. C. R. Hubbard and R. Jenkins, "Design and Preliminary Results of the Second Round Robin to Evaluate Search/Match Methods for Quantitative Powder Diffractometry" (Abstract), 26th Annual Conference on Applications of X-ray Analysis, Denver, CO, p. 42, August 1977.

4. C. R. Hubbard and R. Jenkins, "Design and Results of the Second Round Robin to Evaluate Search/Match Methods for Quantitative Powder Diffractometry," this volume.

5. G. G. Johnson, Jr., "Resolution of X-ray Powder Patterns," in J. S. Mattson, H. B. Mark, Jr., and H. C. MacDonald, Jr. (Eds.), Laboratory Systems and Spectroscopy, Chapter 3 (1977).

6. G. G. Johnson, Jr., "User Guide. Data Base and Search Manual," Publ. by JCPDS, International Center for Diffraction Data, Swarthmore, PA (available on request).

7. P. F. Dismore, "Computer Searching of the JCPDS Powder Diffraction File," Adv. in X-ray Analysis, Vol. 20, H. F. McMurdie, et al., Eds., Plenum Press, New York (1977).

THE CONTROL AND PROCESSING OF DATA FROM AN AUTOMATED X-RAY POWDER DIFFRACTOMETER

Chester L. Mallory and Robert L. Snyder

N.Y.S. College of Ceramics

Alfred University, Alfred, NY 14802

ABSTRACT

A program system design which will facilitate the exchange of automated diffractometer control programs is presented. The various procedures for collecting data and finding peaks, not involving profile fitting, were programmed, tested and evaluated. The optimum strategy evolving from this work involves a decision making algorithim which locates and removes the experimental threshold from the digital diffraction data, removes the $K_{\alpha 2}$ component, smooths and locates diffraction peaks via a second derivative procedure.

INTRODUCTION

The quality of X-ray powder diffraction data is the principal factor limiting the further development of powder techniques. A recent study (1) has shown that the average $\Delta 2\theta$ error of the high quality cubic pattern determined by the National Bureau of Standards is .015° while the average for all cubic patterns in the powder diffraction file is .091°. Neither of these values appears to be improving with time. The smallest $\overline{\Delta 2\theta}$ error available using manual methods is approximately .01°. We believe that the computer control of a powder diffractometer will ultimately permit $\overline{\Delta 2\theta}$ errors on the order of .001°. This, coupled with the removal of preferred orientation effects from the observed intensities by spray drying (2), will permit the revolutionizing of powder diffraction as an analytical tool.

The rapid technological advances in laboratory scale computers have allowed many laboratories to automate their powder diffraction equipment. But the development of algorithims and programming systems which exploit this new potential is still in its infancy. The purpose of this paper is to review and evaluate the various procedures which have been suggested or are currently in use, for the control of, and the collection and reduction of data from automated powder diffractometers. In addition we suggest a basic program system design which will promote the exchange of software between laboratories.

AUTOMATED X-RAY POWDER DIFFRACTION

A schematic picture of our automated X-ray powder diffractometer is shown in Figure 1. This type of system may be obtained in any of three ways:

1) completely designed and built in the laboratory (3)
2) purchase of an interface to adapt existing equipment

or 3) purchase of an entire automated system.

Whatever the source the essential components of the system will not differ significantly from the example in Figure 1. The common ground for all automated systems is the control of the 2θ position of the detector via a stepping motor and the ability to control and read the results from a scaler-timer. Since all systems may be generalized in this manner it is reasonable to propose some rules for the design of control programs which will facilitate exchange.

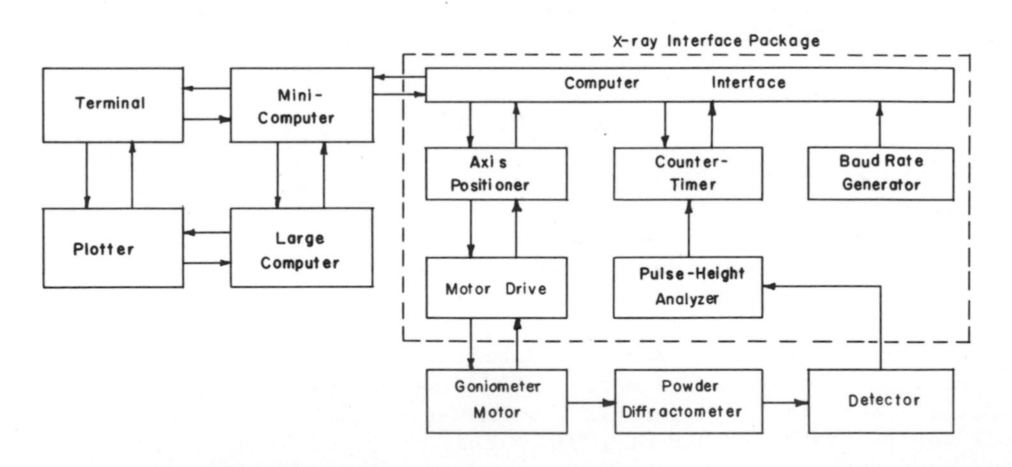

Figure 1. Automated X-ray Powder Diffraction System

A program system design for an automated powder diffractometer is schematically presented in Figure 2. There are three levels of control: Level I consists of user written interactive control algorithims which perform specific tasks such as quantitative and qualitative analysis, preferred orientation analysis, diffractometer alignment, etc. Level II programs are series of Fortran translation subroutines which convert control instructions to the instrument (i.e. move 2θ, count, integrate scan, etc.) into specific commands for the user's interface. Level III is comprised of assembly language interface drivers which carry out the communication between the computer and the 2θ stepping motor and scaler-timer. Some systems may perform these functions with hardware or firmware modules in place of the assembly langugage routines.

Once the level II and III routines are written or obtained from a manufacturer the user may begin to write the level I programs of interest to his laboratory. Since these programs often require months to years of development, the ability to exchange software is of concern to all. If the programming is considered as shown in Figure 2, level I programs with minor modifications can be exchanged by users. The two essential rules are:

1) All code should be written in ANSI standard FORTRAN.
2) All commands to the motor controller and scaler-timer should be accomplished by calls to subroutines with arguments in standard units.

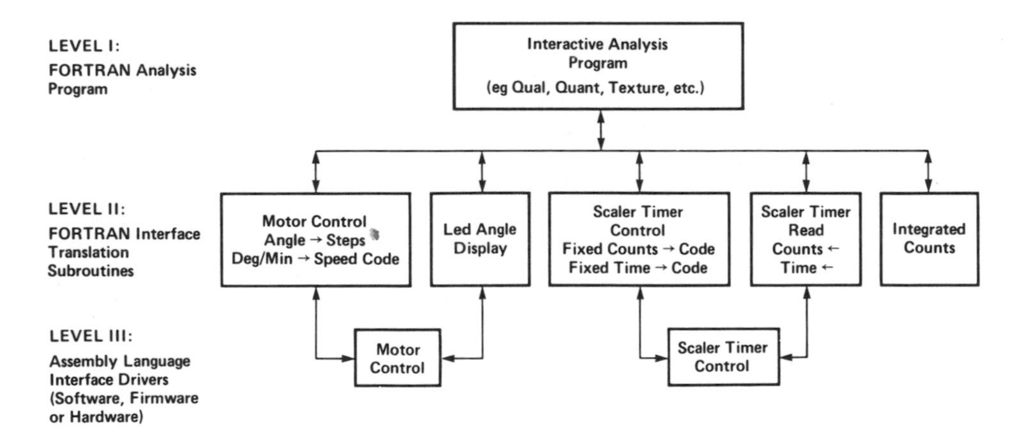

Figure 2. Program System Design for an Automated Powder Diffractometer

For example, the units to the level II motor control routines should be degrees for 2θ angle and degrees/min. for speed. The machine specific subroutine will then translate these standard units to number of motor pulses required to achieve the desired angle and the number of delay loops between pulses to control the the speed of the detector for a particular interface. These very machine specific values may then be passed on to the interface via the level III drivers.

The most important category of level I programs are those which perform the most elementary tasks - those of data collection and peak finding. To meet our goals of improving the quality of diffraction data by an order of magnitude, control and peak finding algorithims must be carefully evaluated to find the optimum strategy.

CONTROL ALGORITHIMS

Level I control algorithims can be divided into two categories - non-decision making or dumb and decision making or smart. The dumb algorithim simply moves the diffractometer to a specified 2θ position and initiates a counting sequence of the incoming intensity signal from the detector. It then moves the detector by a fixed 2θ increment and counts again. This cycle is repeated until the entire 2θ range specified by the user has been scanned. Upon completion the program has stored the digital diffraction pattern which is then analyzed for peak positions and intensities. Several examples of this type of algorithim appear in the literature. (4-6) It is very easy to program but requires large amounts of computer storage either in the main computer memory or on attached peripherals (discs, magnetic or paper tape, etc.). To reduce the amount of storage, the algorithim may perform some data reduction on the fly as it is collecting the pattern. In addition, it does not make intelligent use of time in that it spends as much time on a background point as it does on the top of a peak. On the other hand, the smart or decision making algorithim is able to vary its sampling technique (2θ step width, count time) throughout the specified 2θ range based upon logical and mathematical analyses of the previously collected data.

The speed of the computer in performing Real-Time decision making gives the user increased control over the diffractometer since both types of algorithims may be operated in a Real-Time or in an Off-Line mode. In the case of the dumb algorithim, data collection may be broken into several small regions and analyzed as the data groups are collected, thus saving on memory requirements and more effectively utilizing the computer. The smart algorithim needs only to store enough data to perform the peak search procedure. In more powerful computer systems, foreground/background operation offers the advantage of even more effective use of the computer. In foreground the data is intelligently collected and stored while in background the data is simultaneously analyzed. To fully utilize the abilities of the computer a smart algorithim running in a Real-Time mode performing data reduction on the fly should be implemented. Examples (7-8) are available illustrating such a system.

DATA REDUCTION

The analysis of experimental diffraction data can be handled in many ways. There are basically two different approaches to data reduction. One method is to perform a line profile analysis on the pattern. In this procedure mathematical functions are used to synthesize the actual experimental line profile. These functions describe contributions from the spectral dispersion of the X-ray source, aberrations arising from the instrument and diffraction process, and the contribution from the sample itself. This technique has been shown (9-10) to be superior to other approaches of of peak searching in that it can resolve heavily overlapped peaks, perform spectral stripping, and does not require data smoothing procedures. It also has the advantage of requiring less actual data points from the pattern resulting in faster Real-Time data collection runs. (11) For this work we have not attempted to evaluate this procedure beyond the literature accounts. Our focus instead is on the evaluation of the more numerous traditional peak finding algorithims not involving profile fitting. Other more traditional approaches to peak position determinations involves several discrete steps:

Background Determination

Various techniques for dtermining the background level in diffraction patterns have been suggested. Of these, two stand out as being superior both in generality and effectiveness. The first method was applied by Sonneveld and Vissar (4) to digital patterns obtained from Guinier films with an automated densitometer. In this technique every twentieth data point is used as a first approximation of the background. Since some of these sample points will occur on peaks an iterative process is used to reduce the data to the background level. Each data point is not allowed to be more than a given value above the mean of its adjacent neighbor.

The second method (5,12) is a similar procedure but a linear least squares line is used to describe sections of the pattern. Normally, two hundred data points are fit to a straight line. All points above this line are rejected from the data set and a new line is fit. The first 150 points are then corrected for background by subtracting the background as given by this second line from each data point. The procedure is continued by selecting the next two hundred points from 151 to 350 until the entire pattern has been corrected for background. This procedure has been shown to be very effective in removing the background due to amorphous scattering.

These techniques establish an approximation of the background in the pattern. Once the line is determined the background is subtracted from each original data point. The result is a linearized pattern with all the noise and statistical fluxuation remaining.

Noise Elimination

In a traditional peak search algorithim the process of discriminating between statistical noise and a weak peak is performed by establishing a noise level above which data may be considered to be part of a peak group. The typical procedure is as follows:

(1) Select a sample of N data points from the background corrected pattern ($N \sim 500$).

(2) Calculate the mean (μ) and standard deviation (σ) of the N points.

(3) Reject points above $\mu + n\sigma$ (where n varies from 2.5 to 4).

(4) Recalculate the mean and standard deviation. If the difference between the current σ and the last σ is large go to (2).

(5) The mean, μ, and the standard deviation, σ, are used to describe the noise level at a value of $\mu + n\sigma$.

The difficulty with setting the noise level at a value of $\mu + n\sigma$ is deciding upon the value of n. This value may range from 2.5 to over 4 (7) in some instances.

Another approach to the problem of threshold level determination is to extract the threshold directly from the observed data.(13) The algorithim to perform this is as follows:

(1) Extract one maximum point from each $1/4^{\circ}$ segment of the first 20° of the 2θ range scanned. Ignore any points above a certain cutoff (e.g. 100 cps).

(2) Fit a second or third order least sqaures polynomial to the extracted maxima.

(3) Reject those points which are above the fitted polynomial.

(4) Fit final least squares polynomial to the remaining maximum points.

(5) Evaluate the final polynomial for the first 10° of the 20° segment and subtract this value from the raw data.

(6) Select the next 2θ range from 10 to 30° and repeat until all 2θ ranges have been corrected.

The resultant data is free of any background noise and has at the same time been linearized. Any values which remain at this stage may be considered significant data from real peaks.

Spectral Stripping

In data reduction of powder diffraction results it may be useful to remove the $K_{\alpha 2}$ contribution from the experimental patterns. Three methods have been proposed for the separation of the α_1-α_2 doublet: the substitution correction (frequently referred to as a Rachinger correction (14), the fourier correction (15), and another technique based on a modified Rachinger correction (16).

The first two methods make several assumptions:

(1) The line profile of the α_2 is the same as the line profile of the α_1.

(2) The intensity ratio of the α_1 and α_2 component is accurately known.

(3) The doublet separation is accurately known.

Both the Rachinger and the fourier correction method produce oscillations in the corrected data at the high angle side of the residual α_1 profile. These oscillations may be reduced but not eliminated in the fourier method if an angular dependency of the α_1-α_2 separation is accounted for within a given profile. (17) The fourier correction suffers from an added constraint in that the doublet separation must be constant and the data must be equally spaced. This can only occur on a $\sin\theta$ scale. Since most data collection procedures are performed on a constant 2θ scale, these data must be transformed into an equally spaced $\sin\theta$ scale and the intensities must be interpolated between 2θ values.

A third technique offers results which are considerably better than those from the above methods. However, the process has a 2 $\sin\theta$ scale constraint and, therefore, the data must be transformed into this frame. Basically, the algorithim works as follows:

(1) Transform data into 2 sinθ units.

(2) Compute α_2 intensity components from each data point.

(3) Multiply the intensity values by various weighting factors (either 3, 5 or 7).

(4) Subtract α_2 intensity components (either 3, 5 or 7 of them) from the raw intensity values located at a calculable distance (in 2 sinθ) from the current position.

This algorithim is easily applied to a Real-Time data collection scheme to produce α_2 free data making analysis by conventional methods simpler.

Data Smoothing

Of the various data smoothing techniques proposed the most effective is one first introduced by Savitzky and Golay (18). The process involves a least squares fit of a polynomial to a number of sequential experimental data points. We will use seven to illustrate. The polynomial is evaluated at the mid point and the fourth data point from the real pattern is replaced with the evaluated point. The next set of data is chosen as points 2 through 8 and the process is repeated replacing the fifth experimental point. In so doing, counting statistics are smoothed over many data points and the short range fluctuations are removed. Algorithms which perform peak finding based upon derivatives of the data must first smooth the data. The above smoothing method conveiently allows for the evaluation of the first and second derivatives at each data point from the smoothing polynomial.

This process will degrade the resolution of the pattern if too large a span of points is used for the data smoothing. The most important parameter in data smoothing is the length of the range chosen to be smoothed. It has been recommended (19) that a smoothing range of approximately 0.7 times the full width at half-maximum of the narrowest line profile be used. Another important parameter to be considered in the procedure is the order of the polynomial used to fit the data. We have found a second degree polynomial to be adequate when closely spaced data points ($.01^{o}$ - $.0025^{o}$) are used.

Peak Finding

Routines for locating peaks may be divided into two categories: those which can detect shoulders and those which can not. Under the heading of those which can not locate shoulders are many methods. Some techniques are listed below:

(1) 3-point parabolic fit to maximum point. (8,20)

(2) Least squares parabolic fit to all data in a peak envelope.

(3) 'Tee-Pee' method of fitting linear least squares lines to the opposite sides of the peak and calculating peak location at the intersection of the two lines. (21)

(4) Midchord bisection of full width at half-maximum. (22)

(5) First derivative techniques will not detect shoulders unless a change in sign occurs. (22)

(6) Smart data collection algorithims which search for maximum intensity points after a minimum intensity location. (23)

These six peak search methods account for most of the published techniques. Since shoulders can not be detected with any of these techniques they are not worth considering further.

The techniques which will locate shoulders fall into two categories: a second derivative determination of the smoothed data and a polynomial regression fit to the data in a peak. We have found the latter technique suffers from stability problems in the mathematical fit. The dissymmetry of some peaks will significantly shift the location of the maximum of the polynomial rendering the method invalid. The process of evaluating the second derivative of the polynomial fit to the raw data in the Savitzky-Golay smoothing process yields a plot of two-theta versus the second derivative which indicates peaks at the maximum negative points. (4) The intensity and position may then be estimated by use of the maximum negative point and the roots of the second derivative on either side of that point. Techniques using the procedure yield accurate data for both intensity and position of the peaks. (4,24) Overlapped peaks may also be determined by this method.

CONCLUSION

Programming and testing each of the data collection and peak finding algorithims found in the literature or developed by us which meet the criterion of high inherent peak resolution leads to an optimum strategy. This strategy involves the use of a smart algorithim which takes advantage of the full capabilities of the computer to simultaneously collect and analyze data. Data should be analyzed by first determining and then removing the background threshold. (13) K_{α_2} peaks should be removed by Ladell's (16) modified Rachinger method. Finally, the data should be smoothed and peaks found using a Savitzky-Golay (4) second derivative procedure.

These procedures give the best results of any suggested in the current literature. They have been incorporated into an algorithim in our laboratory which has been written consistent with the program system design presented here. Although changes will almost certainly occur in this program in time, the current version is available by writing to the second author.

REFERENCES

1. R. L. Snyder, Q. C. Johnson, E. Kahara, G. S. Smith and M. C. Nichols, "An Analysis of the Powder Diffraction File," Lawrence Livermore Laboratory Report UCRL-52505, June 22, 1978.
2. S. T. Smith, R. L. Snyder, and W. E. Brownell, "Minimization of Preferred Orientation in Powders by Spray Drying," Advances in X-ray Analysis, this volume.
3. F. Chapman and R. L. Snyder, "Microcomputer Automation of a Powder X-ray Diffractometer," Abstr. Am. Ceram. Soc., Mtg. Detroit (1978).
4. E. J. Sonneveld and J. W. Visser, "Automatic Collection of Powder Data from Photographs," J. Appl. Cryst. 8, 1 (1975).
5. A. Segmüller and H. Cole, "Procedures to Run an Automated Micro-Densitometer on a Shared Computer System," Advances in X-ray Analysis, Vol. 14, p. 338-351, Plenum Press (1970).
6. A. Segmüller, "Automated X-ray Diffraction Laboratory System," Advances in X-ray Analysis, Vol. 15, p. 114-122, Plenum Press (1971).
7. R. Jenkins, D. J. Haas, and F. R. Paolini, "A New Concept in Automated X-ray Powder Diffractometry," Norelco Reporter 18 (2) (1971).
8. M. R. James and J. B. Cohen, "Study of the Precision of X-ray Stress Analysis," Advances in X-ray Analysis, Vol. 20, p. 291-307, Plenum Press (1976).
9. D. Taupin, "Automatic Peak Determination in X-ray Powder Patterns," J. Appl. Cryst. 6, 266 (1973).
10. T. C. Huang and W. Parrish, "Accurate and Rapid Reduction of Experimental X-ray Data," App. Phys. Lett. 27 (3), 123 (1975).
11. W. Parrish, G. L. Ayers and T. C. Huang, "Rapid Recording and Reduction of X-ray Diffractometer Data," Paper D6, Amer. Cryst. Assoc. Summer Meeting Abstr. 5 (2) (1977).
12. R. P. Goehner, "Background Subtract Subroutine for Spectral Data," Anal. Chem. 50 (8) 1223-4 (1978).
13. C. L. Mallory and R. L. Snyder, "A Method for Determining the Threshold of Significance from Digital Powder Diffraction Data," J. Appl. Cryst. submitted (1978).

14. W. A. Rachinger, "A Correction for the $\alpha_1\alpha_2$ Doublet in the Measurement of Widths of X-ray Diffraction Lines," J. Sci. Inst. 25, 254 (1948).
15. A. Gangulee, "Separation of the α_1-α_2 Doublet in X-ray Diffraction Profiles," J. Appl. Cryst. 3, 272 (1970).
16. J. Ladell, A. Zagofsky, and S. Pearlman, "$CuK_{\alpha 2}$ Elimination Algorithim," J. Appl. Cryst. 8, 499 (1975).
17. R. Delhez and E. J. Mittemeyer, "A Comparison of Two Computer Methods for Separation of the α_1-α_2 Doublet," Acta. Cryst. A31, Part S3 (1975).
18. A. Savitzky and M. J. E. Golay, "Smoothing and Differentiation of Data by Simplified Least Squares Procedures," Anal. Chem. 36, 1627 (1964).
19. T. H. Edwards and P. D. Wilson," Digital Least Squares Smoothing of Spectra," Appl. Spec. 28 (6), 541-5 (1974).
20. D. Kirk and P. B. Caulfied, "Location of Diffractometer Profiles in X-ray Stress Analysis," Advances in X-ray Analysis, Vol. 20, p. 283-89, Plenum Press (1976).
21. A. W. Westerberg, "Detection and Resolution of Overlapped Peaks for an On-line Computer System for Gas Chromatographs," Anal. Chem. 41, 1770 (1969).
22. A. Segmüller, "Automated Lattice Parameter Determination on Single Crystals," Advances in X-ray Analysis, Vol. 13, p. 455, Plenum Press (1969).
23. R. Jenkins and R. G. Westberg, "Use of an Automated Powder Diffractometer for the Analysis of Rock Samples," Advances in X-ray Analysis, Vol. 16, p. 310, Plenum Press (1972).
24. R. P. Goehner, General Electric, Schenectady, N.Y. 12301; private communication.

A PRELIMINARY REPORT ON THE DESIGN AND RESULTS OF THE SECOND ROUND ROBIN TO EVALUATE SEARCH/MATCH METHODS FOR QUALITATIVE POWDER DIFFRACTOMETRY.†

Ron Jenkins*, and Camden R. Hubbard**

Philips Electronic Instruments, Inc., Mahwah, NJ

National Bureau of Standards, Washington, D. C. 20234

ABSTRACT

Six computer synthesized data sets, each representing a mixture, and one physical mixture were prepared and widely distributed in order to study the various search/match methods and factors which affect their success. A total of 67 returns were received representing eight countries and three search methods. The participants were primarily from industrial laboratories. The average score exceeded 90%. The Hanawalt search method yielded the best overall score. Use of the Frequently Encountered Phases subfile decreased the search time by about 40% and marginally increased the success rate. For the physical mixture the $\Delta d/d$ and $\Delta I/I$ values were measured to about 2/1000 and 40% respectively. Use of an internal standard improved the d-values by a factor of 2 and resulted in better search/match performance.

INTRODUCTION

This paper briefly outlines the format and results of a round robin to evaluate search/match methods for qualitative powder

---

* Chairman, Computer Subcommittee, JCPDS-International Centre for Diffraction Data

** Chairman, Education Subcommittee, JCPDS-International Centre for Diffraction Data

† Contribution of the National Bureau of Standards. Not Subject to copyright

diffractometry. This round robin is a more extensive follow-up of a similar round robin reported in these proceedings in 1977.[1] This first round robin came to the following tentative conclusions:

- Hand-searching and computer-searching techniques are equally effective for inorganic and mineral analyses.
- Hand-searching is superior to computer-searching for organic analyses.
- In computer-searching techniques more accurate d-spacings (∿ 1 part in 1000 at d=1.5Å) are required than in hand-searching.
- Use of elemental data in computer-searching permits use of lower quality d-values.
- Typical accuracy of the experimental d-values lies in the range 1 to 3 parts in 1000.

The second round robin was designed to test and extend these tentative conclusions.

ROUND ROBIN DESIGN GOALS AND PROBLEM SYNTHESES

Three basic design goals were built into this test and these are listed in Table 1.

TABLE 1

ROUND ROBIN DESIGN GOALS

1. Compare methods
   a) success rate
   b) time
2. Examine factors affecting success
   a) subfiles
   b) accuracy of d and I values
   c) elemental data
   d) common systematic errors in data
3. Evaluate quality of d and I values used for identification

TABLE 2

SYNTHESIZED MIXTURES

Problem	Phases	PDF#	I^{rel}_{wt}	Factors	Set A	Set B
1	Fe_3S_4	16-0713	40	elemental info	no	no
	FeS_2	6-0710	10	$\sigma_{2\theta}$	0.02°	0.05°
	CuS	6-0464	20	σ_I	2.5%	5%
	$CuFe_2O_4$	6-0545	30			
2	$PbBr_2$	5-0608	40	elemental info	no	no
	$PbCl_2$	5-0416	10	$\sigma_{2\theta}$	0.05°	0.02°
	PbO $PbSO_4$	18-0702	30	σ_I	5%	2.5%
	$Pb_3O_2Cl_2$	25-0443	20			
3	$BaSO_4$	24-1035	30	elemental info	yes	no
	$BaCO_3$	5-0378	40			
	$CaSO_4$	6-0226	15	$\sigma_{2\theta}$	0.05°	0.05°
	$SrSO_4$	5-0593	15	σ_I	5%	5%
4	$CaCO_3$(calcite)	5-0586	30	elemental info	no	yes
	$CaCO_3$(aragonite)	5-0453	30	$\sigma_{2\theta}$	0.05°	0.05°
	$CaCO_3$(vaterite)	13-0192	20	σ_I	5%	5%
	$ZnCO_3$	8-0449	20			
5	$Ca_3Si_2O_7$	22-0539	40	elemental info	yes	no
	Fe_2SiO_4	20-1139	30	$\sigma_{2\theta}$	0.05°	0.05°
	α-SiO_2	5-0490	20	σ_I	5%	5%
	$MgSiO_3$	19-0768	10	systematic error in 2θ	0.05°	0.05°
6	meta-dinitrobenzene	11-0855	40	elemental info	yes	yes
	2,4-dinitrophenol	23-1670	40	$\sigma_{2\theta}$	0.05°	0.05°
	2,6-dinitrophenol	26-1826	20	σ_I	5%	5%

In order to achieve these goals it was decided to base the majority of the test data on synthesized patterns since this, a) allowed controlled modification of each problem, b) removed measurement variability from the study of search match methods, c) controlled elemental information and d) allowed meaningful averaging of returns. To this end, a series of six patterns, each representing 3-5 phases, were made up from starred JCPDS cards using a random number generator to add various errors to the data sets. Normal errors with a standard deviation of either 0.02° or 0.05° in 2θ and either 2.5% or 5.0% in intensity were added to each line to obtain high quality and average quality data respectively. Additionally a systematic error of 0.05° 2θ error was added to one problem representing a 2θ zero offset and sample displacement error. Finally, a resolution condition was applied in the pattern synthesis such that lines with $\Delta(2\theta)<0.1°$ 2θ be averaged. Each problem was divided into two parts, an A set or B set of data, designed to reduce perturbations due to analyst experience. Table 2 lists the phases of the synthesized mixtures, their Powder Diffraction File[2] number (PDF #), their weights and the relevant factors for the A and B sets. Each participant received all six problems for set A or set B.

Roughly 120 data sets were circulated and 67 returns were accepted. Two returns were rejected because of the extremely poor quality of response. Table 3 breaks down the participants by country, search method, and organization.

TABLE 3

SECOND ROUND ROBIN PARTICIPATION

Country		Method (Problem 1)		Organization	
		Hand			
Australia	3	Hanawalt	42	Industry	37
Canada	3	Fink	8	Academic	18
Denmark	1	Computer*		Government	7
Germany, Fed. Rep.	3	Johnson/Vand	12	Other	5
Japan	2	Nichols	1		
Netherlands	3	ZRD	1		
U.K.	2	3M	1		
U.S.A.	50				

*See Note Added in Proof

RESULTS

Table 4 summarizes the results of the hand searching methods for the inorganic and mineral mixtures. The rating system used was identical to that used in the first round robin.[1] Score sheets permitted listing a phase as "positively" or "possibly" identified. For identification of a correct phase a score of +5 or +3 was assigned for positive or possible identification. For an incorrect phase a score of -3 or -1 was assigned for positive or possible, respectively. The score for correctly indicating that not all phases were identified was +1. For the four phase inorganic and mineral mixtures the maximum rating was 20. For the three phase organic mixture the maximum was 15. These data show that the Hanawalt method is the preferred method, and those using it spent less time than with the Fink method. Also, those using the Hanawalt method obtained somewhat better results,probably due to the good quality of the intensities in the synthesized patterns which favors the Hanawalt search method compared to the Fink search method.

The computer aided methods were separated into two categories: Johnson/Vand method and various others. Table 5 summarized the data on these computer aided methods for the inorganic and mineral mixtures.

TABLE 4

HAND METHODS

Inorganic and Mineral Mixtures

Feature	Hanawalt	Fink
# of returns	226	48
average rating	19.1	18.1
average time (hr.)	2.3	2.9

TABLE 5

COMPUTER AIDED METHODS

Inorganic and Mineral Mixtures

	Johnson/Vand	Other
# of returns	58	17
average rating	16.8	20.0
average time (hr.)	1.7	0.7
# with rating = 20	33	17

In Table 5 one sees that the Johnson/Vand code is widely used and in over half the cases gave the complete and correct answer. However, as indicated by the average time of 1.7 hrs. the output from this code needs considerable analysis. Although it appears less successful on the average, it must be remembered that some of the other codes involved use of a much smaller data base and are usually being employed by the program designer. Compared with the hand searching data given in Table 4, it is seen that, on the average, the Johnson/Vand code was less successful but required less time than the hand searching methods. In the case of the organic mixture #6, hand searching was faster than computer searching -- the time/rating for the Hanawalt hand search was 1.7 hrs./13.2 compared to the Johnson/Vand computer search averages of 1.9 hrs./9.0. Neither the Fink method or other computer codes were applied to the organic problem. The maximum rating for this sample was 15.

For both inorganic and mineral mixtures, use of the FEP (Frequently Encountered Phases) subfile as a first search decreased the search time by about 40% as well as marginally increasing the success rate. Another interesting sidelight came out of this study, namely that a maximum optimum time was found to occur for hand searching. A plot of success rate as a function of analysis time revealed that if a perfect result were not found within about 2-3 hours it was unlikely that a perfect result would be found at all. Similarly, in those cases where a perfect score was not obtained, analysis times in excess of 2-3 hours did not significantly enhance the success rate.

THE PHYSICAL MIXTURE

In addition to the six synthesized data sets most participants were also provided with an actual four phase mixture. This consisted, by weight, of 50% $BaCl_2 \cdot 2H_2O$, 15% KI, 20% ZnO, plus 15% Si. Samples 7A were identical to samples 7B, except that in the case of sample 7B, the Si was identified as an added d-spacing standard.[3] The major purpose of this specimen was to evaluate measured d and I quality as well as estimating the value of an internal standard for routine qualitative analysis. Six test lines (d_i, I_i i=1,2...6) covering the d-spacing range from 5.45 to 1.648Å were selected as a judge of d quality using a quality factor of the form:

$$1/6 \sum \left([d_i - d_i(\mathrm{PDF})] \;/\; d_i(\mathrm{PDF}) \right)$$

where d_i(PDF) is the d-spacing of the i-th line as listed in the Powder Diffraction File.[2]

Three of these lines were taken as a judge of I quality using a further line as the reference intensity. Again a quality factor was used, in this case being:

$$1/3 \sum \left([I_i/I_{ref} - I_i(NBS)]/I_i(NBS)\right)$$

where I_i(NBS) is the relative intensity of the i-th line determined by averaging ten observations from six samples.

Three breakdowns of data were then used to examine the quality in terms of equipment (Table 6), of specimen preparation method (Table 7), and for use of internal standard (Table 8). In several cases the number of lines or number of analyses are quite small. This limits the conclusions that can be drawn from the data in Tables 6 and 7.

The data included in Table 6 indicate that there is little difference in the quality of d-spacings measured on automated or manual diffractometers, or with Guinier cameras. The Debye-Scherrer data is worse by a factor of 3-4. Unfortunately, only one Debye-Scherrer data set is involved in this analysis so not too much significance can be placed on this single result. The authors were surprised to note that Debye-Scherrer and Guinier intensity data quality was marginally better than both automated and manual diffractometer data.

By far the best d and I data were obtained by rear-packing of the specimen, as shown in Table 7. The most popular specimen techniques were "pressing" (front loading) and "smearing" a slurry on glass but, of all methods compared, pressing gave the worst d and I quality. "Side drifting" (side loading) gave data of average quality. The differences between the ratings may not be significant due to the limited number of returns for many categories.

We were also surprised to observe that less than one half of the participants who were supplied with specimen 7B actually made use of the internal standard data contained therein. As is clearly seen in Table 8, use of internal standard data to correct 2θ values improved d quality by a factor of 2 and resulted in better search/ match rating.

CONCLUSIONS

Although a detailed analysis of all the second round robin data has yet to be completed the findings of this brief review are indicated in Table 9. It is the intention to publish a detailed study including the original data and a list of participants as an NBS Report in the very near future.

TABLE 6

QUALITY VERSUS EQUIPMENT

	# Lines	d-Quality	# Lines	I-Quality	#	Rating
CAMERAS						
Debye-Scherrer	6	0.0054	3	0.26	1	20.0
Guinier	28	0.0014	15	0.37	4	18.0
DIFFRACTOMETERS						
Manual	137	0.0016	66	0.46	24	17.7
Automated	51	0.0013	22	0.43	8	17.9

TABLE 7

QUALITY VERSUS MOUNTING

Sample Mount	# Lines	d-Quality	# Lines	I-Quality	#	Rating
Fiber	6	0.0054	3	0.26	1	20.0
Dusted Slide	-	--	3	0.68	1	20.0
Side Drifted	12	0.0015	9	0.29	2	17.0
Thin Smear	81	0.0014	34	0.37	13	18.0
Pressed	94	0.0017	42	0.57	16	17.0
Rear Packed	17	0.0011	9	0.13	2	20.0

TABLE 8

QUALITY VERSUS USE OF INTERNAL STANDARD

Use of Standard	# Lines	d-Quality	#	Rating
None	157	0.0017	24	17.4
Visual Check on 2θ-values	28	0.0019	4	17.5
Correct 2θ-values	49	0.0007	9	19.0

TABLE 9

BRIEF REVIEW OF FINDINGS OF ROUND ROBIN

1. High number of correct results.

2. Time/success curves indicate that 2-3 hours is the "optimum" time, for manual searching.

3. Hanawalt system preferred for hand searching.

4. Those using their own computer search/match program had good scores. Many using the Johnson/Vand program distributed by the JCPDS also had good scores, but some obtained noticeably worse results.

5. Computer search/match poor for organic.

6. Manual searching less sensitive to quality of d and I data than computer searching.

7. FEP subfile helped in searching success and reduced search time.

8. $\Delta d/d$ values were measured to around 2/1000. $\Delta I/I$ values were measured to around 40%.

9. Diffractometer and Guinier Camera gave about the same quality of result. Debye-Scherrer Camera 3-4 times worse.

10. Rear packing was the best specimen preparation technique.

11. Internal standard data provided was generally not used. When used it lead to significantly better d-values and improved success.

We would like to thank all of those who participated in this round robin, and we particularly acknowledge the help of our colleagues at the JCPDS-International Centre for Diffraction Data.

REFERENCES

1. Jenkins, R., "A Round Robin Test To Evaluate Computer Search/Match Methods for Qualitative Powder Diffractometry," Adv. X-Ray Anal., 20 (1977) p 125-137.

2. JCPDS-International Centre for Diffraction Data, 1601 Park Lane, Swarthmore, PA 19081 U.S.A.

3. The Si (SRM-640) was kindly supplied by the Office of Standard Reference Materials, National Bureau of Standards, Washington, D.C. 20234 U.S.A.

NOTE ADDED IN PROOF

Listed below are computer codes which were used for analysis of one or more round robin problems. Users of the NICHOLS and ZRD codes both obtained the correct answers to the organic problem. Two users of the Johnson/Vand code also obtained perfect scores on this problem.

Code	Contact/reference	Data Base	Problems
JOHNSON/VAND	G. G. Johnson, Jr. Penn State Univ. University Park, PA 16801	PDF	1-6
NICHOLS	M. C. Nichols Sandia Laboratory Livermore, CA 94550	PDF	1-6
ZRD	L. K. Frevel J. Appl. Cryst, 9 199-204 (1976)	Dow File	1-6
3M	W. E. Thatcher 3M Co., P.O. Box 33221 St. Paul, MN 55133	PDF	1-5
XRD SEARCH	D. B. McIntyre Pomona College Claremont, CA 91711	Minerals	4,5
WAIT	B. H. O'Connor W. Australia Inst. of Tech. S. Bently, W. Aust.	Minerals	4

APPLICATION OF GUINIER CAMERA, MICROCOMPUTER CONTROLLED FILM DENSITOMETRY, AND PATTERN SEARCH-MATCH PROCEDURES TO RAPID ROUTINE X-RAY POWDER DIFFRACTION ANALYSIS

J. W. Edmonds
Dow Chemical Company
Midland, MI 48640

and

W. W. Henslee
Dow Chemical Company
Freeport, TX 77541

INTRODUCTION

In most routine chemical analyses, a trade-off is made between quality of data and time required to obtain and analyze the data. In X-ray powder diffraction, identifications are normally made by Debye-Scherrer film methods or by medium speed (1-2° 2Θ/min.) diffractometry, with or without an internal standard. With one notable exception [1], the inherent precision of the Guinier camera geometry has been virtually ignored as too expensive or time consuming for routine work, or relegated to special projects [2,3]. The accessibility of microcomputers, however, not only makes it economically and realistically feasible to automate the equipment previously used for special Guinier projects, but to extend the overall precision of observed d-spacings into the area of routine analysis. Search-match procedures benefit from the increased data precision to such an extent that they can be used routinely to propose the identity of major pure phases and release the analyst to concentrate on minor components and impure phases which may be subject to lattice constant shifts.

INSTRUMENTATION AND DATA MANIPULATION

Guinier data of high quality are more dependent upon camera quality and stability, sample preparation technique, and the use of precise internal d-spacing standards than they are upon computational hardware and film reading devices, provided the latter are able to resolve linear distances on the film corresponding to 0.005 °2Θ (∿ 0.010 mm on a film from a Guinier camera of 114.59 mm radius). The instrumentation used for this work includes a conventional microcomputer/FORTRAN IV operating an older emission spectroscopy photometer [2], and a desk-top calculator/unique language operating a rebuilt single crystal film unit. Other than amplifiers, the specific interfaces are stock items from the computer manufacturers.

Films are digitized directly onto mass storage devices at rates exceeding 25 mm/min. ($\geq$ 13,568 I_o data per film). Data reduction software locates peaks, identifies internal reference lines, linearly interpolates between reference lines to obtain d_{obs}, and computes integrated I_{obs}. The hardware and software sequence of data collection/reduction also permits inexpensive attachment of the diffractometer ratemeter output and provides the same data manipulation capabilities for diffractometer data.

DATA QUALITY

The quality of data that can be obtained by the two systems is shown in the comparison of d-spacings obtained for an identical sample (JCPDS/NBS Round Robin 7) prepared in the individual laboratories and processed from a single photometer pass. (Table 1. The error in d due to mechanical reproducibility is in the least significant digit reported.)

Discrepancies will be due to errors in peak location caused by signal noise (the software currently interpolates peak position at peak maximum), nonlinearily in film travel, diffraction line flaring due to sample preparation, and distortion of line shape for intense or over-exposed lines.

Most of these errors can be minimized for more demanding work by averaging d-spacings obtained from scanning multiple exposures obtained from the same

Table 1
INTERLABORATORY COMPARISON OF HIGH RESOLUTION D-SPACINGS SEMI-AUTOMATED X-RAY DIFFRACTION
SAMPLE = NBS ROUND ROBIN 7A, B

Dow Freeport, Å	Dow Midland, Å
5.977	5.984
5.733	5.732
5.465	5.462
4.945	4.944
4.853	4.849
4.503	4.503
4.432	4.432
4.336	4.336
4.081	4.082
3.662	3.663
3.623	3.626
3.533	3.535
3.390	3.391
2.4101	2.4091
2.3871	2.3875
2.3652	2.3656
2.2543	2.2552
2.2281	2.2282
2.2086	2.2095
2.1486	2.1498

sample. The improved internal agreement is demonstrated in the following data for $BaSO_4$ (the Hannawalt mineral), which has been averaged from four separate films (Tables 2,3).

Table 2
$BaSO_4$ (Hannawalt Mineral)
Least squares cell constants*

a:	8.8613 (2) Å	α:	90.°
b:	5.4515 (1) Å	β:	90.°
c:	7.1446 (1) Å	γ:	90.°

*Angles fixed. 57 (hkl) and d's for 4.43 Å $\geq$ d $\geq$ 1.1932 Å
Calculated σ is indicated in parentheses.

Table 3

$BaSO_4$ (Hanawalt Mineral)

Data agreement for a selection of (hkl)

(hkl)	d_{obs}, Å	d_{calc}, Å	Δd, Å	Δ2Θ($CuK_{\alpha 1}$)
200	4.430	4.431	-0.001	0.003
111	3.8946	3.8932	0.0014	-0.008
201	3.7657	3.7654	0.0003	-0.002
102	3.3145	3.3132	0.0013	-0.011
220	2.3215	2.3216	-0.0001	0.001
221	2.2079	2.2079	0.0000	0.000
*113	2.1188	2.1191	-0.0003	0.005
*401	2.1158	2.1159	-0.0001	0.003
*122	2.1048	2.1050	-0.0002	0.003
*312	2.1005	2.1006	-0.0001	0.002
230	1.6813	1.6813	0.0000	0.000
+421	1.6716	1.6714	0.0002	-0.006
+114	1.6674	1.6671	0.0003	-0.012
+204	1.6565	1.6566	-0.0001	0.004
224	1.4160	1.4157	0.0003	-0.018
702	1.1932	1.1932	0.0000	0.000

* = Quartet area used as resolution test
+ = Triplet area used as d-spacing check

While the average Δ2Θ (d_{obs} -d_{calc}) for typical high quality NBS data ranges from 0.02° to 0.04° 2Θ, the worst Δ2Θ for $BaSO_4$ is 0.018° and the average (Δ2Θ) is 0.005°.

Internal data agreement of this magnitude places increased demands upon the quality of camera resolution and the use of well characterized internal d-spacing standards. Since commercial Guinier cameras/monochromators offer the required resolution, care in sample preparation and routine use of internal standards is the minimum necessary to obtain high quality data on film. Only a convenient means of registering the linear precision from the film is required to obtain data of this quality on a routine basis.

SAMPLE PREPARATION

The Guinier camera operates in the sample transmission mode for the forward reflection geometry, and this requires the preparation of a thin sample. Too thick a sample causes defocusing of diffracted lines, and attempts to compress the sample may distort the sample/support from the focusing condition. Since technique can vary, it is advisable to prepare and retain a specific sample as a camera resolution/alignment check. (See below).

The sample support is usually a thin amorphous film (e.g., mylar) or a foil which can also serve as an internal standard, (e.g. Al, see below). An adhesive medium is used to hold the powdered sample on the support (e.g., vaseline, cellosolve), and must also be very thin ("painting" a thin smear with a camel hair paint brush works adequately, and does not wrinkle the support). The sample/internal standard must be well ground, since large particles will orient and cause "streaked" diffraction lines whose position is difficult to define accurately. The sample is placed on the support/adhesive (within the die opening, if required) and the sample holder gently tapped, causing the powder to move across the adhesive("dusting").

Excess sample is poured off the support, and the sample holder tapped to release poorly held sample (loose sample will contaminate adjacent sample areas, if the camera is equipped to handle multiple samples at one exposure).

INTERNAL STANDARDS

Internal standards of accurately known (and stable) d-spacings are necessary to obtain data of high quality. For lattice parameter determination, a material such as Si (NBS Silicon Standard 640) is advisable, plus a standard for high d-values, such as As_2O_3. Well crystallized As_2O_3 (grown by sublimation in sealed tubes) is cubic (a_o=11.0743 Å) and provides lines at 6.39375 Å (111) and 3.19592 Å (220). An additional standard can be pure α-SiO_2 (e.g. Min-U-sil 5, Pennsylvania Glass Sand Corporation, 3 Penn Center, Pittsburg, PA 15235.

Agreement between Guinier data from three independent laboratories is given in Table 4.

For routine work where identification is the goal, the sample can be dusted on Al foil with As_2O_3 mixed in the vaseline adhesive ($\sim$ 7.5% by weight). The sample and internal standards are not homogeneously co-mixed, but the loss in quality is not significant for routine purposes, as the data can easily be used for search/match procedures (See below).

Table 4
α-SiO_2 Quartz

(hkl)	Midland*	Texas*	A. Brown+	$d_{lit, obs}$
100	4.2545	4.2557	4.2528	4.26
101	-	3.3426	3.3392	3.343
110	2.4567	2.4565	2.4566	2.458
102	2.2811	2.2811	2.2815	2.282
111	2.2365	2.2362	2.2361	2.237
200	2.1273	2.1273	2.1268	2.128
201	1.9795	1.9794	1.9798	1.980
112	1.8177	1.8174	1.8181	1.817
	a = 4.9129 (1)		4.9125 (1)	4.913
	c = 5.4045 (2)		5.4045 (3)	5.405

σ is reported in parenthesis for least squares results
*Min-U-sil 5 (5 μm nominal particle size)
+Belgian sand. See Reference 1.

RESOLUTION STANDARD

A stable standard with closely spaced lines is recommended to be kept to check on camera alignment. A good material is $BaSO_4$ which has been shown to possess a resolved quartet (see Table 3), i.e., only a solid solution containing ∿ 1% Sr or other atomic species will show the splitting, not pure $BaSO_4$. Resolution of this quartet is also highly dependent upon thermal history/crystallite size, and excellent patterns are obtained by calcining the mineral at 800°C for 18 hours, and dusting unground sample on the support. Sample preparation (thickness) is also critical for resolution, and serves to develop good technique. The successful preparation should be retained, if possible, for reference.

SEARCH/MATCH PROCEDURES

Routinely obtained d-spacings ("heterogeneous internal standard", i.e., Al supporting foil and As_2O_3 in the vaseline) are of such increased precision as to permit the application of search match procedures on a sample to sample basis to propose candidate matches on the major phases present. The Frevel ZRD search match program [4,6] can operate inhouse and be executed serially after data collection/reduction with no

Table 5

NBS/JCPDS ROUND ROBIN 7B

	SAMPLE		Si PHASE 1		KI PHASE 2		ZnO PHASE 3		$BaCl_2 \cdot 2H_2O$ PHASE 4		I(N)-
N	D(N)	I(N)	D/A	I	D/A	I	D/A	I	D/A	I	SUM I
1	5.720	19							5.725	9	10
2	5.452	66							5.440	66	0
3	4.938	54							4.938	33	21
4	4.844	19							4.845	9	10
5	4.496	29							4.497	23	6
6	4.427	61							4.427	49	12
7	4.333	15							4.335	9	6
8	4.138	12 – VASELINE									12
9	4.079	38			4.080	31					7
10	3.661	29							3.660	23	6
11	3.622	13							3.622	10	~~13~~ 3
12	3.533	100			3.530	73			3.565	6	~~27~~ 21
13	3.389	29							3.389	19	10
14	3.360	15							3.360	9	6
15	3.135	62	3.138	54							8
16	2.949	29							2.949	23	6
17	2.929	20							2.928	26	-6
18	2.908	43							2.908	46	-3
19	2.862	24							2.859	26	-2
20	2.815	40					2.816	43			-3
21	2.711	31							2.710	26	5
22	2.604	35					2.602	34			1
23	2.547	36							2.545	33	3
24	2.528	8							2.527	9	-1
25	2.498	46			2.498	51					-5
26	2.476	72					2.476	61	2.468	5	6
27	2.409	23							2.407	16	7
28	2.387	16							2.386	6	10
29	2.365	14							2.365	9	5
30	2.255	23							2.252	6	17
31	2.227	11							2.227	9	2
32	2.209	7							2.208	9	-2
33	2.131	24			2.131	21					3
34	2.118	13							2.118	6	7
35	2.087	23							2.085	16	7
36	2.062	6							2.061	6	0
37	2.040	20			2.039	19			2.042	6	-5
38	1.999	15							1.998	6	9
39	1.956	7									7
40	1.921	30	1.920	32							-2
41	1.911	17					1.911	17			0
42	1.767	11			1.767	11					0
43	1.638	29	1.638	19							10
44	1.625	26			1.621	5	1.626	24			~~2~~ 3
45	1.579	24			1.580	17					7
46	1.557	9									9
47	1.500	9									9
48	1.478	25					1.477	21			~~25~~ 4
49	1.442	9			1.442	10					-1
50	1.379	30					1.379	17			13
51	1.358	22	1.357	4	1.360	2	1.359	8			~~22~~ ~~14~~ ~~12~~ 8
52	1.307	7									7
53	1.247	15	1.246	7	1.249	1					~~15~~ ~~14~~ 7
54	1.109	7	1.108	9							~~7~~ -2
55	1.093	11					1.093	6			5
56	1.046	35	1.045	4							~~35~~ 31
57	1.042	6					1.042	6			0
58	1.016	6					1.016	3			3

Handwritten entries are from final editing by the analyst

operator intervention once film scanning begins. The JCPDS/NBS Round Robin sample 7 (Table 1) was run in this manner, and resulted in the correct identification of all four phases present. (The Frevel ZRD program uses the mini search file approach, and all four phases were in this standard file.) Table 5 shows the final ZRD output for the matched phases, and the ΔI residual for each peak. D-spacings are output from ZRD only to ± 0.001 Å due to the internal program format (output from data reduction is reported to ± 0.0001 Å, however). D-spacings indicated by hand are those which were not matched due to the narrowness of the match windows.

For the complete analysis, the operator spent 10 minutes in sample preparation and film development, and less than 2 minutes in defining computer variables. Data collection, reduction, search/match were performed without intervention and required 21 minutes, most of which was for teletype output (See Table 5).

CONCLUSIONS

The high quality of data inherent in the Guinier geometry camera can be routinely extracted by the use of internal standards and precision film densitometry. This data can be obtained for routine identification purposes and greatly increase the utility of computer search/match procedures, releasing the analyst from many routine tasks.

REFERENCES

1. Brown, A., X-ray Powder Diffraction with Guinier-Hägg Focusing Cameras, Aktiebologet Atomenergi, Pub. AE-409, Studsvitz, Nyköping, Sweden Rev. 1977.
2. Frevel, L. K., Anal. Chem. 38(13), 1914-20 (1966). Frevel, L. K., Advan. in X-ray Anal., Vol. 20, 15-25, Plenum Press (1977).
3. Clark, D. E., Scott, G. J., and Henich, L. L., Ceramurgia Int., 1(1), 19-22 (1975).
4. Frevel, L. K., Adams, C. E., and Ruhberg, L. R., J. Appl. Cryst., 9 199-204 (1976).
5. Nichols, M., Snyder, R., Discussions at the August, 1977, meeting of the Amer. Cryst. Assoc., and March, 1978 JCPDS meeting.
6. Edmonds, J. W., in preparation.

COMPUTER AUTOMATION OF X-RAY POWDER DIFFRACTION

J.S.Lindsey, C.P.Christenson, and W.W.Henslee

Central Laboratory, Dow Chemical U.S.A.

Freeport, Texas 77541

ABSTRACT

The interfacing of a Hewlett-Packard 9825 calculator to an Enraf-Nonius microdensitometer and a Philips powder diffractometer provides automated data collection and manipulation on a routine basis. Data is acquired and stored on a floppy disk; background corrections are made; diffraction peak maxima are found, normalized and plotted; a table of approximate d-spacing (±0.01Å) is printed. The analyst designates the internal standard peaks. Corrected d-spacings are then computed, tabulated and printed above the appropriate peak. An accurate d-spacing scale is drawn on the plot. The total time for a scan is approximately twenty minutes with only about five minutes of operator time required. The program is conversational and requires little prior knowledge of the software by the operator. Data interpretation is further aided by supporting programs to calculate patterns from literature, determine lattice constants for observed phases, help index patterns, deconvolute overlapping peaks, carry out crystallite size determinations and perform search/match procedures. Such supporting programs are possible because, even though the system can be used in a "dedicated" manner, it is still a stand-alone computer. Thus, the desired degree of automation in data collection has been obtained without sacrificing flexibility.

INTRODUCTION

The automation of X-ray powder diffraction has been developing for some time. The early approach was to acquire data off-line and reduce it on a large computer via batch mode. Although this is a great advance over manual systems, its application is limited and time consuming. More recently, automated powder diffractometers using microprocessors have become available. However, software for these dedicated computers has been limited. The need today is for an on-line system which can handle the large number and variety of samples typical of an industrial laboratory, without sacrificing accuracy or speed. Graphical output of data should be available to facilitate interpretation. The data should be stored in digital form to allow secondary transformations of the data such as curve smoothing, peak deconvolution, lattice parameter determinations, and phase identification by search-match procedures.

These goals have been accomplished using a Hewlett-Packard 9825 desk-top programmable calculator interfaced to an Enraf-Nonius microdensitometer. The data is reduced in a short period (20 min. overall time; 5 min. operator time) and presented as both an anotated graph of the diffraction pattern and a tabulated summary of the data.

HARDWARE

The Hüber-Guinier type X-ray camera produces films of excellent quality. To collect this information in a digital form without losing resolution or accuracy requires taking a large number of precise voltage readings from the microdensitometer in a short time period. This necessitated making several changes to the Enraf-Nonius densitometer. A synchronous drive motor and gear train available from the manufacturer was installed to move the film past the detector at precise speeds. Slits ranging in size from 20 to 500 microns were fabricated. The smaller slits are used when maximum resolution of well crystallized samples is desired. The broader slits are used to gain an improved signal-to-noise ratio by optically averaging the background at the expense of resolution. To obtain a noise free signal at .1 to 14V for the digital voltmeter, a "Pin" photo diode, fast response amplifier, and DC power to the illumination lamp were added. With these modifications to the slits, optics and electronics of the microdensitometer, an output accurately reflecting the quality of the film was obtained.

The signal from the microdensitometer is digitized using a Hewlett-Packard 3437A system voltmeter. The 3437A is a microprocessor controlled $3\frac{1}{2}$ digit successive approximation voltmeter capable of sampling rates in excess of 5000 readings per second with input voltage ranges of 0.1, 1, and 10 volts full scale. Up to 9999 readings can be taken without resetting the internal counter. Range select, sampling frequency and internal counter are software controlled, using a Hewlett-Packard 9825A calculator with 24K bytes of core.

Data collection is initiated by a signal from the calculator to the voltmeter. The DVM serves as an A to D converter and timing device to measure and transmit signals to the 9825 at a preset frequency (normally 100 readings/second) under priority interrupt control. Twenty-eight thousand voltage readings are taken over the 47° θ span of the film for a point-to-point separation of 0.0017° θ. The data is stored on disk by the 9825 using a flip-flop buffer routine. Data is read into one buffer until the buffer (1000 data points) is filled. Acquisition is then transferred to the second buffer while the data from the first buffer is written to the floppy disk. The buffers then reverse roles in the input/output scheme. After the film is scanned (approx. 4½ min.), software routines are used to reduce the data. The spectrum is reproduced graphically using the HP 9872 plotter and the results are summarized in tabular form using the HP 9866B thermal printer.

SOFTWARE

The software is conversational in nature and asks the operator to supply the needed information by answering a series of questions. After the film has been positioned on the microdensitometer and the light intensity adjusted to match the overall film darkening, the drive motor is engaged. The motor can be started using a remote switch located at the calculator. To run the program, the operator turns the calculator power switch on. This automatically loads a file from tape that lists the programs available and asks which one to run. After the appropriate program number has been entered, initial dialogue commences. During this phase the operator answers questions regarding the running of the sample (sample description, data, name, radiation used, etc.). This takes one to two minutes. The program then instructs the operator to throw the remote start switch for the densitometer and simultaneously press the "continue" key on the calculator. This initiates data acquisition. At this point the operator may leave. In approximately twenty minutes, the data will have been acquired, reduced, plotted, and diffraction peaks automatically found.

All of the information necessary to control the DVM is provided by the software. Data acquisition continues until the film has been scanned. The program then retrieves and processes the data. Due to core limitations, the data is handled as 1000 data point blocks. The twenty-eight blocks are serially accessed under software control. Disk storage is optimized by encoding the voltage readings into two eight bit bytes as opposed to the standard 8 bytes needed to store a decimal number. This necessitates using a binary bit manipulation routine to convert the reading to decimal equivalents prior to any data reduction. Using this packing routine, data for up to six films can be stored on a floppy disk. However, since reading a film requires less than five minutes, the film itself is the primary permanent record of the data. Usually, only the data from the last film read resides on disk. This allows the programs which control the system to be stored on disk, speeding execution time by eliminating slower tape transfers.

Data reduction is accomplished in several stages. The initial process involves scanning the 28 data blocks to establish their local minima and maxima. In examining the data, only every fifth data point is used. The data points are unpacked to convert them to the decimal equivalent of the original voltage. The log inverse of the voltage (log 1/v) is taken to generate the actual data point. This compensates for the logarithmic nature of the film darkening and converts transmittance to optical density. The minima of the data blocks are used to generate a background correction function. The curved background is approximated by a series of straight lines based upon the minimum values of the data in each of the 28 blocks. The local maxima of the blocks are corrected for background and a net system maximum established. This is used as a scaling factor for normalizing the plot on standard 11 x 17 inch (centimeter grid) graph paper. Descriptive information concerning the sample is printed at the top of the graph and the plotting regions are outlined (See Figure 1). This is followed by the plotting of the data and the concurrent automatic detection of lines. Only every fifth or tenth point is plotted. No additional detail is apparent when plotting every point versus every tenth one and to plot all the data would significantly increase execution time.

The peak detection routine involves generating a least squares 9-point running-smoothed second order derivative function of the corrected data. By using the method of Savitsky and Golay (1), this function is easily computed by use of a look-up table of appropriate coefficients. This method is particularly good at finding shoulders on peaks due to the nature of the second order derivative function. It is not especially sensitive to the broad peaks which are observed in back reflection mode and for poorly crystallized samples. However, by increasing the step size from every fifth data point to every tenth data point, and decreasing the threshold value used to distinguish a peak from noise, most broad peaks can be automatically found. The operator can specify either the broad or sharp peak mode for front reflection shots. The broad peak search is the default used for back reflections. In addition, the threshold value for peak detection can be changed during plotting without interruption of the program. Since the derivative function only examines every fifth or tenth point, only the approximate location is known for the peaks found. To locate the exact maximum, the program branches to another routine which searches every data point in the vicinity of this approximate location. Most peaks possess a unique maximum. When the exact peak position is found, a tic mark is placed over the peak on the plot and a sequential peak number, approximate d-spacing and intensity are printed on the printer. This process continues until the pattern is plotted and the peaks located and marked.

At this point the program halts and asks the operator if additional manual peak finding is needed. If necessary, the operator

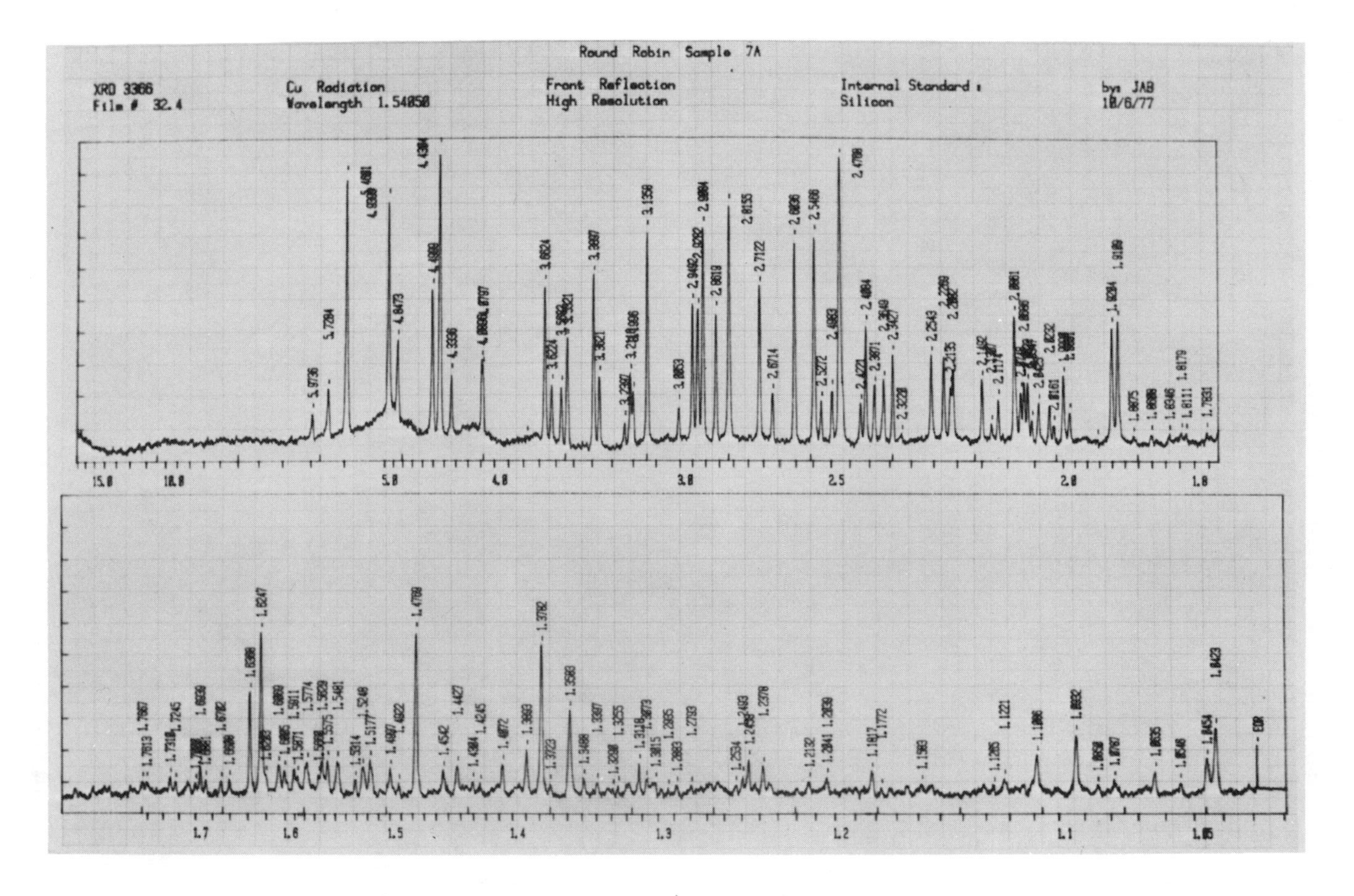

FIGURE 1: Computer Plot of X-Ray Diffraction Pattern of Round Robin Sample 7A from Guinier Film

can use the manual peak finding routine to position the plotter point at a new peak and automatically place a tic mark over it, adding it to the list of peaks found. To aid in locating the maximum of a peak, the position and intensity of the data point under the pen are displayed on the LED panel of the 9825. The operator can use step sizes of up to 1000 data points to step through the data until the pen is in the vicinity of the peak and then can examine every data point until a maximum is found.

When all of the peaks are located, the operator presses a key to load the second part of the program which uses a fit of the internal standard lines to calculate accurate d-spacing, prints the summarized results and the sample running conditions in formatted $8\frac{1}{2}$ x 11 pages and labels the observed peaks with their corrected d-spacings. To perform the regression analysis of the standard lines, the operator is asked to press a key corresponding to the internal standard used. The d-spacings for common standards calculated from cell constants are stored in the program (silicon, using either copper or iron radiation, front or back reflection, alumina and NaCl). A standard key labeled "other" can be used to enter the d-spacing of any known lines in the spectrum (up to 10 lines) to be used in the fit. The operator then enters all appropriate line numbers (from the list printed out when the peaks were found) for the standard used. When all the lines are entered, the fit of the observed line position to literature two-theta values is made. A list of the observed d-spacing, calculated d-spacing and difference is printed. The operator has the option to delete a line from the fit, or enter alternate line numbers and recalculate. When the operator is satisfied that the best fit has been obtained, he presses a key to get the final printout. Corrected d-spacings for all of the observed lines in the spectrum are calculated and printed in tabular form along with the sample description, running conditions and goodness-of-fit data (See Figure 2). An accurate d-spacing scale based on the standard fit, radiation and geometry used is added to the plot. The peaks are then labeled with d-spacings. The operator can specify any peaks which he does not want labeled or printed in the table. This eliminates clutter due to extraneous peaks. When the printout and labeling are completed, the program reloads the library program and the system is ready to be used again. The flow chart of these programs is shown in Figure 3.

An additional option from the library program permits data collection from a Philips powder diffractometer. The signal is transmitted from a 0-10 volts BNC connector on the recorder to the DVM through a standard coaxial cable. The program is identical to that from film with minor modifications. The user can specify scan rate and the region of 2θ to be covered. The response of the DVM is such that scans can be made at significantly higher speeds. Figure 4 shows data collected at 2°, 4° and 70° 2θ/minute. The

Calculated d-spacings

XRD 3366
10/6/77

Film no. 32.4
by: JAB

Round Robin Sample 7A

Cu radiation
Wavelength 1.54050

High resolution
Front reflection

Internal Standard Silicon

Line no.	d lit	d calc	diff x10**4
0	3.1353	3.1358	5.0
1	1.9200	1.9204	3.9
2	1.6374	1.6368	-5.2
3 *	1.3576	1.3583	6.6
4	1.2458	1.2458	-0.8
5	1.1085	1.1086	1.1
6 *	1.0451	1.0454	3.2

* not used in least squares fit

Least Squares Coeff	
Slope	Intercept
-0.66462	48.77262

calculated d-spacings

line #	pos	int	d
1	62.2375	11	5.9736
2	61.7575	19	5.7284
3	61.1825	94	5.4601
4	59.8875	85	4.9399
5	59.6275	36	4.8473
6	58.5550	54	4.4999
7	58.3200	103	4.4304
8	57.9800	24	4.3336
9	57.0500	29	4.0896
10	57.0100	27	4.0797
11	55.1175	55	3.6624
12	54.9125	20	3.6224
13	54.6275	20	3.5682
14	54.4325	36	3.5321
15	53.6225	58	3.3897
16	53.4575	23	3.3621
17	52.6900	8	3.2397
18	52.5050	24	3.2116
19	52.4250	18	3.1996
20	51.9900	74	3.1358
21	51.0400	13	3.0053
22	50.6050	48	2.9492
23	50.4375	46	2.9282
24	50.2775	75	2.9084
25	49.8925	46	2.8619
26	49.4950	83	2.8155
27	48.5600	54	2.7122
28	48.1700	17	2.6714
29	47.4925	70	2.6036
30	46.8950	72	2.5466
31	46.6850	15	2.5272
32	46.3650	18	2.4983
33	46.1225	99	2.4768
34	45.4850	14	2.4221
35	45.3200	41	2.4084
36	45.0600	20	2.3871
37	44.7850	22	2.3649
38	44.5025	31	2.3427
39	44.2350	4	2.3220
40	43.3225	30	2.2543
41	42.9375	30	2.2269
42	42.7450	20	2.2135
43	42.6675	26	2.2082
44	41.7850	22	2.1492
45	41.4975	7	2.1307
46	41.2875	16	2.1174
47	40.7825	44	2.0861
48	40.5950	18	2.0748
49	40.4925	22	2.0686
50	40.3850	21	2.0622
51	40.2575	9	2.0547
52	40.0500	14	2.0425
53	39.7150	14	2.0232
54	39.5900	7	2.0161
55	39.2850	23	1.9990
56	39.1050	11	1.9891
57	37.8050	41	1.9204
58	37.6175	43	1.9109

FIGURE 2: Computer Printout for Guinier Film of Round Robin Sample 7A.

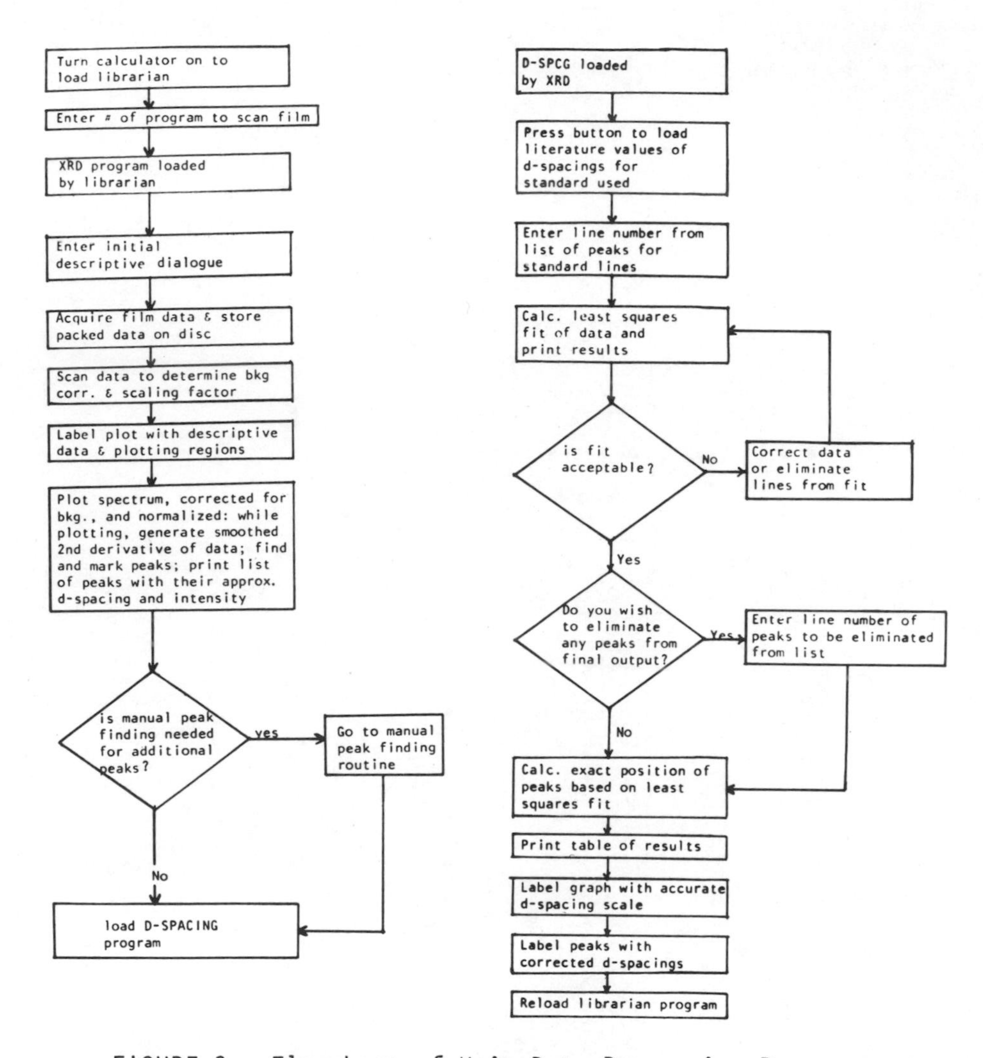

FIGURE 3: Flowchart of Main Data Processing Programs

70°/minute scan was made using the slewing speed on the diffractometer. D-spacings are given in Table 1 for these scans but labelling was suppressed in Figure 4 for purposes of clarity. At 70°/minute weak peaks are not always found and extra noise spikes are sometimes detected, depending on the parameters chosen for finding peaks. Routine identifications are now made in our laboratory from data collected at 8°/min. with minimal loss in the quality of the data. This was accomplished by changing the frequency of the stepper motor driving circuit.

TABLE 1: D-spacings (Å) for a mixture of SiO_2, Al_2O_3 * and Si from diffractometer scans at various rates. Accepted values were calculated from cell constants which were determined for each substance independently.

	Accepted Value	Slew 70°/min.	$\|\Delta d/d\|$ x 10^4	4°/min.	$\|\Delta d/d\|$ x 10^4	2°/min.	$\|\Delta d/d\|$ x 10^4
1	4.2548	N.O.		4.259	9	4.258	7
2	3.4832	3.4821	3	3.4840	2	3.4795	11
3	3.3432	3.3418	4	3.3442	3	3.3430	1
4	3.1353	3.1344	3	3.1356	1	3.1352	1
5	2.5529	2.5520	4	2.5532	2	2.5529	0
6	2.4565	2.4571	2	2.4580	6	2.4549	7
7	2.3819	2.3824	2	2.3821	1	2.3804	6
8	2.2812	2.2800	5	2.2808	2	2.2828	7
9	2.2364	2.2351	6	2.2359	2	2.2377	6
10	2.1274	N.O.		2.1276	1	2.1254	9
11	2.0873	2.0875	1	2.0874	0	2.0863	5
12	1.9796	N.O.		1.9784	6	1.9787	5
13	1.9200	1.9201	1	1.9199	1	1.9195	3
14	1.8178	1.8185	4	1.8180	1	1.8173	3
15	1.7416	1.7428	7	1.7410	3	1.7407	5
16	1.6716	N.O.		1.6707	5	1.6708	5
17	1.6374	1.6378	2	1.6384	6	1.6380	4
18	1.6027	1.6039	7	1.6030	2	1.6031	2
19	1.5414	N.O.		1.5423	6	1.5424	6
20	1.5118	1.5125	4	1.5124	4	1.5136	12
21	1.4059	1.4060	1	1.4052	5	1.4054	4
22	1.3820	N.O.		1.3828	6	1.3819	1
23	1.3749	1.3760	8	1.3747	1	1.3748	1
24	1.3576	1.3569	5	1.3577	1	1.3577	1
25	1.2458	1.2456	2	1.2452	5	1.2455	2
26	1.1085	1.1087	2	1.1086	1	1.1086	1
Mean absolute $\Delta d/d$ x 10^4			4		2		4

Lines 4, 13, 17, 24, 25, and 26 (Si) used in data fit.

Lines 1, 10, 12, 16, 19, and 22 omitted from mean $\Delta d/d$ calculation.

*Cell constants differ from literature due to impurity.

Reground Al-Si-O_2, 2°/min.

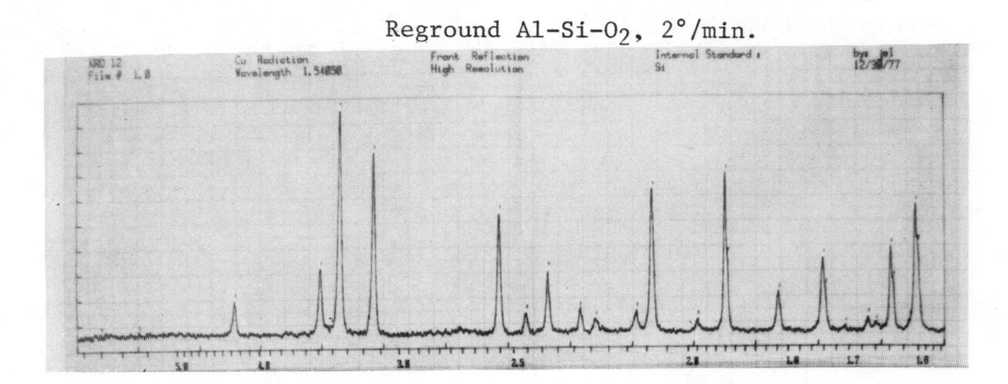

Reground Al-Si-O_2, 4°/min.

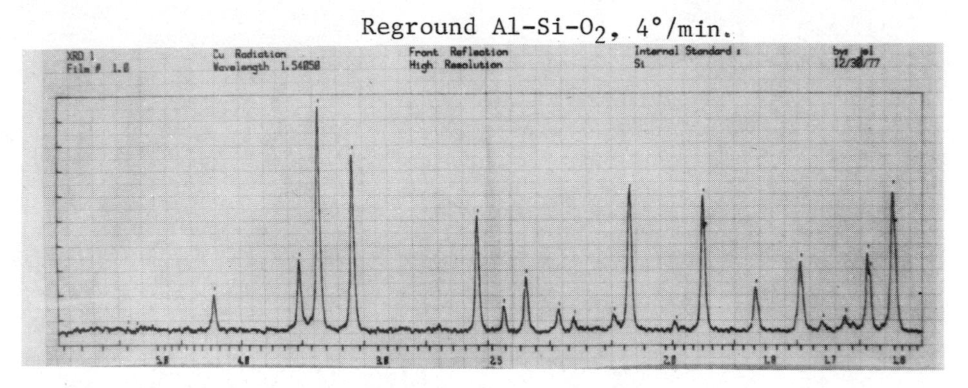

Reground Al-Si-O_2, 70°/min.

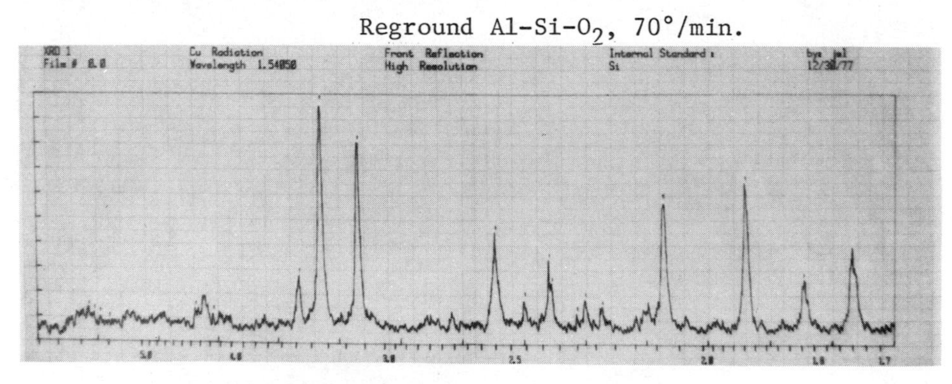

FIGURE 4: Partial Diffractometer Scans of a Mixture of Corundum, Quartz and Silicon Obtained at 2, 4, and 70 Degrees Two Theta Per Minute.

In addition to the basic programs for data collection, several supporting programs are available. These programs utilize the final d-spacing information stored on disk by the basic program, eliminating tedious re-entry of data.

A lattice parameter program (2) utilizes a least squares approach to obtain refined cell constants for a given phase. The output includes direct and reciprocal cell data, sigma, observed and calculated values of d and 2θ, Δd and $\Delta 2\theta$ (See Figure 5). The program is used as an aid to identification, especially in cases of solid solutions, by testing the internal consistency of a tentative match. The interactive nature of the program allows data to be added, changed or deleted easily and cell parameters recalculated.

An auto-assignment program uses cell constants and lattice type to assign Miller indices in a list of d-spacings. The range of calculation and error window, $\Delta d/d$, are user specified. This program is useful in cases where a literature pattern does not contain all of the observable lines. It is also used in combination with the lattice parameter program to aid in solid solution identifications.

Profile fittings of diffraction peaks is currently an area of intense interest (3,4). One of the most extensively used supporting programs deconvolutes up to six overlapping diffraction peaks in a given region (5). The data may be obtained from films or diffractometer. Diffractometer data is normally collected at 4° 2θ/min. over an 8° to 10° span. A unique feature of this program is that a user defined image string is used rather than making any theoretical assumptions about peak shape.

A single peak of an appropriate standard is used to build an image string. The image is chosen to reflect the absorption of the sample, the two theta region and the instrument conditions under which the sample is to be run. Several image strings can be stored on disk for most commonly encountered situations. Additional image strings can be constructed as needed within a few minutes. For example, several overlapping peaks at ≃3.4Å from a mixture of aluminosilicates might be deconvoluted using the 3.48Å Al_2O_3 line or the 3.34Å SiO_2 line. The output of this program includes accurate d-spacings (calculated based on a designated standard line), peak height, peak width at half height, peak area and area percent (See Figure 6). Uses of these data include crystallite size calculations, percent crystallinity of polymers, quantitative phase comparisons and studies of solid solutions.

Linking of our lab computer system to large remote computers via a telephone line further expands the capabilities. Current uses include search/match programs and calculation of intensity data from atomic positions.

Round Robin 7A Barium Chloride dihydrate phase 2/14/78 JSLSilicon 3366f

	Reciprocal	Direct	Sig(dir)
A	0.14879	6.72226	0.00061
B	0.09168	10.90696	0.00086
C	0.14023	7.13227	0.00054
ALPHA	90.00000	90.00000	0.00000
BETA	88.89400	91.10600	0.00787
GAMMA	90.00000	90.00000	0.00000
VOLUME	0.00191	522.83588	

Round Robin 7A Barium Chloride dihydrate phase 2/14/78 JSLSilicon 3366f

Line #	h	k	l	d obs	d calc	diff x10↑4	2 theta obs	2 theta calc	diff x10↑3
1	0	1	1	5.9734	5.9685	48.9	14.817	14.830	-12.2
2	1	1	0	5.7269	5.7219	50.2	15.459	15.473	-13.6
3	0	2	0	5.4577	5.4535	42.2	16.227	16.239	-12.6
4	-1	0	1	4.9398	4.9388	10.2	17.941	17.945	-3.7
5	1	0	1	4.8437	4.8445	-8.2	18.300	18.297	3.1
6	-1	1	1	4.4982	4.4990	-8.4	19.719	19.716	3.7
7	1	1	1	4.4288	4.4274	13.7	20.031	20.038	-6.3
8	0	2	1	4.3335	4.3319	16.1	20.477	20.484	-7.7
9	-1	2	1	3.6608	3.6607	0.8	24.292	24.293	-0.5
10	1	2	1	3.6223	3.6218	4.7	24.554	24.558	-3.2
11	0	0	2	3.5676	3.5655	21.3	24.937	24.952	-15.1
12	0	1	2	3.3896	3.3890	6.1	26.269	26.274	-4.8
13	2	0	0	3.3608	3.3605	3.0	26.498	26.501	-2.4
14	0	3	1	3.2396	3.2390	6.2	27.509	27.514	-5.4
15	2	1	0	3.2115	3.2115	-0.2	27.754	27.754	0.2
16	1	3	0	3.1987	3.1978	9.3	27.868	27.876	-8.2
17	1	1	2	3.0032	3.0040	-7.8	29.722	29.714	7.9
18	-2	1	1	2.9494	2.9487	7.1	30.277	30.285	-7.4
19	-1	3	1	2.9275	2.9279	-3.8	30.509	30.505	4.1
20	2	2	0	2.8615	2.8609	5.6	31.231	31.237	-6.3
21	1	2	2	2.7116	2.7113	3.2	33.005	33.009	-4.0
22	-2	2	1	2.6711	2.6704	6.8	33.520	33.529	-8.7
23	1	4	0	2.5269	2.5267	1.9	35.495	35.497	-2.7
24	2	0	2	2.4223	2.4223	0.4	37.082	37.083	-0.6
25	-2	1	2	2.4083	2.4084	-1.4	37.305	37.303	2.2
26	-1	4	1	2.3870	2.3871	-0.9	37.651	37.649	1.4
27	2	1	2	2.3649	2.3646	2.5	38.016	38.020	-4.2
28	-2	3	1	2.3426	2.3424	2.4	38.392	38.396	-4.0
29	-1	0	3	2.2546	2.2547	-0.8	39.953	39.952	1.5
30	1	0	3	2.2269	2.2275	-5.8	40.472	40.461	11.1
31	2	2	2	2.2135	2.2137	-2.2	40.727	40.723	4.2
32	-1	1	3	2.2076	2.2080	-4.0	40.841	40.833	7.7
33	-3	0	1	2.1488	2.1492	-4.3	42.011	42.002	8.8
34	2	4	0	2.1173	2.1174	-0.9	42.666	42.664	1.9
35	-1	4	2	2.0680	2.0686	-6.3	43.735	43.721	14.1
36	1	2	3	2.0614	2.0621	-7.0	43.882	43.867	15.7
37	1	4	2	2.0549	2.0545	3.8	44.028	44.037	-8.6
38	2	4	1	2.0230	2.0231	-1.0	44.760	44.758	2.3
39	-3	2	1	1.9993	1.9996	-2.5	45.320	45.314	6.0
40	0	5	2	1.8606	1.8608	-1.6	48.910	48.906	4.6
41	0	6	0	1.8176	1.8178	-2.3	50.146	50.139	6.7
42	0	0	4	1.7832	1.7827	4.7	51.183	51.197	-14.3

FIGURE 5: Least Squares Lattice Parameters for the Barium Chloride Dihydrate Phase of Round Robin Sample 7A

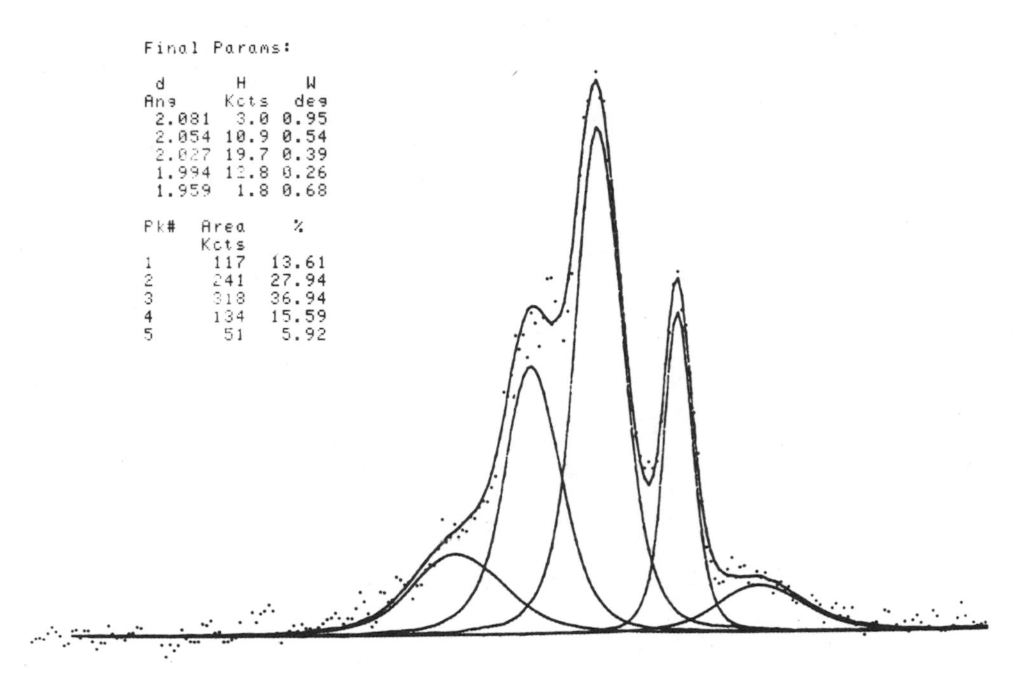

6/6/78 AH SALT
5215C1
Final Params:
d H W
Ang Kcts deg
2.081 3.0 0.95
2.054 10.9 0.54
2.027 19.7 0.39
1.994 12.8 0.26
1.959 1.8 0.68
Pk# Area %
Kcts
1 117 13.61
2 241 27.94
3 318 36.94
4 134 15.59
5 51 5.92

FIGURE 6: Results of Deconvolution Program Showing Original Data (dots), Overall Fit and Individual Peaks

REFERENCES

1. A. Savitsky, M. Golay, "Smoothing and Differentiation of Data by Simplified Least Squares Procedures", Analytical Chemistry, 36, No. 8, 1627-1639 (July, 1964).

2. R. E. Davis, University of Texas at Austin, private communication (1977).

3. T. C. Herang and W. Parrish, "Qualitative Analysis of Complicated Mixtures by Profile Fitting X-Ray Diffractometer Patterns", Advances in X-Ray Analysis, Vol. 21 (1977).

4. A. Brown and O. Rosdahl, "X-Ray Characterization of Non-Equilibrium Solid Solutions", Atomic Energy Commission of Sweden, Report AE-507 (1975).

5. M. A. Kelly, Surface Science Laboratories, Palo Alto, CA., private communication (1977).

A MICROCOMPUTER CONTROLLED DIFFRACTOMETER

R. P. Goehner and W. T. Hatfield

General Electric Company

Schenectady, New York 12301

Automated powder diffractometers are beginning to come into general use. The principal reason for automating or purchasing an automated powder diffractometer is to increase the speed and accuracy of the analysis. Three principal methods are used. The first is that a multisample changer increases the through-put on the instrument since the unit can be operated unattended during the evening (1,2). The second is the diffractometer can be scanned at faster rates when digital data are collected, as opposed to the analog method of using a strip chart recorder (3,4,5). The third is the elimination of reading the peak positions and intensities from the strip charts. These readings would typically have to be tabulated manually for later interpretation. Another reason for having an automated powder diffractometer is for experiments where digital data are necessary to make the analysis practical. These experiments include routine quantitative analysis, radial distribution functions, Fourier profile analysis, and signal averaging to bring out very weak diffraction peaks.

A microcomputer interface was designed principally because it provides the most versatile hardware at the smallest cost. This type of interface also makes it possible to do distributive processing. The microcomputer can be used to control the diffractometer and to acquire the data in its RAM (random access memory) or on a digital magnetic tape. Upon completion of the experiment, the data can then be sent to a host computer for processing. The host computer can be a minicomputer or a timesharing computer. This makes it possible for one host to service a large number of instruments. It also makes it possible to continue to utilize the instruments when the host is down for maintenance or just overloaded. The microcomputer can be equipped with a higher level language such as BASIC. This then allows the operator to do some simple data reduction independent of the host computer. Since a higher level language is available both on the microcomputer and host, software development becomes much easier than if only machine language was available (6).

This is the philosophy that was followed when a microcomputer interface, designated laboratory automation module (LAM), was designed for controlling a Siemens D500 powder diffractometer. The D500 is equipped with two independent stepping motors for theta (θ) and 2 theta (2θ) motions. The LAM allows independent control of θ, 2θ, and θ-2θ coupled. In addition, the LAM senses the shutter and drives a stepping motor on a digital strip chart recorder. Allowances have been made for control of a multisample changer. The LAM is based on Intel's single board computer (SBC) system using an SBC 80/20. The system contains: 6K of ROM (read only memory) which is used for the Intel monitor, utility routines and D500 command system; 7K of additional ROM, which contains the BASIC interpreter; and 16K of dynamic RAM. Two serial ports, two parallel ports, a cassette tape deck, a line frequency clock, a 32-character display, and the interface board, which has a built-in timer-scalar, complete the system. The LAM command software allows the execution of a single command, as well as a sequence of commands. The commands can be typed in on a terminal connected to the LAM, or sent through the second serial port by a host computer. The LAM will buffer the data resulting from a command sequence in its RAM memory. The data can then be written out on its cassette tape for later retrieval, or sent to the host computer, or to the terminal.

The LAM is presently being used to control the D500 diffractometer and to buffer the digital data in its RAM memory. These data are then transmitted to a time-sharing host, which in this case is a Honeywell 600. The software on the host is written in FORTRAN, consists of archival storage and retrieval programs for data management, and a data reduction and display program. The data sent by the LAM are stored in binary on the disk of the host, along with the starting 2θ angle, step size, preset time, and 48 characters of general title information. The name of this binary file, along with the title information, is stored in a directory file which can be searched by sample number, date of run, or any character string in the title. SPECPLOT is the name of the data reduction and display program. This program will do smoothing, background subtraction, peak search, and interactive display of the diffraction patterns. Some of the interactions the operator can perform on the data are plotting a windowed portion of the spectrum, a parabolic fit to the top of a peak, integration of a specified peak with the background supplied by the operator, d-spacing of a peak, *etc*.

A 7K ROM version of ALTAIR BASIC interpreter is incorporated into the LAM. Plans have been made to use the BASIC to do some simple data reduction and intelligent control of the D500 diffractometer.

REFERENCES

1. Jenkins, R., Haas, D., and Paoline, F. R., Norelco Reporter, 18, 12-27, (1972).

2. Jenkins, R., and Westberg, R. G., Advances in X-Ray Analysis, 16, 310-321 (1973).

3. Holland, H. J., and Medrud, R. C., J. Appl. Cryst., 10, 386-389, (1977).

4. King, P. J. and Smith, W. L., J. Appl. Cryst., 7, 603-608 (1974).

5. Parrish, W., Huang, T. C., and Ayers, G. L., Trans. ACA 12, 55-73 (1976).

6. Kelly, C. J. and Gagliardi, C. A., Proc. IEEE 63, 1426-1431 (1975).

X-RAY POWDER DIFFRACTION INVESTIGATION OF THE MONUMENTAL STONE OF THE CASTEL dell'OVO, NAPLES ITALY

S. Z. LEWIN and A.E.CHAROLA

Department of Chemistry, New York University

4 Washington Place, New York, N.Y. 10003

The Castel dell'Ovo on the Bay of Naples, Italy, is an important historic structure dating from the 12th century. It is composed of a soft, volcanic tuff that was employed because it was locally available and easy to quarry and work. The exposed stone of this monument shows extensive decay, principally in the form of the crumbling away of the surface, and this has progressed to the greatest extent in the lower parts of the structure (1). Restoration and preservative intervention are now being considered. The essential prerequisites to the formulation of a safe and effective strategy of treatment of the monument are the understanding of the nature and properties of the stone, and of the factors responsible for its decay. The present investigation was undertaken to provide that information. In the course of these studies certain novel and significant features of the crystal chemistry of one of the common zeolite minerals have been observed, and are reported here.

EXPERIMENTAL

Specimens were obtained of the decayed stone from the monument itself, and of the pristine stone from the original quarry site. These were examined in thin section by visual microscopy; in fracture cross-section by SEM; and ground samples were analyzed by XRD.

I. Mineral Components

At 20X to 40X magnification with the optical stereo microscope, the gently crushed stone from the original quarry appeared to consist of six visually distinguishable types of particles. Each of these types was separated out by hand picking to yield a sufficient quantity of homogeneous sample for identification and characterization. These constituents proved to consist of four different mineral phases, as follows:

1. The principal constituent, which composes the pale brown ground mass in which the other minerals are embedded (Fig. 1A), is the zeolite, chabazite. It occurs in two very different habits (Fig. 1B-D): (a) solid, irregular sheets, square prisms, and pseudo-cubic crystals, and (b) hollow tubular, spherical, or sheet-like spongy masses. The latter form predominates, and contributes a very large internal porosity and specific surface area to the stone. Similar structures have been observed in the growth of synthetic zeolites (2) by aging silica-alumina mixed gels at 100°C.

Standard wet chemical analyses of the chabazite component showed it to correspond to the following stoichiometry, based upon two moles of SiO_2 per formal unit:

$$(Na/K)_{0.58}Mg_{0.04}Ca_{0.19}(Al/Fe)_{1.04}Si_2O_{6.08}\cdot 2.91H_2O$$

or, in terms of oxide proportions:

$$(Na_2O/K_2O)_{0.29}(MgO)_{0.04}(CaO)_{0.19}(Al_2O_3/Fe_2O_3)_{0.52}(SiO_2)_2\cdot 2.91H_2O$$

This composition corresponds to an average type of chabazite (3), with close to the "ideal" ratio of Al to Si (vide infra). The x-ray diffraction pattern is similar to that reported by Gude and Sheppard for a specimen from Colorado (4). The refractive index determined by immersion was 1.488 to 1.490 at 25°C for various particles. Some specimens of the fines showed extra weak diffraction lines at d=7.0 and 2.97; these could correspond to the presence of a minor amount of the zeolite mineral, potassium natrolite (5).

2. A small proportion of the zeolite ground mass (ca. 0.5 to 1% by volume) of the stone consists of the mineral analcime; it is present as discrete white, opaque masses about 0.1 to 5 mm in dimensions. These are very friable; they consist of small, equant grains (ca. 0.05 - 1 mm) with well-developed crystal faces, but all of the grains are so permeated by cracks that they crumble under slight pressure. The x-ray diffraction pattern is very similar to that reported by Coombs for a specimen from Australia (6); an extra small peak is present at d=3.33 which may be due to the presence of a minor amount of sanidine (vide infra).

3. About 1 to 2% of the stone consists of phenocrysts of the feldspar mineral, sanidine (Fig. 2A-D). This occurs as (a) well-developed crystals with flat faces and sharp angles, either colorless and transparent, or black and clear; and (b) irregular, pumice-like masses showing little external evidence of crystallinity. Both types

Fig. 1A. Fracture section of Castel del'Ovo stone, showing two phenocrysts of sanidine in a porous matrix of chabazite.

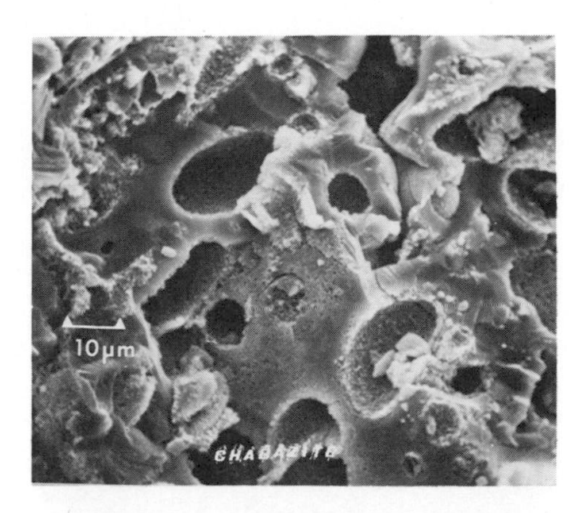

Fig. 1B. Chabazite in the form of sheets that are partly solid, partly sponge-like.

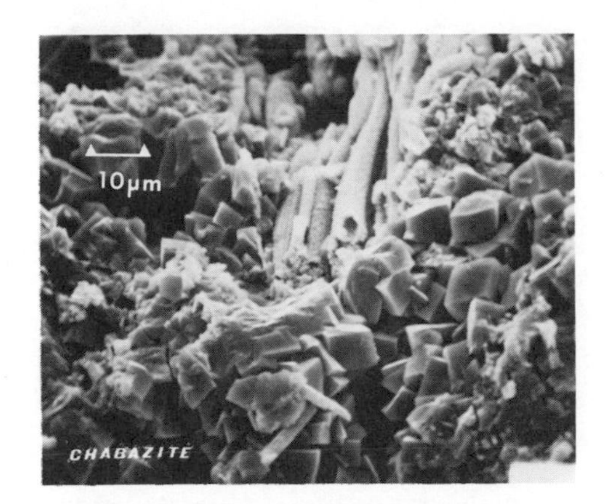

Fig. 1C. The two habits of chabazite: solid, pseudo-cubic crystals, and highly porous, sponge-like structures in the form of tubes, spheres and irregular masses.

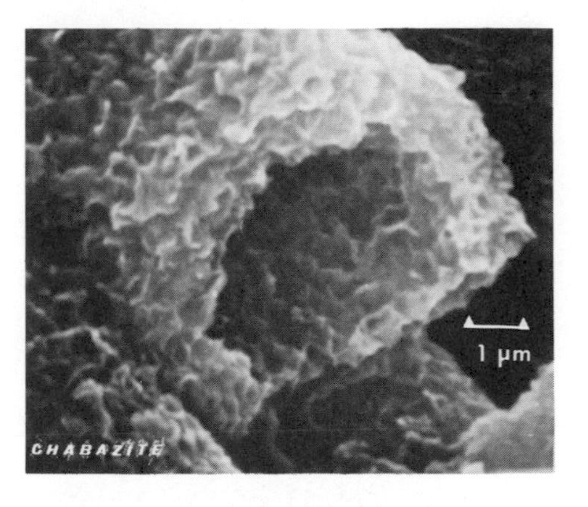

Fig. 1D. High magnification micrograph of one of the sponge-like tubes of chabazite. These structures contribute a tremendous specific surface area to the stone.

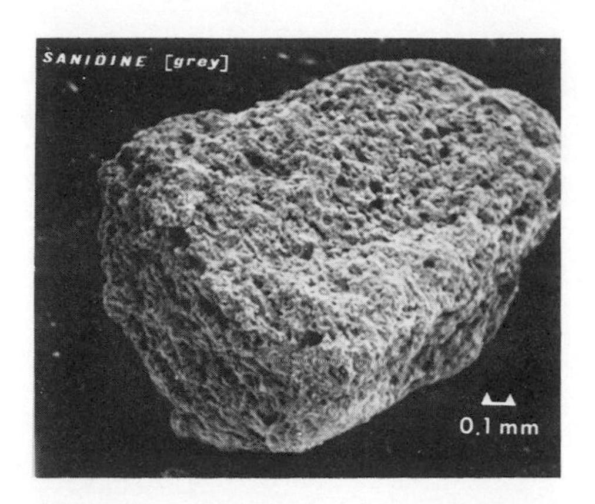

Fig. 2A. Inclusions of the feldspar mineral sanidine occur in the form of grey to black pumice-like masses, and ...

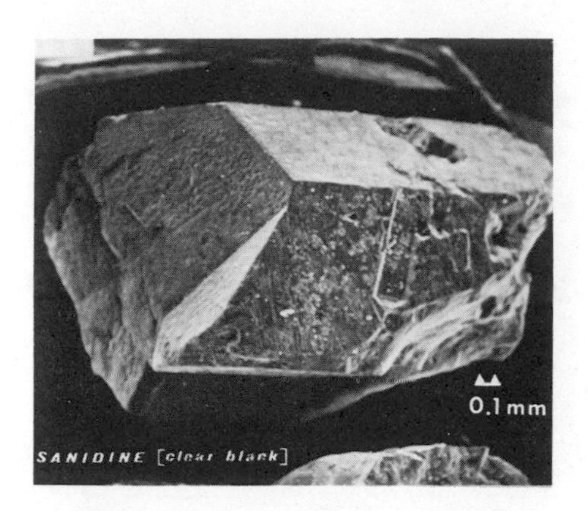

Fig. 2B. ...well-developed crystals with flat faces and sharp interfacial angles.

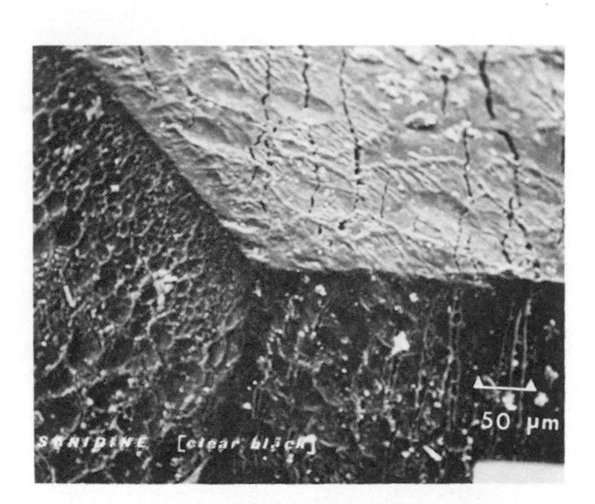

Fig. 2C. Detail of the phenocryst of Fig. 2B, illustrating that most of the well-developed crystals are permeated by cracks, channels, and imperfections.

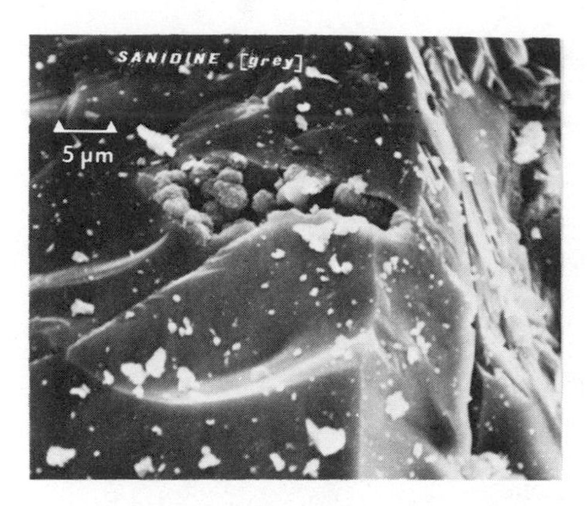

Fig. 2D. At high magnification, the pumine-like particles have a glassy surface, yet give the x-ray diffraction pattern of sanidine.

gave identical x-ray diffraction patterns that agree within experiental reproducibility with the reported pattern (7). Although many of the particles had high concentrations of cracks and imperfections, they were hard and resistant to crushing. The crystals showed refractive indices ranging from 1.520 to 1.528, the higher values occurring in the black crystals. These variations in optical properties, due to different iron contents, were not associated with detectable alterations in the diffraction patterns (within ± 0.01 in d).

4. About 2 to 4% of the stone consists of a black or dark grey glassy component, which is non-crystalline. This would appear to be the residuum of the original volcanic glass that has undergone geochemical transformation into the present zeolitic ground mass as a result of hydrolytic reactions and hydrothermal crystal growth.

This mineral suite is what would be formed as the result of the _in situ_ hydrothermal transformation of a tuff of volcanic ash (8). Chabazite is a hydrated, low density aluminosilicate, and forms first at low temperatures and pressures. With increasing age and weight of the overburden on the sedimentary deposit, gradual transformation of the chabazite to the denser, less hydrous analcime takes place. When the burial temperatures are higher than 150°C, the zeolites are replaced by feldspars, viz., sanidine in the present case (9).

II. Chabazite Properties

This stone can be characterized as a porphyritic rock containing phenocrysts of feldspar in a ground mass of zeolite. Since the monument is situated in direct contact with salt water and spray, it is to be expected that some of the calcium in the chabazite at the exposed surfaces would in the course of time be exchanged by sodium ions (10). In order to determine whether this exchange produces strains in the crystal lattice that could result in fracturing of crystal grains and consequent crumbling of the stone, a study was made of the effect of ion exchange on the lattice parameters of chabazite.

A manual step scanning technique was employed to determine the precise locations of the stronger diffraction maxima. The goniometer of the diffractometer was advanced in 0.01° two-theta steps; at each setting the time necessary to accumulate 6400 counts was measured, so that the random error of all the count rate data is constant. Each peak investigated was scanned twice over an angular range sufficient to yield the full contour of the diffraction line. The location of the peak maximum could be established with a precision of $\pm 0.01°$.

Specimens were prepared by shaking portions of the powdered original chabazite with concentrated solutions (1-5 F) of NaCl, $CaCl_2$

and $BaCl_2$ respectively, for periods of time ranging from 1 day to 1 week. It was found that a maximum of only about 80% of the calcium in the original chabazite was exchangeable by either sodium or barium ions; about 95% of the alkali ion content in the original chabazite was exchangeable by either calcium or barium ions. These results are consistent with the general behavior of mineral zeolites (11). Specimens in which these maximum extents of exchange had taken place were employed for the step scanning measurements.

Figure 3 illustrates the nature of the effect observed in the diffraction peak due to the (003) planes. The d-value is smaller for the calcium-enriched chabazite by 0.031±0.006 A compared to the sodium-enriched form; i.e., replacing $2Na^+$ by Ca^{++} in the lattice results in a compression along the c-axis of 0.62±0.12%. The reality of this effect is demonstrated by the strictly additive result obtained in the step scan of the physical mixture of these two samples.

Figure 4 shows that the effect of replacing $2Na^+$ by Ca^{++} is in the opposite sense for the (110) planes; i.e., the lattice expands by 0.30±0.15% in the a and b directions. The peaks at 7.1A in this Figure do not correspond to any of the predicted reflections based on the crystal structure of "ideal" chabazite (4). A similar weak peak is observed at d=5.58-9. These could be due to a minor component that escapes detection in ordinary diffractometry; it is probably a zeolite, and could be wairakite (12), which is the Ca-analog of analcime.

The results of the step scanning measurements for the Na-, Ca- and Ba-exchanged chabazites are collected in Table I. In addition to these crystallographic measurements, some experiments were performed to test the chemical stability of the chabazite in this stone. It was found that prolonged exposure to water causes the zeolite to hydrolyze, leaving amorphous hydrous iron oxide, alumina, and silica as the insoluble residue. Contact with dilute acids produces a similar result, the hydrolytic decomposition being more rapid and extensive, the more concentrated the acid. Hydrochloric acid (1 M) preferentially leaches out first the alkalis, alkaline earths and iron, and then the alumina, leaving a residue that is increasingly

Table I. Effect of Exchangeable Ion on Crystal Plane Spacings

Plane	Na-Exchanged	Ca-Exchanged	Ba-Exchanged
110	6.873±0.005	6.894±0.005	6.910±0.005
201	5.543±0.004	5.543±0.004	5.567±0.004
003	5.019±0.003	4.999±0.003	5.014±0.003
401	2.924±0.001	2.928±0.001	2.934±0.001
214	2.891±0.001	2.881±0.001	2.893±0.001
317	1.8100±0.0002	1.8012±0.0002	1.8079±0.0004

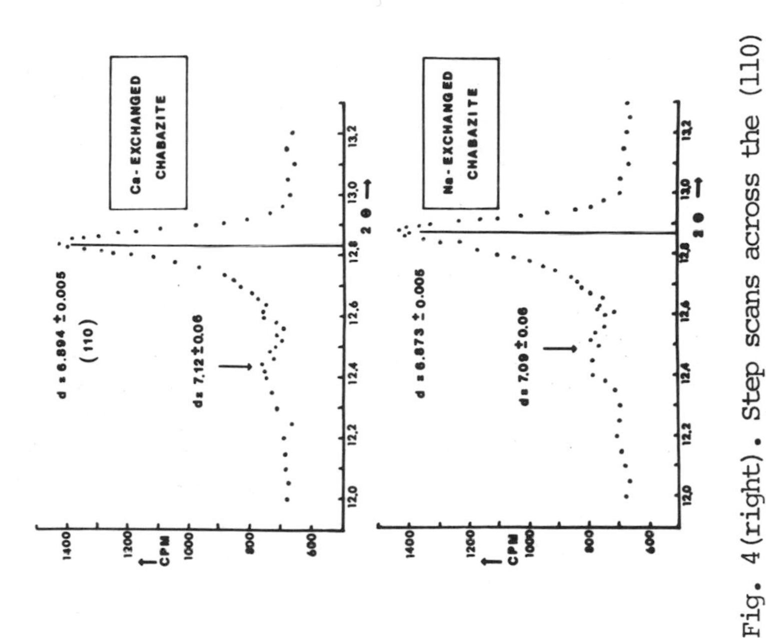

Fig. 3 (left). Step scans in increments of 0.01° across the (003) diffraction peak of the Castel del'Ovo chabazite in the Ca- and Na-exchanged states.

Fig. 4 (right). Step scans across the (110) peak. Replacement of Ca by Na shifts the d-value in the opposite sense from that for the (003) diffraction.

rich in silica as the reaction proceeds. This is consistent with the generalization that zeolites with a Si/Al ratio of 1.5 or less are converted into a gel by acid hydrolysis, whereas those with a ratio greater than 1.5 (as in the present case) decompose with formation of a precipitate of hydrous silica (13).

DISCUSSION

If, following Gude and Sheppard (4), we take the "ideal" or "normal" crystal lattice of chabazite as corresponding to the zeolite entirely in the calcium form, i.e., $Ca_2Al_4Si_8O_{24}\cdot 12H_2O$, then the various d-values observed for the Na-, Ca- and Ba-exchanged samples of the Castel del'Ovo chabazite can be employed to estimate the relative differences between their unit cell parameters and those of the ideal. For this purpose, Δd_{hkl}^{exp} is defined as the difference between the observed d_{hkl} and that calculated from the formula for a hexagonal crystal with $\underline{a}_o$=13.786 and $\underline{c}_o$=15.065, the parameters of ideal chabazite. Then $(\Delta d_{hkl}^{exp}/d_{hkl})$ is compared with the values calculated from:

$$\left[\frac{\Delta d_{hkl}}{d_{hkl}}\right]_{calc.} = \left[\frac{4(h^2+hk+k^2)}{3\underline{a}_o^{\,2}} + \frac{l^2}{\underline{c}_o^{\,2}}\right]^{-1}\left[\frac{4(h^2+hk+k^2)}{3\underline{a}_o^{\,2}}\cdot\frac{\Delta\underline{a}_o}{\underline{a}_o} + \frac{l^2}{\underline{c}_o^{\,2}}\cdot\frac{\Delta\underline{c}_o}{\underline{c}_o}\right]$$

and the values of $(\Delta\underline{a}_o/\underline{a}_o)$ and $(\Delta\underline{c}_o/\underline{c}_o)$ are adjusted to give the best fit of calculated and experimental values. Table II shows the results of the calculation, and Table III gives the derived lattice parameters of the Castel dell'Ovo specimens compared with those of "ideal" chabazite, and literature data.

Table II. Comparison of Observed and Calculated Lattice Plane Shifts

hkl	Na-Exchanged		Ca-Exchanged		Ba-Exchanged	
	$\Delta d^{exp}/d$	$(\Delta d/d)_{calc}$	$\Delta d^{exp}/d$	$(\Delta d/d)_{calc}$	$\Delta d^{exp}/d$	$(\Delta d/d)_{calc}$
110	-0.29%	-0.20%	+0.02%	+0.01%	+0.25%	+0.35%
003	-0.06	+0.25	-0.46	-0.47	-0.16	-0.10
201	-0.13	-0.14	-0.13	-0.06	+0.31	+0.09
401	-0.14	-0.18	0.00	-0.01	+0.20	+0.34
214	-0.03	+0.07	-0.38	-0.27	+0.03	+0.08
317	+0.30	+0.12	-0.18	-0.33	+0.61	+0.03

Table III. Lattice Parameters of Exchanged and Ideal Chabazites

	Castel dell'Ovo	Nova Scotia (14)	Ideal (4)
		Na-Exchanged	
a_o	13.758±0.005	13.9_0	13.799±0.001
c_o	15.103±0.005	15.2_5	15.102±0.002
		Ca-Exchanged	
a_o	13.787±0.005	13.8_5	13.786±0.002
c_o	14.994±0.005	15.0_0	15.065±0.004
		Ba-Exchanged	
a_o	13.834±0.005	13.8_6	
c_o	15.050±0.005	15.2_0	

It has generally been assumed that the effects of ion exchange on zeolites such as chabazite are a function of ion size, and consist of simple expansions or contractions, if indeed there are any effects at all. Thus, the reports of Gude and Sheppard (4) and Barrer and Sammon (14) suggest either no effect, or a slight contraction along all three axes when 1 Ca^{++} (octahedral crystal radius=0.99 unhydrated; 4.12A hydrated) (15) replaces 2 Na^+ (each Na^+ with r=0.95 unhydrated; or 3.58 hydrated). The Ba^{++} ion is larger (r=1.35) unhydrated, but smaller hydrated (r=4.04) compared with Ca^{++}. According to Barrer and Sammon when Ba^{++} replaced 2 Na^+ they found no detectable effect on the unit cell parameters.

However, the present work shows that there are in fact specific non-isotropic distortions of the crystal lattice due to the exchange reaction. The Castel dell'Ovo chabazite in the Na-exchanged form has a significantly smaller a_o and a larger c_o than the "ideal" chabazite. Upon exchanging the Na^+ in the zeolite by Ca^{++}, the lattice expands 0.2% along the a and b directions, and contracts 0.7% along the c direction. When Ba^{++} takes the place of 2 Na^+, the expansion in the a and b directions is 0.6%, and the contraction along c is 0.4%

According to Hoss and Roy (16), the Na- and Ca-exchanged chabazites have the same total water content under equivalent conditions of temperature and humidity, and this reflects the equivalence in the crystal of $Ca \cdot 10H_2O$ and $2(Na \cdot 5H_2O)$.

Hence, it is evident that the distortion of the crystal lattice accompanying exchange is not merely a bulk effect. At least half of the exchangeable ions reside inside the hexagonal prisms formed

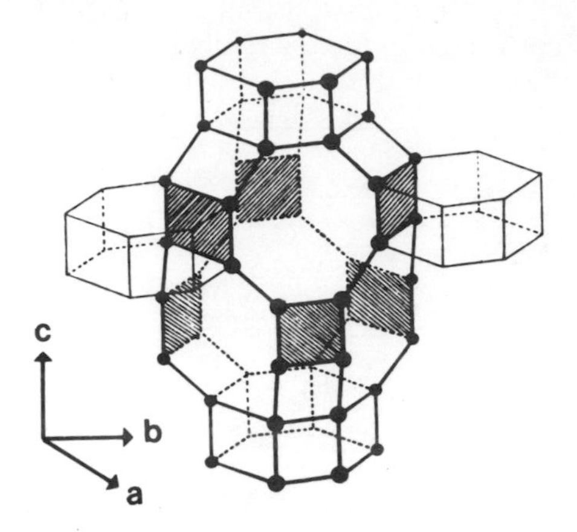

Fig. 5. Crystal structure of "ideal" chabazite (19).

by the six-membered aluminosilicate rings (17) and most of the remainder are in or near the 8-membered rings (18) shown in the diagram of Fig. 5 (19).

The doubly charged ions, representing a positive field that is more concentrated near the centers of the prisms and the 8-membered rings than is the field of the two Na^+'s, pull the negatively charged oxygens on the inner peripheries of the rings toward the center, thereby compressing the unit cell along the c-axis (cf. Fig. 5). The increased closeness of the rings along this axis increases their mutual repulsion, and this is relieved in part by expansion in the a, b directions.

The unit cell volumes are equal (within ±0.1%) in the Na- and Ca-exchanged forms, further supporting the deduction that the principal consequence of exchange is a distortion of the unit cell due to electric field effects, rather than a steric effect due to differences in bulk. The unit cell volume of the Ba-exchanged form is 0.9±0.1% larger than that of the Na- or Ca-exchanged forms. This would be the consequence of a smaller electrostrictive effect due to the smaller ionic potential (e/r) of the Ba^{++} field.

CONCLUSION

The exchange of calcium by sodium in the chabazite matrix of the Castel dell'Ovo stone (and probably in all chabazite rocks) results in strains that are at least of the order of several tenths of a percent, and this is sufficient to generate cracks in a brittle substance. In this connection it is significant to note the general experience of several investigators summed up by Hoss and Roy (16) that "repeated exchange of cations tends to decrease the crystallinity of all five zeolites until they finally become x-ray amorphous." Hence it is probable that a significant part of the decay observed in the Castel dell'Ovo is due to the ion exchange process resulting from daily contact with the salt waters of the Bay of Naples. In addition, the susceptibility of the zeolitic ground mass to hydrolysis, and the accelerating effect on this reaction of acidic substances such as those provided by ambient polluted air, contribute to the total decay.

It follows that the stone of the Castel dell'Ovo is inherently unstable in its local environment, and can be fully stabilized only by either a chemical treatment that modifies its internal structure, or a physical intervention (e.g., impregnation) that isolates it from contact with salt and water.

REFERENCES

1. R. Rossi-Manaresi, "Causes of Decay and Conservation Treatments of the Tuff of Castel dell'Ovo in Naples," in 2nd International Symposium on the Deterioration of Building Stones, National Technical University, 42 Patission St., Athens T.T.147, Greece, pp. 151-68 (1976).

2. A.K.Patel and L.B.Sand, "Mechanisms of Synthesizing Pseudomorphic Zeolite Particulates Using High Concentration Gradients," in Molecular Sieves, II, ACS Symposium Series No. 40, Amer. Chem. Soc., Wash., D.C., pp. 207-17 (1977).

3. D.S.Coombs, et al., "The Zeolite Facies, with Comments on the Interpretation of Hydrothermal Syntheses," Geochim. et Cosmichim. Acta, 17, 53-107 (1959).

4. A.J.Gude and R.A.Sheppard, "Silica-Rich Chabazite from the Barstow Formation, San Bernardino County, Southern California," Amer. Mineral., 51, 909-15 (1966).

5. Powder Diffraction File 3-0062, Joint Committee on Powder Diffraction Standards, Swarthmore, Pa.

6. D.S.Coombs, "X-Ray Investigation of Wairakite and Non-Cubic Analcime," Mineral. Mag., 30, 699 (1955).

7. Powder Diffraction File 19-1227.

8. D.W.Breck, Zeolite Molecular Sieves, p. 202, Wiley (1974).

9. H. Hoss and R. Roy, "Zeolite Studies III: On Natural Phillipsite, Gismondite, Harmotome, Chabazite, and Gmelinite," Beitr.

Mineral. u. Petrogr., 7, 404 (1960).

10. A.Cremers, "Ion Exchange in Zeolites," in Molecular Sieves, II, loc. cit., pp. 182-4.

11. M.H.Hey, "Studies on the Zeolites," Mineral. Mag., 22, 422-37, (1930); 24, 225-53 (1936).

12. A.Steiner, "Wairakite, Ca-Analogue of Analcime, A New Zeolite Mineral," Mineral Mag., 30, 691 (1955).

13. D.W.Breck, loc. cit., p. 503.

14. R.M.Barrer and D.C.Sammon, "Exchange Equilibria in Crystals of Chabazite," J. Chem. Soc., 2838-49 (1955).

15. E.R.Nightingale, "Phenomenological Theory of Ion Solvation. Effective Radii of Hydrated Ions," J. Phys. Chem., 63, 1381-7 (1959).

16. H.Hoss and R.Roy, loc. cit., p. 399.

17. G.T.Kokotailo, S.L.Lawton and S.Sawruk, "Inter- and Intraparticle Diffusion of Ions in Zeolites," in Molecular Sieves, II, loc. cit., pp. 439-50. "When the calcium ion loses its water of hydration, it wants to surround itself with oxygen ions and therefore buries itself into the anionic framework with a preference for small cages, such as the double six-membered rings." p. 448.

18. D.W.Breck, loc. cit., pp. 108-9.

19. L.S.Dent and J.V.Smith, "Crystal Structures of Chabazite," Nature, 181, 1794-6 (1958); J.V.Smith, F.Rinaldi and L.S.Dent-Glasser, Acta Crystallogr., 16, 45 (1963).

QUANTITATIVE PHASE ANALYSIS OF DEVONIAN SHALES BY COMPUTER CONTROLLED X-RAY DIFFRACTION OF SPRAY DRIED SAMPLES

Steven T. Smith, Robert L. Snyder and W. E. Brownell

NYS College of Ceramics

Alfred, N.Y. 14802

ABSTRACT

Spray drying is shown to be an effective and rapid method for preparing samples for quantitative analysis by x-ray powder diffraction. Previously intractable problems like the simultaneous analysis of multiple phases in orientation prone systems can be carried out. Using this method, and a computer controlled diffractometer, five and six phase analyses of Devonian shales can be accomplished in approximately forty minutes. A rapid and convenient method for using the absorption diffraction technique for x-ray quantitative analysis is described.

INTRODUCTION

X-ray powder diffraction is the principal tool used for the quantitative analysis of crystalline phases in mixtures. In spite of its wide spread use, the unreliability of intensity measurements, principally due to preferred orientation, continues to severely limit its application. Quantitative analyses of orientation prone materials, like the clay minerals, have been the cause of a considerable number of studies over the years. The methods arising from these studies usually involve the preparation of a large number of reference patterns of oriented specimens along with tedious and problematic particle size separations (1-2).

Spray drying powders to form spherical shaped agglomerates has been shown to minimize preferred orientation (3). This technique should allow x-ray quantitative analysis of previously intractable systems. This study was undertaken for two reasons. One was to establish a rapid and routine method for quantitative phase analysis as part of a large scale project for characterizing the eastern United States Devonian oil and gas bearing shales. The second reason was to evaluate the effectiveness of spray drying as a general sample preparation technique for use in the quantitative analysis of orientation prone materials. The complex clay containing Devonian shales are well suited to this purpose.

Experimental

To assess the validity, accuracy and precision of quantitative analysis using spray dried samples, two mixtures of minerals were prepared. One was a 60-40 weight percent mixture of kaolinite and quartz. The other was a mixture of illite, quartz, feldspar and chlorite in proportions similar to that found in the Devonian shales. In addition the procedure to be described was applied to 29 samples of shales from West Virginia and New York State.

Since full chemical analysis of each of the shales to be analyzed was available from other work being done in conjunction with this study, the x-ray absorption coefficients could be calculated. This permitted the use of the Diffraction Absorption Technique of Analysis (4-5) based on the relationship:

$$\text{weight fraction of phase j} = \frac{I_{hk\ell \text{ j mixture}} \left(\frac{\mu}{\rho}\right)_{\text{j mixture}}}{I_{hk\ell \text{ j pure}} \left(\frac{\mu}{\rho}\right)_{\text{j pure}}}$$

This equation has been applied in a number of studies where authors usually determine $(\mu/\rho)_{\text{mixture}}$ by direct measurement. Most of these studies have commented on the errors introduced by both orientation efforts and the errors in measuring (μ/ρ). Both of these sources of error should be minimal in this study.

Qualitative analysis by x-ray powder diffraction showed that the shales typically contain as principal phases: illite, quartz, feldspar, chlorite or kaolinite, pyrite and sometimes siderite. In order to obtain the intensities of lines in the pure materials, our mineral collection was surveyed. The sample of each mineral with diffraction pattern closest to the patterns found in the shales was selected as a standard. Due to the high degree of peak overlap in the shale powder patterns, integrated intensities could not be obtained so background corrected peak heights were used. In order to avoid redetermining the standards before each analysis, a method was devised which eliminated day to day variations in incident beam intensity and line voltage instabilities in the detector system. This method involved the fabrication, from brass, of an x-ray diffractometer powder sample holder with a narrower than usual sample chamber. A constant portion of this sample holder was in the x-ray beam during each analysis. Brass was chosen because it had a non-overlapped line at 49.075° 2θ and because its other lines did not obscure those to be used in the mineral analyses. The ratio of the intensity of the analysis line of each mineral standard to that of brass was then determined. Table 1 gives the 2θ angles used for counting the peaks and background for each material. Using typical count rates for illite, a major phase, we calculated (6) that counting times of 300 seconds on peaks and 100 seconds on background would produce results with counting errors on the order of 1 percent. These times were used to measure the intensity ratios given in Table 1. The standard deviations were computed from counting statistics.

All analyses were carried out using a Norelco Vertical diffractometer with a diffracted beam graphite monochromator using Cu $K\alpha$ radiation. The diffractometer had a 1° divergence slit and a 0.003" recieving slit. It was aligned for high resolution such that the $K\alpha_1$ - $K\alpha_2$ doublet would resolve at 28° 2θ, and therefore relatively low count rates were observed. The scientillation detector and a 2θ controlling stepping motor were connected to a Canberra interface which was controlled by a PDP 11 minicomputer. The computer is configured with 56 K bytes of memory and two floppy disk drives.

Computer Controlled Analysis

A computer program was written in FORTRAN IV for the PDP 11 using the RT 11 operating system. The program is interactive, requesting operating parameters from the user. Routine information, which will not normally change from run to run, is stored on a disk file. In fact, Table 1 is a listing of this file. With the exception of the first record for brass, each record on the file contains the information necessary for the computer to measure and compute the weight percent of a particular mineral

TABLE 1

Data File for Computer Controlled Shale Analysis

	Bkg Angle	Peak Angle	μ/ρ	I Standard / I Brass	σ
Brass	51.800	49.075	[100.]	[300.]	
Illite	17.000	19.840	43.0	1.070	0.087
Quartz	17.000	20.875	34.4	7.200	0.330
Feldspar	28.900	27.950	37.9	5.740	0.510
Chlorite	13.200	12.515	94.2	1.630	0.240
Pyrite	33.600	33.085	191.0	2.400	0.500
Kaolin	11.400	12.380	30.0	2.775	0.041
Siderite	28.900	32.100	154.0	2.580	0.140

phase in any unknown sample. The first record provides the peak and background angles for the brass line and the background and peak counting times in seconds to be used in the analysis of each phase (these have been placed in brackets).

The program has two modes of operation. The first mode is to measure data for a new standard and add this to the existing file, or create a new file if one does not exist. In this mode the diffractometer is directed to oscillate, a user specified number of times, between determination of the background corrected intensity for the brass peak and the peak chosen for the new standard. When finished an average value for I standard/I brass is computed along with its standard deviation. These values with the other information in Table 1 are then added to the end of the computer file. As new phases were encountered in the shales a run of a pure standard mineral in this mode would update the file and permit its analysis in any future specimen.

The second mode of operation is the normal analysis mode. Here the computer directs the movement of the detector to the 2θ position of the brass peak, found in the first record on the file and counts for 300 sec. Then the detector is positioned to count background and the corrected brass intensity is computed. Next the record corresponding to the first of the mineral phases to be analyzed is read and the intensities are measured. The mineral to brass intensity ratio is then calculated. Finally this value is converted into a weight percent using the mass absorption coefficient and the intensity ratio of the pure phase also found on the file. The analysis result for this phase is printed on the operators console and analysis of

Table 2. Sample run of quantitative analysis computer program.

```
AUTOMATED QUANTITATIVE ANALYSIS USING
THE INTENSITY RATIO METHOD. VER 6

ENTER CURRENT TWO-THETA VALUE:  4.0

ENTER MU ON RHO FOR THIS SPECIMEN:  48.5

ENTER TITLE FOR RUN:
MIDDLESEX SHALE 10 AM APRIL 20

IS THIS A NORMAL RUN?  (ENTER A ZERO) OR A LOOP RUN OR STANDARD?
(ENTER NUMBER OF LOOPS):
0

DO YOU WANT THE REF. PEAK DETERMINED ONLY ONCE?  (ENTER 0), OR
BEFORE EACH UNK PEAK?  (ENTER 1):
0

HOW MANY STANDARDS DO YOU WANT TO DETERMINE?
LISTING OF STANDARDS      = (-1)
USE ALL STANDARDS IN RUN = (0)
SPECIFIC NUMBER OF STDS  = (#):  5

ENTER THE NUMBER OF STANDARDS TO BE RUN:  2,3,4,5,6

MOUNT THE SPECIMEN IN THE DIFFRACTOMETER.
WHEN READY; HIT RETURN


        ANALYSIS RESULTS FOR-MIDDLESEX SHALE 10 AM APR 20
```

NAME	PEAK-CNT	BKG	NET-CPS	SIGMA	RATIO	SIGMA	CONC	SIGMA
BRASS	38174.	1402.	113.23	0.75				
ILLITE	22862.	1946.	56.75	0.67	0.50	0.01	52.83	4.30
QUARTZ	63861.	1992.	192.94	0.95	1.58	0.02	31.03	1.42
FELDSPAR	15343.	2230.	28.84	0.63	0.23	0.01	5.15	0.46
CHLORITE	17318.	2394.	33.79	0.66	0.26	0.01	8.18	1.20
PYRITE	9675.	2280.	9.45	0.58	0.07	0.00	0.72	0.15

```
                          TOTAL    97.91 +, -      7.53

ANOTHER SAMPLE FOR THESE STANDARDS?  YES = (0); NO = (1):  1

ANOTHER NEW SAMPLE?  YES = (0); NO = (1):  1

EXIT
```

the next phase is begun. An example run of this mode of operation is shown in Table 2. The underlined items are the user's response to the questions asked by the program. This example requested that standards 2 through 6 (as given in Table 1) be analyzed. This five-phase determination was then carried out with no further user intervention in approximately forty minutes.

Materials

Illite: this mica-like clay mineral has a variable composition due to extensive solid solution. Our illite standard material was purified from a sample of the American Petroleum Institute reference clay No. 36 (7) obtained from Wards (8). This illite found at Morris, Illinois, contains quartz as the major impurity and probably a trace of kaolinite. Preparation of this material involved grinding, suspension in an aqueous medium (about 10% by weight illite) with Calgon added as a deflocculant. After 24 hours of settling, the top part of the solution was withdrawn, centrifuged and dried. This purified material, which was subsequently spray dried, contained only a trace of quartz, with a total impurity level of probably 3%. Chemical analysis showed Si = 21.76%, Al 12.98%, Fe 2.88%, Mg 1.49%, Ca 0.42%, P 0.47%, Ti 0.50%, Na 0.25%, K 3.24%. The most intense peak free from serious overlap was chosen for analysis of each phase. For spray dried illite this was the 111 line near 4.46Å. It is interesting to note that this line is more intense than the basal (001) reflection when the specimen is spray dried.

Quartz: a pure sample of < 200 mesh was ground for one hour and spray dried. Since the most intense quartz line interferes with the illite 003 we chose the second most intense line at 4.26Å for analysis.

Feldspar: a sample of oligoclase from Mitchell County, North Carolina was closest in diffraction pattern to the shale feldspars. This material was analyzed as Si 28.80%, Al 12.38%, Fe 0.27%, Ca 3.44%, Na 6.25%, K 0.42%. The sample was ground to < 325 mesh and spray dried. The most intense line near 3.20Å was chosen for analysis.

Chlorite: a sample of ripidolite was used as a standard for this platy disilicate material. A Wards sample from Goscheneralpe Switzerland was ground to <325 mesh and spray dried. Chemical analysis showed Si 11.5%, Al 10.96%, Fe 24.06%, Mg 6.03%, Ti 0.02%. The 002 line near 12.5° was used for analysis.

Pyrite: a Wards sample from Rico, Colorado was ground to <325 mesh and spray dried. The 200 line which is the most intense peak in both the calculated and spray dried patterns was used. The powder Diffraction File pattern (6-710) incorrectly lists the 311 as the 100% line.

Kaolinite: the sample, kaolex D-6, a Georgia kaolin from the J. M. Huber Corp (9) was sufficiently fine to be spray dried directly . The partical size is reported (9) as being 62-68% < 2μm and 12-15% >5 μm. The most intense line, 002 near 12.4^o was chosen for analysis. This line is in the same location as the line used for chlorite. We were forced to use these conflicting lines because of their high relative intensities and the low concentration of these two minerals. To differentiate chlorite from kaolinite, oriented slides of non-spray dried shales were prepared and treated with DMSO (10). The analysis of these diffraction patterns showed that when chlorite was present kaolin was not and when kaolin was present there were only trace amounts of chlorite. Thus with the introduction of a small amount of error the 12.4^o line was measured using either the chlorite or kaolin standard depending on the result of the DMSO test.

Siderite: a Wards sample from Roxbury, Connecticut,was ground to <325 mesh and spray dried. This sample contained about 5% quartz which was ignored. The 104 reflection was used in the analysis.

Shales: the shales of unknown composition were broken into pieces of about 5 cm and ground first in a roll crusher. Representative samples were obtained using a riffle-type splitter. A 15g sample was then ground to <325 mesh in a spex mill or mortar and spray dried. Mass absorption coeffecients were obtained by calculation from the known chemical analysis.

Two Phase Test Mixture: a 60% kaolinite 40% quartz mixture was weighed,ground together and spray dried. The mass absorption coefficient was calculated as 31.8.

Four Phase Test Mixture: a mixture of 47.2% illite, 34.1% quartz 11.4% oligoclase and 7.3% ripidolite was weighed, ground and spray dried. These proportions are similar to those found in the shales. The mass absorption coefficient was calculated to be 45.0.

Results and Discussion

Figure 1 is a scanning electron micrograph of ripidolite which had been ground to <325 mesh and spray dried. This is a

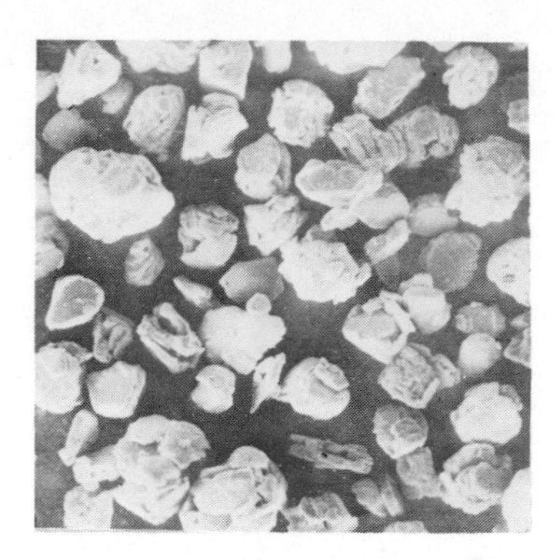

Fig. 1. Spray dried ripidolite before fine grinding (200X)

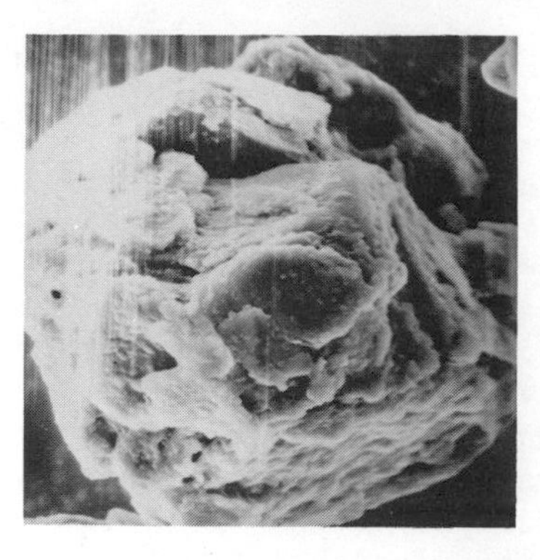

Fig. 2. Spray dried ripidolite before fine grinding (1200X)

Fig. 3. Spray dried ripidolite after fine grinding (200X)

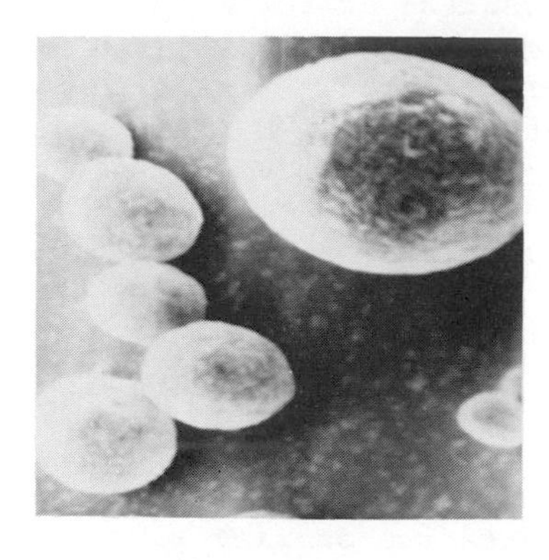

Fig. 4. Spray dried illite (640X)

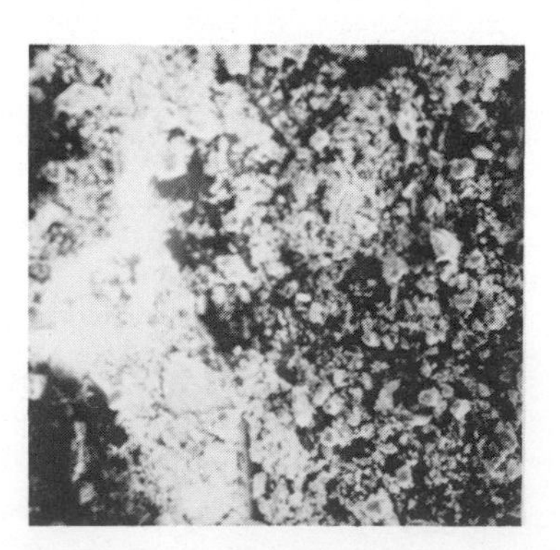

Fig. 5. Middlesex shale before spray drying (200X)

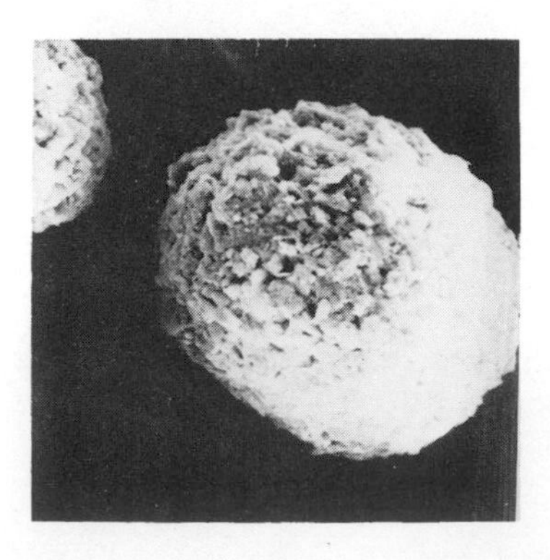

Fig. 6. Middlesex shale after spray drying (720X)

good illustration of the sphere distorting effects of too large a ratio of crystallite size to agglomerate size. Figure 2 shows a close up of one of these agglomerates, clearly indicating the controlling effect which relatilvely large crystallites have on the microstructure of the sphere. After regrinding the ripidolite powder for 40 more minutes and spray drying, the spherical agglomerates shown in Fig. 3 were obtained. The finest particle size material we had available was the illite powder obtained by Stokes Law settling. On spray drying this powder, the smooth eliptically distorted spheres shown in Fig.4 were obtained. The spray drying of the finely ground shales, such as the Middlesex, NY shale shown in Fig. 5 presents no special problems. The spray dried shale shown in Fig. 6 formed reasonably spherical agglomerates with no evidence of either orientation or particle segregation.

The quantitative analysis of the prepared kaolin-quartz mixture,averaged over three computer runs of approximately one half hour each,gave:

	Measured	Actual
Kaolin	57.7% ± 4.1	60.0%
Quartz	41.3% ± 3.6	40.0%

The analysis results for the prepared mixture of minerals with proportions similar to those found in Devonian shales are:

	Measured	Prepared
Illite	45.1% ± 1.0	47.2%
Quartz	31.1% ± 1.9	34.1%
Feldspar	11.6% ± 1.7	11.4
Chlorite	7.4% ± 0.6	7.3

These data were also averaged over three computer runs of approximately 40 minutes each.

These results show accuracy and precision typical of quantitative x-ray powder diffraction analyses on synthetic materials. Quantitative mineral analyses are often only reported to the nearest five percent. On considering such factors as the mineral nature of these materials, that analyses are made from a single pure standard and the numerous sources of error in the method of analysis, these results are impressive. They further confirm the validity and desirability of spray drying as a preparation procedure for quantitative analysis.

The quantitative analysis of the five or six phases identified in the 29 Devonian shales summed to give a total analyzed mineral content between 90% and 105%. A typical computer run on the Middlesex NY shale shown in Figs. 5 and 6 is given in Table 2. The second from last column at the end of this table gives the concentration of the phase in weight percent. These runs usually take the computer about forty minutes to carry out.

Since the results on the synthetic samples show deviations of as much as 3% and the total mineral analysis of the shales vary over a 15% range, some consideration of the sources of error is in order. First let's consider errors in the general procedure not related to the mineral nature of the samples. The problems of microabsorption and of the peak counting method are well established and have some effect in our procedure. A more serious problem which arises only with spray dried samples is the tendency of our agglomerates to roll. This problem gets more serious at high angles where the sample holder angle is significant. The rolling of the specimen introduces a sample displacement error which will significantly shift a peak's position. Particularly when using the peak counting method this can introduce significant errors. This effect can be controlled by two methods. One is to spray a clear amorphous lacquer over the sample in the holder to keep it in place. The other is to add a peak location routine into the computer algorithm to find the top of the peak before counting. The traditional errors due to orientation and sample inhomogeneity appear not to be significant in this method.

The principal source of error in our analyses of the shales is the difference between the mineral phase found in the shale and that used as a standard. A number of serious assumptions have to be made in selecting standards. The extremely solid solution prone chlorites, feldspars, illites and siderite found in the Devonian shales showed considerable variation in their diffraction patterns. For example the choice of oligoclase as our feldspar standard does not imply that plagioclase and the alkali feldspars were not also present. In addition to this large error source,the presence of impurity phases in our standards and the presence of small amounts of kaolinite in the chlorite shales and vice versa introduced more error. In general the shales show a high background with numerous unidentifiable small peaks; relative changes in these under the peaks to be analysed, is another source of error. On considering all of these difficulties the variations in the shale analyses seem reasonable. A more accurate analysis can be performed on any one shale by a careful matching of standards to the mineral phases. For the survey we are conducting,this improvement would not justify the extra time required.

CONCLUSION

Spray drying has been shown to be a simple and effective method for minimizing orientation effects for quantitative x-ray diffraction analysis. Use of this technique provides a method for the previous intractable problem of simultaneous multiphase analysis in orientation prone materials. The complex mixtures found in Devonian shales can be analyzed by this method.

ACKNOWLEDGMENTS

The authors wish to express their gratitude to C. L. Mallory for writing the interface control subroutines. This work was supported in part by U. S. Department of Energy Contract No. EY-76-C-05-5207 and W-7405-ENG-48.

REFERENCES

1. Sudo, Ouinuma and Kobayaski, "Mineralogical Problems Concerning Rapid Clay Mineral Analysis of Sedimentary Rocks," Acta Universitatis Carolinae Geologica Supplementum I, 189-219 (1961).
2. Schultz, "Quantitative Interpretation of Mineralogical Composition from X-ray and Chemical Data from the Pierre Shale," Geological Survey Professional Paper 391-C.
3. S. T. Smith, R. L. Snyder and W. E. Brownell, "Minimization of Preferred Orientation in Powders by Spray Drying," Adv. in X-ray Anal. this volume.
4. H. P. Klug and L. E. Alexander, "X-Ray Diffraction Procedures," 2nd ed., Wiley, New York (1974).
5. J. Leroux, D. H. Lennox and K. Kay, "Direct Quantitative X-Ray Analysis by Diffraction Absorption Technique," Anal. Chem 25, 740-743 (1953).
6. R. Jenkins and J. L. deVrie, "Practical X-ray Spectrometry," Springer Verlag, New York (1969).
7. Reference Clay Mineral, American Petroleum Institute Research Project 49, Columbia University (1951).
8. Wards Natural Science Establishment, Inc. P. O. Box 1712, Rochester, NY 14603
9. "Kaolin Clays and Their Industrial Uses," J. M. Huber Corp., New York, NY (1955).
10. S. G. Garcia and M. S. Camagano, "Differentiation of Kaolinite form Chlorite by treatment with Dimethyl Sulfoxide," Clay Miner. 7,447 (1968).

X-RAY DIFFRACTION STUDIES OF STABILITIES OF ZEOLITES TO TEMPERATURE CHANGES AND SOLUTION TREATMENT

Hossein Salek, Harold Vincent and Melinda Patton

The Anaconda Company

P. O. Box 27007, Tucson, Arizona 85726

ABSTRACT

Natural zeolite 1010A, mostly clinoptilolite, was found to be stable to prolonged heat treatment up to 600° C. Stability to treatment by acids and bases is dependent on solution strength, treatment time and temperature. Variations in intensities of diffraction lines correlate well with changes in exchange capacities for ammonium ions and other observed physico-chemical changes.

INTRODUCTION

This paper discusses the X-ray diffraction studies of heat and solution treated zeolite 1010A. The studies were part of a project to evaluate a zeolite from a particular orebody for its ability to retain its properties under certain physical and chemical stresses. The natural zeolite 1010A, a newly commercial material marketed by The Anaconda Company, is an aggregate consisting primarily of clinoptilolite with a compositional formula:

$$[(Ca, Na, Mg, K)\ (Al, Fe)_2\ Si_3\ O_{18} \cdot 6H_2O]$$

The X-ray diffraction and optical studies of the ore samples indicate that the ore material contains up to plus 95% clinoptilolite with accessory, quartz, feldspar, carbonate, clays, mica, iron and manganese oxides, glass, and other minerals. A trace to minor amount of heulandite was also noted in a few areas of the orebody. Fluorite is seldom found in the rock ore, and then only in trace amounts.

The ion-exchange properties of this zeolite are used in a number of applications for the concentrating and separating of cations. Conditioning, elution, and regeneration of the ion-exchanger is usually accomplished by the application of strong acid or base solutions or strong salt solutions. Variation of temperature may be a part of these processes. It is necessary to know the stability of the zeolite under these various conditions.

The molecular sieve properties of this zeolite may be used for dehydrating, purifying, and isolating certain gaseous components. Gas absorption properties are strongly temperature dependent. Regeneration and activation are usually carried out by elevating the temperature of the material by several hundred degrees. Knowledge of its temperature stability is necessary in order to plan for cyclic use of the material.

EXPERIMENTAL - THERMAL TREATMENT

Zeolite 1010A samples were heated in air to temperature levels up to 1000° C for various time periods. The products of that treatment were examined with respect to physical characteristics such as hardness and color and with respect to maximum exchange capacity for ammonium ions. Measurement of the ammonium ion absorption is appropriate because of the relatively high selectivity for this ion and because it does not occur in nature with the zeolite. X-ray diffraction patterns were obtained for each of these products. Tests were carried out on samples sized to -20 mesh +48 mesh and on samples -200 mesh.

RESULTS OF THERMAL TREATMENT

The hardness of the Zeolite 1010A increases with increasing temperature. For the untreated aggregate particles the hardness is about 3 and the hardness changes to about 5 when the sample is heated to 750° C. The color changes to pink at 600° C, red at 700° C and dark red to reddish brown at 800° C. More fine particles are evident in the material heated to 800° C than for that heated only to 700° C.

The X-ray diffraction studies of heat treated zeolite of two size fractions (-20 +48 mesh and -200 mesh) indicate that there are no changes in the line intensities and no pattern line shifting for each heated sample from 100° to 700° C. The reactions and their rates are apparently not dependent on the aggregate particle size. X-ray patterns for some heat treated samples above 300° C from different areas of the 1010A zeolite orebody show minor decreases in line intensities. This change is due to the heulandite impurities.

Heulandite is another zeolite mineral which has an X-ray pattern that is similar to one for the clinoptilolite and its structure collapses when the mineral is heated to about 350° C.

As shown in Figure 1, the structure line intensities for the clinoptilolite decrease as temperature is increased above 700° C. During the temperature increase, an increase in the intensity of the scattered lines for the amorphous material is noted, shown by the hump appearing in the middle of the diffraction pattern. These changes correlate with the measured ion exchange capacity for ammonium ion as shown in Table #1. Very little change in this capacity is observed up to 600° C. Slight decreases at 650° C and 700° C are shown. Above 700° C, the capacities diminish until none is detected for samples heated to 800° C and 850° C. The diffraction patterns indicate that some clinoptilolite remains in the samples at 800° C. This diminished capacity may be due to the blocking or damaging the cages of zeolite so that the ammonium ion is too large in size to go inside the cages.

TABLE I Ammonia Absorption Capacity for Heat Treated Zeilite 1010A (for Period of Four Hours)

Temp. C	% of Zeolite Structure Collapsed (by XRD)	Meq NH_4^+/gram
Untreated Sample	--------	1.95
100^0	No change	2.03
200^0	No change	2.10
300^0	No change	2.18
400^0	No change	2.14
500^0	No change	2.08
600^0	No change	2.01
650^0	Possible trace	1.89
700^0	Possible trace	1.84
725^0	20-25	1.52
750^0	30-35	1.50
775^0	40-45	0.98
800^0	70-75	0.00
825^0	85-90	0.00
850^0	95-97	0.00
900^0	100	0.00
1000^0	100	0.00

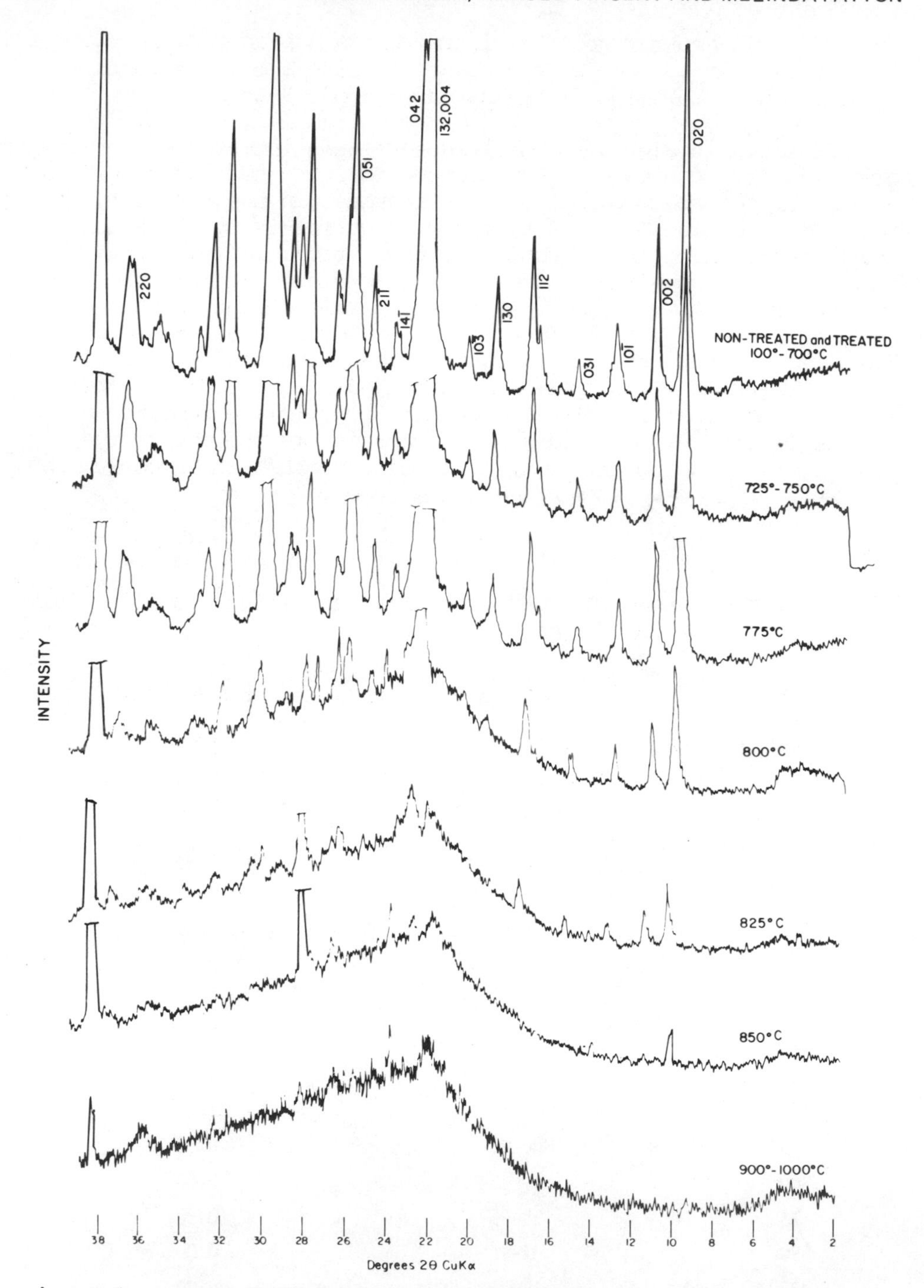

Figure 1 - X-ray diffraction patterns for zeolite 1010A samples heated to various temperatures for 4 hours in air.

EXPERIMENTAL - SOLUTION TREATMENT

Zeolite 1010A samples were placed for various periods of time in NaOH solutions ranging in concentration from 10^{-5} normal to one normal. Other 1010A samples were treated for various time periods in HCl solutions ranging from 0.01 N to 12 N. Most of the solution experiments were carried out at ambient temperature and ranged in exposure times up to several weeks. A series of experiments at 60° C were also carried out with 6N HCl.

Zeolite 1010A in the -20 + 48 mesh range were used for most experiments. Some tests on -200 mesh material were carried out in order to describe effects of particle size and diffusion. Chemical analyses were obtained for products and corresponding solutions. X-ray diffraction patterns were obtained for the treatment residues. Ion exchange capacities for ammonium ions were measured for these materials.

RESULTS - SOLUTION TREATMENT

The data in Table #2 show that the zeolite has deteriorated in the 0.1 normal HCl and stronger solutions. The effect on the ammonia absorption capacity of treating the zeolite with 6 normal HCl solution at 60°C for 24 hours is the same as ambient temperature treatment for 22 days. Some capacity remains after this treatment corresponding to lines shown on the diffraction patterns of the residues. Boiling with 6N HCl gives a product with no capacity for ammonia absorption even though the diffraction pattern (Figure 2)

TABLE II Ammonia Absorption Capacities for Zeolite 1010A After HCL Treatment

Soln	Time	Temperature C	Meq NH_4^+/gram	% of Zeolite Structure Collapsed (by XRD)
H_2O	10 days	Ambient	1.94	No change
0.01 N HCl	10 days	Ambient	1.90	No change
0.10	10 days	Ambient	1.61	No change
1.0	10 days	Ambient	1.35	No change
3.0	10 days	Ambient	1.26	No change
6.0	10 days	Ambient	0.71	20-25
6.0	14 days	Ambient	0.61	40-45
6.0	22 days	Ambient	0.54	20-25
6.0	4 hours	60°	1.00	15-20
6.0	8 hours	60°	0.81	15-20
6.0	16 hours	60°	0.72	35-40
6.0	24 hours	60°	0.56	55-60
6.0	15 minutes	Boil	0.00	45-50

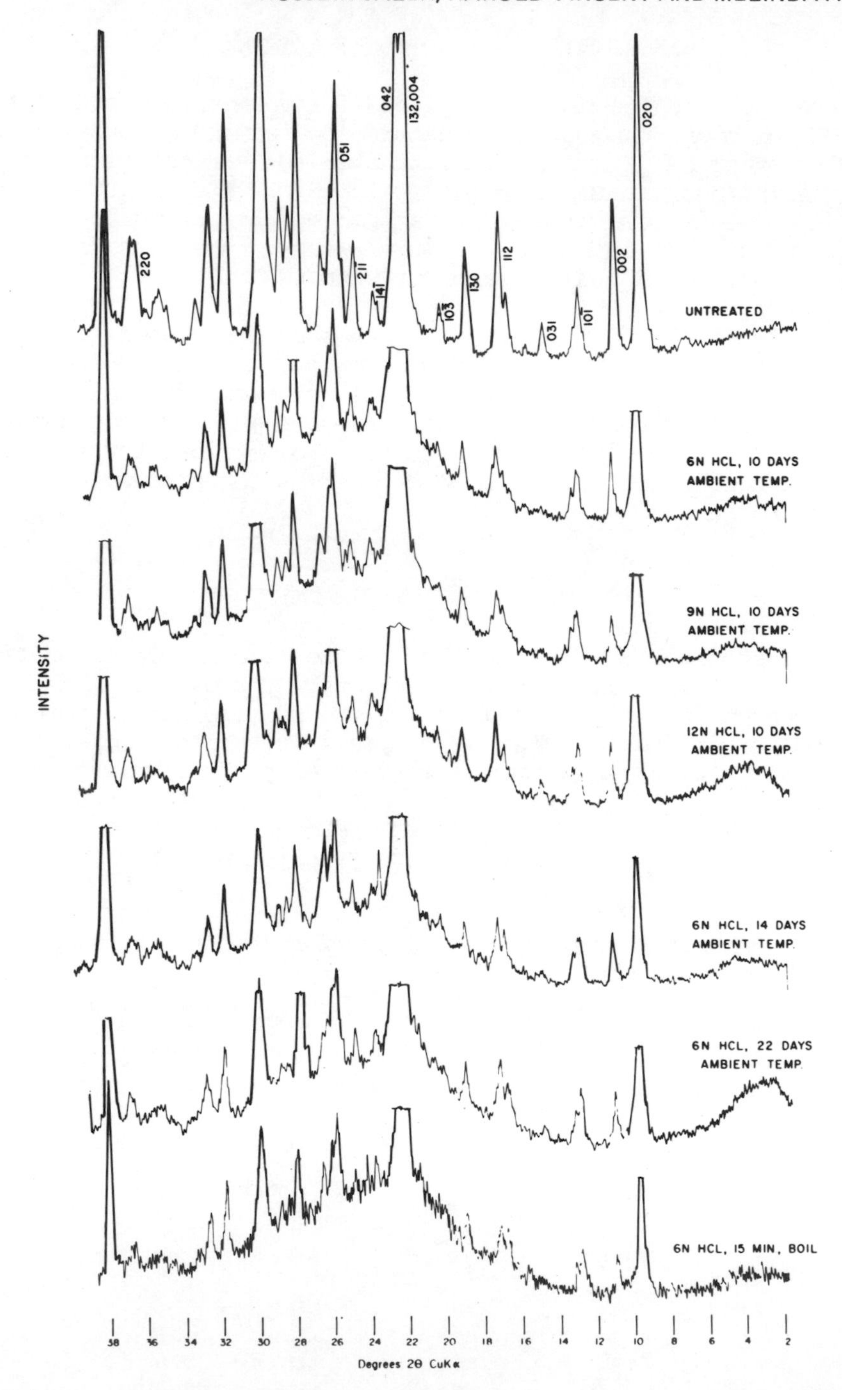

Figure 2 - X-ray diffraction patterns for zeolite 1010A samples after solution treatments.

TABLE III Acid Solution Equilibrium Components (MEQ/Gram) Zeolite 1010A (Ambient Temperature, 21 Days)

	Na	K	Mg	Ca	Fe	Si
H_2O	0.15	.005	.001	0.00	.005	0.40
0.01 N HCl	0.62	.041	.004	.001	.00	0.79
0.1 N HCl	1.08	.29	.076	.02	.11	1.41
1.0 N HCl	1.16	.97	.14	.03	.33	1.14
3.0 N HCl	1.18	1.14	.15	.04	.41	0.00
6.0 N HCl	1.25	1.21	.20	.08	.37	0.00

indicates that nearly half the original amount of clinoptilolite still remains. A buildup is shown of the scatter from amorphous material represented by the broad hump in the middle of the pattern. This is similar to that noted for thermal treatment although the products may be different. Table #3 shows the appearance in the equilibrium solutions of the various components. Sodium changes to the greatest extent with the weaker acid application with potassium showing with stronger solutions. This is in agreement with the relative cation affinities considering displacement by hydrogen ions. Some of the potassium and the magnesium may come from gangue mineralization dissolving in the solutions stronger than 1.0 normal. No cation exchangeable iron occurs in the zeolite as determined by ion exchange tests so that the appearance of iron in solution may represent breakdown of either zeolite or gangue minerals or both.

Basic solutions at ambient temperature affect zeolite breakdown when the solution is 10^{-3} normal or stronger. X-ray diffraction patterns of the residues show solubilization of the clinoptilolite.

CONCLUSION

X-ray diffraction intensities related to clinoptilolite in the natural zeolite 1010A are affected by changes in the aggregate material brought about by heating or treating the material with strong acid or base solutions. Diffraction intensities correlate well with ammonium ion exchange capacities except for the products of stronger thermal and chemical treatments where some clinoptilolite structures are evident but ion exchange capacities are not. Further work is needed to explain this.

APPLICATIONS OF THE GANDOLFI X-RAY CAMERA IN MINERALS INDUSTRY RESEARCH

Ronald K. Corbett

Colorado School of Mines Research Institute

P. O. Box 112, Golden, Colorado 80401

ABSTRACT

The Gandolfi camera, which produces simulated powder diffraction patterns from single crystals, has been found to be a very useful tool in applied mineralogical research. For many micro-diffraction problems, the advantages of the method far outweigh its weaknesses, and Gandolfi films are usually quite adequate for identification purposes. A simple modification to the small Gandolfi camera permits the recording of low-angle reflections corresponding to d-spacings of nearly 14 angstroms for CuK_α radiation. Applications of the method include identification of radioactive minerals, other heavy minerals, contaminants, and glass stones.

INTRODUCTION

Although the Gandolfi camera has become widely available in the decade following Gandolfi's publication (1) on his method, only a limited amount of information has appeared concerning applications of the method. The camera produces a powder-like diffraction pattern from a single crystal by simultaneous rotation of the crystal about two axes. Extensive use of the Gandolfi device as a supplement to other x-ray diffraction techniques in a wide variety of materials analysis problems has shown it to be a very valuable tool when the nature of the sample requires a micro-diffraction technique.

SOME STRENGTHS AND WEAKNESSES OF THE METHOD

The technique has certain advantages over other x-ray diffraction methods in some situations. A tiny sample, which may be a single crystal as small as 30 microns, can be easily mounted without regard to orientation. With proper sample mounting procedures, even smaller particles (down to 10 microns or less) can be examined with some degree of success. A pure, single-phase sample can thus often be obtained, and the sample can be recovered intact for subsequent examination by microscopy or other methods. Preferred orientation effects are eliminated, as is the grinding damage which often produces weak, diffuse back-reflection patterns in the normal powder method.

Certain weaknesses of the method, which are readily explainable in terms of the system geometry, seldom present problems in practice. The gaps and nonuniform intensities along some diffraction arcs, commonly observed on Gandolfi films, are due to limitations imposed by camera configuration, crystal symmetry, and the multiplicity factor for the reflection in question. In general, pattern quality deteriorates with decreasing crystal symmetry. The smoothness of individual lines on a given film depends upon the chance orientation given to the crystal and the multiplicity factor for each class of reflection. Since low-angle lines are often those from planes with simple indices and low multiplicity factors, pattern quality tends to show some improvement with increased diffraction angle. If a small aggregate of several diversely oriented crystals can be obtained or made without introducing unwanted phases, or if the sample is by nature finely polycrystalline, the problem of gappy lines is virtually eliminated.

The method may also give rise to estimated intensities which deviate somewhat from published JCPDS data. Some factors affecting the extent of the deviation are: 1) the method used for estimation of intensities, 2) the quality of the Gandolfi film, as influenced by the physical nature and symmetry of the sample, 3) the tendency of the phase to show preferred orientation in the normal powder method, and 4) the extent to which the phase is subject to grinding damage in the normal powder method. Since no sample grinding is involved with the Gandolfi technique, back-reflection patterns are typically sharp and strong compared to powder films. But "abnormal" Gandolfi intensities almost never present identification problems.

LOW-ANGLE MODIFICATION OF THE SMALL CAMERA

A simple modification to the 57.3 mm Gandolfi camera greatly extends its usefulness by permitting the recording of low-angle

a

b

Figure 1. Low-angle modification of small Gandolfi camera.
a. Film cassette in normal configuration.
b. Spacer withdrawing beam stop (right).

reflections corresponding to d-spacings of nearly 14 angstroms for CuK_{α} radiation. This modification, described by Sorem in 1960 (2), involves the placement of a rubber tubing spacer around the beam stop tube. The resulting withdrawal of the tube allows use of an ordinary paper punch to make a smaller exit hole in the film. Placement of the spacer on the beam stop is illustrated in Figure 1. The diameter of the exit hole is reduced from the normal 9.5 mm to about 6.4 mm, thus insuring that spacings in the 8-14 angstrom range (for CuK_{α}) will not be missed. The short exposure times, coupled with the ability to record large d-spacings, make the small camera particularly attractive and quite adequate for most determinative work.

APPLICATIONS OF THE METHOD

The Gandolfi camera is especially well suited to determinative uranium-thorium mineralogy. Used in conjunction with heavy liquid separations, ultraviolet light, and autoradiography, it allows very selective sampling and identification of radioactive phases. Expensive microprobe analyses can often be avoided. Metamict grains may be easily dismounted, heat treated, and remounted. The low-angle modification is very helpful in recording the large d-spacings commonly encountered in hydrated, secondary uranium minerals. The mineral metatorbernite, for example, has a strong characteristic reflection at about 8.7 angstroms. This is seen as the first dark ring right around the small hole in the lower film of Figure 2. Use of the normal 9.5 mm hole punch, as observed on the upper film, causes this key reflection to be lost in and around the hole, however.

Phases of economic or geologic interest in a variety of heavy mineral assemblages can be readily identified and characterized by use of the Gandolfi method. Specific grains can be removed from the surfaces of thin and polished sections and mounted directly in the camera without any need for further sample preparation. Contaminants of various kinds, present in minor to trace amounts as small particles, can often be singled out and identified. Sources of trouble in glass furnaces or glass raw materials, which may give rise to tiny stones in glass, have been tracked down by means of Gandolfi photographs on numerous occasions. Additional applications of the versatile Gandolfi camera seem to present themselves with continued use, and each individual user can be led by his or her own ingenuity and imagination.

ACKNOWLEDGEMENT

I wish to thank Edward H. Leland for taking and preparing the photographs.

a

b

Figure 2. Two Gandolfi films from a small polycrystalline aggregate of metatorbernite. Enlarged comparison of front-reflection regions. a. Using normal large hole punch, 8.7 Å ring is not recorded. b. With beam stop spacer and small hole punch, this reflection is observed. Both exposures 12 hours using Ni-filtered Cu radiation in 57.3 mm camera.

REFERENCES

1. G. Gandolfi, "Discussion upon Methods to Obtain X-Ray 'Powder Patterns' from a Single Crystal," Miner. Petrogr. Acta 13, 67-74 (1967).

2. R. K. Sorem, "X-Ray Diffraction Technique for Small Samples," Am. Mineral. 45, 1104-1108 (1960).

INSTRUMENTAL TECHNIQUES FOR THE ANALYSIS OF PAPER FILLERS AND PIGMENTS

Claude Robert Andrews and Robert Kenneth Mays

J. M. Huber Corporation

P. O. Box 310, Revolution St., Havre de Grace, MD 21078

INTRODUCTION

The chemical analyses of pigments and fillers are important to both the manufacturer and the papermaker. Most standard analytical methods, including those of Technical Association of the Paper and Pulp Industry, are based on the so-called classical gravimetric and volumetric techniques. The major components analyzed are TiO_2, SiO_2 and Al_2O_3. X-ray diffraction methods are available for the identification of crystalline fillers and pigments in finished paper.

QUALITATIVE ANALYSIS

At present, fillers and pigments cannot be analyzed without destroying the paper matrix and the crystalline identity of many of the additive materials. The exceptions to this are the analysis of Ti in paper by radioactive isotopic x-ray fluorescence(1) and the low temperature ashing technique (2). The low temperature ashing technique destroys the paper matrix using an oxygen plasma, but does not destroy the crystalline identity of the additives. Most analyses start with a qualitative investigation for the crystalline and/or mineral fillers using x-ray diffraction methods on the original paper sample (3, 4).

In a diffractogram of a typical paper sample, the peaks are identified and the crystalline material (TiO_2, talc, or kaolin) have the "d" spacings for the minerals calculated using copper $K\alpha$ radiation. The crystalline patterns are usually superimposed upon the cellulose amorphous humps.

A diffraction pattern for the ash of the sample paper is usually the diffraction pattern for TiO_2 (Anatase). The patterns for kaolin, talc and cellulose have been destroyed by the ashing process at 900°C, and this is the normal result of the ashing process. X-ray diffraction cannot detect or produce any useful information on amorphous fillers and pigments. Their presence may be indicated by the appearance of amorphous humps in the diffraction traces of the paper ash.

EXPERIMENTAL DETAILS

The major components of the most widely used paper pigments are usually SiO_2 and Al_2O_3. This is true for kaolins and sodium alumino silicates as well as natural and synthetic silicas and silicates. Their use in most paper applications is in combination with TiO_2. TiO_2, SiO_2 and Al_2O_3 are determined by x-ray fluorescence spectrometry.

The instrumental parameters for the x-ray spectrometer are shown in Table I. Spectral interference is encountered by TiO_2 on Al_2O_3. This occurs because the 3d order reflection of TiO_2 at 36.09° (2θ) is very close to $K\alpha$ line of Al_2O_3 at 36.51° (2θ). Calibration of the TiO_2 line contribution at the $K\alpha$ of Al_2O_3 must be performed and the Al_2O_3 intensities corrected for this interfering TiO_2 contribution.

The samples (standards and paper ash samples) were prepared by a fusion technique (5, 6). In the fusion procedure, 1.000 g of paper ash is mixed with 2.500 g of $Li_2B_4O_7$-Li_2CO_3-NaCl flux and 0.5000 g of Na_2CO_3 in a Pt-5% Au dish. The NaCl is necessary as an anti-wetting and mold releasing agent. The mixture is then fused over a blast burner, swirled regularly to ensure homogeneity and cast into discs by pouring the melt into previously heated (red hot) Pt-5% Au molds. The molds are then air-cooled and the disc removed by inverting the molds. The discs are ready for placement in the spectrometer for analysis.

The standard samples were prepared from a full fraction Georgia filler clay, reagent grade TiO_2, and sodium aluminosilicate. All these materials were ignited separately for 2 hours at 900°C. They were analyzed by multiple gravimetric and volumetric methods for TiO_2, SiO_2 and Al_2O_3.

TABLE I

INSTRUMENTAL CONDITIONS

ELEMENT	Al	Si	Ti
Atomic No. (Z)	13	14	22
Spectral Line	Kα	Kα	Kα (2d order)
Wavelength (λ), Å	8.339	7.126	2.750
2θ line degrees	36.51	31.05	23.82
Background 2θ	39.00	32.00	26.00
X-ray Operating Potential	45 KV	45 KV	45 KV
X-ray Operating Current	30 ma	30 ma	30 ma
Crystal	KAP	KAP	KAP
Radiation Path	Vacuum	Vacuum	Vacuum
Counting Time	20 sec	20 sec	20 sec
Pulse Height Level	3	3	3
Pulse Height Window	12	15	open

These analyses are shown in Table II. All these values are calculated on the anhydrous basis (2 hrs @ 900°C). The x-ray fluorescence standard discs were prepared by compounding different amounts of TiO_2, filler Georgia Clay and Sodium Aluminosilicate and carrying the samples through the fusion process. An instrumental standard disc was prepared from TiO_2 and NBS Flint Clay, and all standard and sample disc intensities were ratioed against the corresponding intensity from the instrument standard disc. This technique was employed to remove instrumental drift and radiation path variances.

TABLE II

ANALYSIS OF IGNITED STANDARD MATERIALS

COMPOUND	TiO_2	Filler Clay	Sodium Aluminosilicate
% SiO_2	---	52.9	75.0
% Al_2O_3	---	44.4	10.7
% TiO_2	99.9	0.58	0.20
% Fe_2O_3	---	0.15	0.07

RESULTS

TiO_2

The intensity ratio of TiKα radiation was found to be linear with percent TiO_2 in the anhydrous samples. The correlation coefficient was 0.9969. In the equation, R' is the ratio of sample intensity to the instrument standard intensity. The linearity was found to extend to 50% TiO_2. The equation for TiO_2 is the following: $\% TiO_2 = 50.835 R'_{TiO_2} - 0.73$

SiO_2

The intensity ratio of SiKα was found to be non-linear with percent SiO_2 because of a severe matrix absorption effect from TiO_2 and Al_2O_3. This matrix effect was removed by the application of the Lucas-Tooth-Price equation (7).

This is a correction assuming that the interfering element intensity ratios are linear with concentration. The corrected intensity ratios of SiKα are linear with percent SiO_2. The overall correlation co-efficient is 0.9995 and the linear range is 0 to 75% SiO_2. The equation for SiO_2 is the following:

$$\%SiO_2 = 0.39 + 23.223 R'_{SiO_2}(0.9871 + 0.1074 R''_{Al_2O_3} + 0.0593 R'_{TiO_2}).$$

Al_2O_3

AlKα radiation displays a spectral interference by TiKα 3d order. Al_2O_3 intensity ratios were measured while varying the TiO_2 content. No Al_2O_3 was present in the discs. The intensity ratio of AlKα was compared against the intensity ratio of TiO_2 at TiKα and found to be linear with a correlation co-efficient of 0.9996. The Al_2O_3 x-ray data were corrected for the TiO_2 contribution by the equation, $R'' = R' - R°$ and a linear relationship was found. $R°_{Al_2O_3} = 0.3330 R'_{TiO_2} - 0.0017$

The corrected data have a correlation co-efficient of 0.9988 and a linear working range of 0 - 45% Al_2O_3.

$$\% Al_2O_3 = 31.955 R''_{Al_2O_3} + 0.32$$

The x-ray fluorescence method just described was used to analyze paper samples. The type of samples analyzed by the x-ray fluorescence method was the ash of actual paper samples. These samples were supplied by our paper laboratory. The data in Table III show the excellent agreement of the oxide composition between the x-ray fluorescence method and the standard method. Note the wide range in the TiO_2, SiO_2, and the Al_2O_3 content. The reproducibility of the x-ray fluorescence method was found to be ± 1%.

TABLE III

Paper Sample Code	% Ash	Oxide	% Oxide Analysis by Classical Methods	XRF Methods
PL-37-77	19.6	TiO_2	46.4	47.0
		SiO_2	29.3	28.6
		Al_2O_3	19.1	18.8
1639	21.6	TiO_2	3.5	3.2
		SiO_2	49.8	50.7
		Al_2O_3	40.8	40.7
1640	21.4	TiO_2	3.0	3.4
		SiO_2	48.6	48.9
		Al_2O_3	42.8	43.0
1641	21.4	TiO_2	1.6	1.5
		SiO_2	50.6	50.7
		Al_2O_3	42.3	41.9

DISCUSSION

Two of the main instrumental techniques which are available to the analytical chemist of today are atomic absorption spectrophotometry and x-ray fluorescence spectrometry. These methods should be considered complementary. X-ray fluorescence techniques are well-established for the analysis of elements having atomic numbers down to 12 (Magnesium) in concentrations as low as a few tenths of a percent.

However, classical analyses are not made obsolete by instrumentation techniques. They perform an important complementary function to instrumental analytical methods. It seems that a manual of instrumental methods published by various technical organizations for their particular industry would be very desirable.

REFERENCES

1. Bauman, H D., Pulp and Paper, May 1972, 120-121
2. Sennett, P., Brodhag, E., Morris, H. H., TAPPI*, 55, #6, 1972, 918-923
3. Garey, C. L., Swanson, J. W., TAPPI*, 43, #10, 1960, 813-818
4. Parham, R. A., Hultman, J. D., TAPPI*, 59, #1, 1976, 152-153

5. Claisse, F., Norelco Reporter, 4, 3-7, 17, 19, 20 (1957); Province of Quebec, Canada, Dept. Mines, 1956
6. Longobucco, R., Anal. Chem., 34, 1263-1267 (1962)
7. Jenkins, R., "An Introduction to X-ray Spectrometry", Heyden, New York, 1974, p. 137, 138

*TAPPI refers to the Journal of the Technical Association of the Paper and Pulp Industry.

INVESTIGATIONS OF DETECTION THRESHOLD VARIATION AND MICROCRYSTALLITE NUCLEATION IN NUCLEAR TRACK DETECTORS USING SMALL ANGLE X-RAY ANALYSIS

A. Aframian

Nuclear Research Center AEOI

P. O. Box 3327, Tehran, Irah

ABSTRACT

The variations in the threshold registration of a number of plastic dielectric nuclear track detectors, DNTD's, have been investigated in terms of range energy relationships and the changes in structure affecting electron density distributions studied, using small angle X-ray diffraction. A similar mechanism has been employed to investigate the presence and formation of microcrystallite nucleation sites in lunar and terrestrial minerals, following proton and charged particle bombardments of 10^{18} particles/cm^2.

INTRODUCTION

The response variations of many polymeric radiation detectors and biological systems to ionizing radiation are described by the delta-ray theory of track formation due to Katz and Kobetich (1,2). The observed charged particle damage in polymeric detectors is the result of a chain of events initiated by secondary and higher generation electrons which interact with the polymer detector. In mineral crystals, electronic interaction is associated with charge exchange followed by atomic collision processes [Fleischer *et al.* (3)] along the particle trajectories, leading to permanent crystal damage which is observable with an electron microscope or an optical microscope after selective chemical etching. In DNTD's, track formation is a critical function of the rate of energy loss by the charged particles. The detection sensitivity and its variations, therefore, due to adverse conditions of background radiation, heat and shock waves, cause variations in the threshold registration sensitivity that are the subject of this study. Small angle X-ray diffraction has been used as a method to investigate the formation and changes of pseudo-lattices in dielectrics as a means of studying variations in electron densities, and for the measurement and behavior of microcrystallites formed as a result of charged particle interactions of cosmic rays in lunar and terrestrial minerals.

EXPERIMENT

Variations of resolving power in cellulose-based DNTD's, polycarbonates, acetates and nitrates, for particles of different energies, is in the main part due to the polymeric chain structures resulting in an ionizing charged particle's detection threshold or sensitivity through linear energy transfer. In general, the paths of the individual charged particles, *e.g.*, He, Ar and Kr of different energies, in DNTD's are revealed by selective chemical etching (25% NaOH 60°C, 30 min). Under normal conditions of registration the particles have an entrance energy that is equal to or higher than the threshold (critical) registration energy, $(dE/dx)_{critical}$, leading to the registration of the particle in the detector medium, thus forming tracks. The detection threshold, relative to the particle's range, is a direct function of the intact detector, the particle's charge Z and energy E.

Diffraction data of pre-irradiation annealed and γ-irradiated samples of plastic detectors, *e.g.*, CA80-15*, LR 115*, Makrofol-E** and Triafol-Tx**(Fig. 1) show that annealing increases crystallinity in most polymers. In cellulose nitrates, however, electron density distributions are reduced through molecular chain rearrangements (Tables 1 and 2), resulting in decreased detection sensitivity and decreased (dE/dx), leading to greater penetration ranges for charged particles (Table 3). Alternatively, thermal annealing of Triafol-Tx at 80°C for one hour increases detection sensitivity and lowers $(dE/dx)_{critical}$, *i.e.*, increases dE/dx. In this respect, both the annealing rate and its duration are effective in stimulating variations of electronic interaction. The apparent decrease in sensitivity of CA80-15 upon annealing is shown by the increase in the average repeat distance, from 3.03Å at the scanning angle of 17.5° for the unannealed detector to 4.24 Å, when annealed at 120°C for 5 hours. The diffuse peak at 12.65° gradually gets more pro-

*CA80-15 and LR 115 (cellulose nitrates) manufactured by Kodak Pathé, France

**Makrofol-E and Triafol-Tx (polycarbonates and acetates) manufactured by Bayer AG, Germany

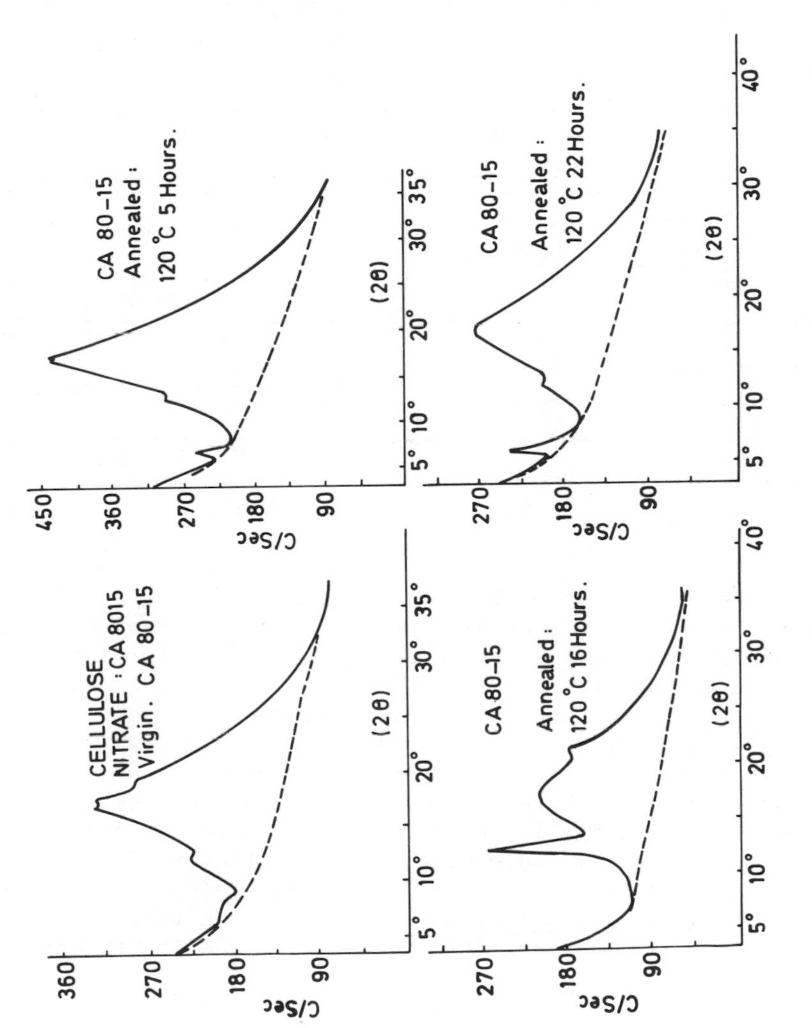

Fig. 1. Small angle X-ray spectrographic analysis of preannealed CA80-15

Table 1

CA80-15

	$(2\theta)°$	Repeat distance d-spacing = $\frac{\lambda}{2 \sin \theta}$	I/I_o	Average repeat distance width at half height $\Delta(2\theta)°$	$\equiv \Delta dx^{\pi}/180(Å)$*
Not annealed	19.85	4.473	0.788	-	-
	18.25	4.86	1.00	5.4	2.86
	17.25	5.14	1.00	5.4	3.2
	12.65	6.998	0.5464	diffuse	-
120°C 5 hours	18.35	4.835	1.00	7.8	4.08
	17.70	5.011	.978	7.8	4.4
	13.28	6.667	.504	-	-
	7.47	11.835	.127	.8	2.54
120°C 16 hours	22.02	4.037	71	-	-
	17.90	4.955	.87	8.2	4.5
	12.78	6.927	1.00	1.1	1.2
120°C 22 hours	17.90	4.955	1.00	9.8	5.38
	12.867	6.88	.567	-	-
	6.98	12.664	.347	1.00	3.62

nounced, resulting in the average repeat distance of 3.62° at 120°C for 22 hours. Whereas thermal annealing does not seem to effect chain spacings in the polycarbonate derivative, γ-irradiation increases the average repeat distance from 3.72Å to 4.4Å, accompanied by a shift in the scanning angle from 16.82° to 16.98°. In this case, however, no detectable sensitivity variations were observed. The observed sensitivity enhancement in Triafol, on the other hand, can be attributed to increased average ionization density due to increased crystallinity in the detector. The variations of the screening parameter K, indicating overall differences in electron density distribution in the equations of Bethe (4) and Block (5), later modified by Barkas (6), account for such differences, by describing the primary ionization density J, and its threshold Jc.

$$J = Z_{eff}^2 \{ \ln(\gamma^2\beta^2) - \beta^2 + K \} / 10^4\beta^2$$

where the effective nuclear charge $Z_{eff}^2 = Z\ (1-e^{-125\beta Z^{-2/3}})$, $\beta = V/C$ and $\gamma^2 = 1/1-\beta^2$. Whereas K varies with the composition and the crystallinity of the stopping medium, J depends on the projectile's charge, mass and energy.

The observations, therefore, seem to indicate that pre-irradiation annealing of polymeric detectors leads to increased

Table 2

Makrofol-E.

	$(2\theta)°$	Repeat distance d-spacing = $\frac{\lambda}{2 \sin \theta}$	I/I_o	Average repeat distance width at half height $\Delta(2\theta)°$	$\equiv \Delta d x^{\pi}/180(\mathring{A})$*
Not annealed	16.817	5.272	1.00	6	3.72
150°C	18.167	4.88	.982	-	-
1 hour	17.733	5.002	1.00	6	3.63
10^9 Rad	17.067	5.195	.945	-	-
γ-ray	16.98	5.222	1.00	7.2	4.4

* $\frac{\lambda}{d} = 2 \sin \theta$ differentiating $\frac{\partial \theta}{\partial d}$ $\frac{\partial \theta}{\partial d} = 2 \text{ Cos } \theta \times \partial \theta = - \frac{\lambda}{d^2} \partial d$

$- \frac{\lambda}{d^2} \frac{\partial d}{\partial \theta} = 2 \text{ Cos } \theta$ $\frac{\partial d}{\partial \theta} = - \frac{2d^2}{\lambda} \text{ Cos } \theta$ But $\Delta d = \frac{\partial d}{\partial \theta} \Delta \theta$

Substitute for λ $\Delta d = \frac{2d^2}{\lambda} \text{ Cos } \theta \quad \Delta \theta$

Substitute for λ $\Delta d = -d \text{ Cot } \theta \ \Delta(2\theta)$

molecular symmetry, causing the normalized ranges to increase. Thus, the rate of energy transfer from secondary electrons to the molecules in the amorphous polymers is enhanced by the presence of distorted bonds, while increased polymer crystallinity reduces critical energy transfer and increases the particle's overall range.

Observations of lunar materials, particularly those of the feldspars and some iron-rich glasses have revealed superficial amorphous layers, nearly 1000Å thick and a high density of visicles, generally attributed to the ionizing effects of solar wind and cosmic ray particles. The heavy radiation damage can be deduced from the present solar flux during quiet periods of solar activity due to proton bombardments amounting to 10^{24} particles/cm^2 for 10^9 years.

Table 3

Track Length in Normal and Annealed (Crystalline) Cellulose Nitrate CA80 15

Ions		Eo MeV/N	(dE/dx) (MeV)/(mg/cm^2)	$(dE/dx)_{crit}$ (MeV)/(mg/cm^2) non-annealed	$(dE/dx)_{crit}$ (MeV)/(mg/cm^2) annealed	track length (μm) non-annealed	track length (μm) annealed
Cf-252 fission	light	0.901	64.4	12 ± 1	22	17.8 ± 0.5	20.4 ± 0.5
fragments	heavy	0.76	86.87	14 ± 1	21	15.2 ± 0.5	17.6 ± 0.5
Eα = 6.01		1.52	0.78	0.55	1.2	36.5 ± 0.5	41.6 ± 0.5
$^*Kr^{30+}$		9.6	31.4	11.8 ± 1	27 ± 1	145.0 ± 2.0	153.0 ± 1.0
$^*Ar^{26+}$		9.6	11.27	10 ± 1	11.5 ± 1	170.0 ± 2.0	177.0 ± 2.0
$^*Fe^{22+}$		9.6	19.74	14 ± 1	21 ± 1	150.0 ± 1.0	156.0 ± 2.0

Eo = energy of the primary particle

* = the charge states generally refer to statistical charges.

The immediate effects of charged particle interaction with such media are ionization and generation of collision cascades, creating equal numbers of interstitials and vacancies. The differential free energy fields created by this interaction are governed by the equilibrium localized zones of vacancy and interstitial rich regions, from the interaction core outwards. The rearrangement into clusters is controlled by repeated and consecutive nuclear collision processes and occur when mobile defects (vacancy or interstitial) are trapped at stable nucleation sites and grow to visible dimensions, forming microcrystallites, observable by H.V. (1 MeV) transmission electron microscopes. In this respect simulations have been carried out by irradiating samples of mica and terrestrial feldspar with protons and heavily charged particles abundant in cosmic rays such as Kr, Ar, Fe, Cr, Cu and Zn ions at various energies under lunar conditions. The present experiments have indicated examples of (radiation) induced annealing of tracks in silicate structures. Thus, the small angle X-ray scattering measurements on such samples showed the formation of small clusters of defects along the path of the incident ions which coagulated into larger, more widely separated clusters that reached a maximum size of about 38Å as the temperature was raised to 279°C, but then shrank and disappeared with increasing temperature. Fig. 2 shows the presence of such microcrystallites in a micron-sized diameter of lunar breccia as observed under a 1 MeV transmission electron microscope.

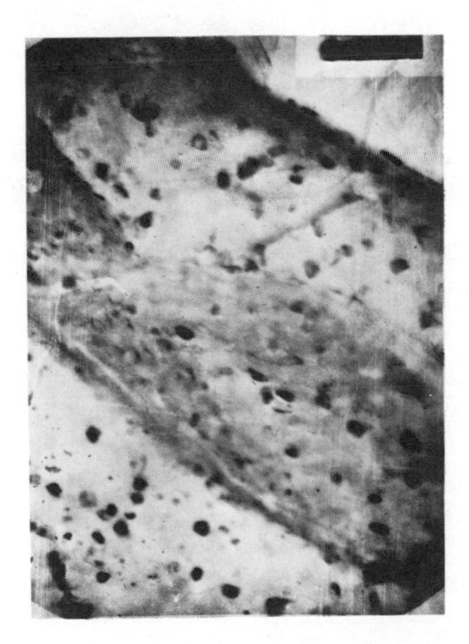

Fig. 2. Transmission electromicrograph of microcrystallite nucleation of charged particle cosmic ray tracks in Appollo 16 lunar breccia, (etched in: $1HF+2H_2SO_4+180H_2O$ for 15 sec.) Mag. 100,000X

DISCUSSION AND CONCLUSION

Pre-irradiation annealing of cellulose based derivates has given rise to polymer crystallinity and/or changes in electron density distribution, thus leading to charged particle range variations. The consequence of such annealing of polymers is a probable increase in the molecular symmetry, indicated by X-ray diffraction data, thus creating one or more cycles in the molecule causing the normalized ranges to vary through changes in bond distortion and strain. Of paramount importance to the electrical and the power cable industries are the changes in morphology of insulating polymers and the associated variations in such parameters as electrical behavior and dielectric constants, electrical breakdown, conduction, interfacial polarization and piezoelectricity, mostly associated with the creation of crystallites and pseudo-lattices. Thus transition from an amorphous polymer to a more symmetrical structure can give rise to a conjugated system of π-bonding, leading to electronic conduction in previously insulating polymers. It is now thought that the major contributing factors to electrical breakdown in transmission cables, after only a few years' service, are the physical and morphological structures present rather than the chemical nature of the polymer. Observations of charge annealing both by track etching and microcrystalline nucleation by high intensity radiation in silicate structures, as observed by X-ray diffraction, are indicative of contributing elements in the stimulation of morphological and mechanical changes affecting electron energy transfer interaction cross sections.

REFERENCES

1. R. Katz and E. J. Kobetich, Formation of etchable tracks in dielectrics, Phys. Rev. 170, 2, 401-405 (1968).

2. R. Katz and E. J. Kobetich, Energy deposition by electron beams and γ-rays, Phys. Rev. 170, 2, 391-396 (1968).

3. R. L. Fleischer, P. B. Price and R. M. Walker, Solid State Track Detectors: Applications to Nuclear Science and Geophysics, Ann. Rev. Nucl. Sci. 15, 1-28 (1965).

4. H. A. Bethe, Theory of the passage of rapid corpuscular rays through matter, Ann. Physik, 5, 325-400 (1930).

5. F. Block, Stopping power of atoms with many electrons, Z. Physik 81, 363-376 (1933).

6. W. H. Barkas, (M. J. Burger, S. T. Seltzer Edit.), Penetration of Proton, α-particles and Mesons, National Research Council, Publication 1133 (1964).

APPLICATION OF MICROBEAM X-RAY TECHNIQUE TO THE EVALUATION OF THE PLASTIC ZONE SIZE OF FATIGUED STEELS

Akimasa Izumiyama, Yasuo Yoshioka and Masao Terasawa

Musashi Institute of Technology

1, Tamazutsumi, Setagaya, Tokyo 158, Japan

ABSTRACT

The microbeam X-ray diffraction technique was used for the evaluation of fracture analysis on the fatigued low carbon steel specimens.

It is possible to evaluate both monotonic and cyclic plastic zone sizes beneath fatigue fracture surface. Thus the stress intensity factor can be estimated.

INTRODUCTION

Recently, the X-ray diffraction technique has been used in Japan as one method of the evaluation of fracture analysis of metals in relation to fracture mechanics. We call this technique "X-ray fractography."

The plastic deformation area is formed at the tip of fatigue crack, and it should be proportional to the square of stress intensity factor K (1). In this study, we measured the plastic zone size beneath the fracture surface in carbon steel specimens by means of microbeam X-rays. We report the relation between plastic zone size and the stress intensity factor K, and discuss the feasibility of the evaluation of K by the X-ray diffraction technique.

PLASTIC ZONE AROUND CRACK

Fig. 1 represents the crack surrounded by the envelope of all zones which during crack growth had been subjected to monotonic plastic deformation under constant amplitude loading. The plastic zone size Wm perpendicular to the crack surface can be expressed as $Wm = C(Kmax/\sigma ys)^2$, where Kmax is the maximum stress intensity factor at crack length of a. In this figure, the hatched area shows the plastic zone area caused by cyclic loading, and Wc can be expressed as $Wc = C(\Delta K/2\sigma ys)^2$, where ΔK is the stress intensity factor range. We attempted to measure these plastic zone sizes.

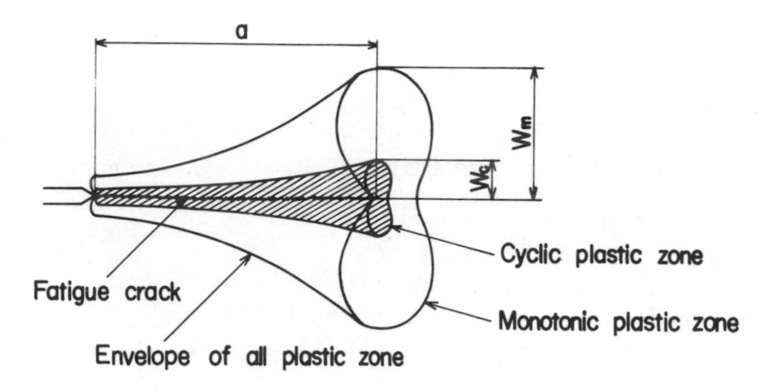

Fig. 1. Development of plastic zone envelope around a fatigue crack

EXPERIMENTAL PROCEDURES

The specimen has the planar dimensions of an ASTM standard 1 inch thickness compact tension specimen, but the thickness employed was nominally 12 mm. The material used was 0.16%C carbon steel which was annealed at 850°C for a half hour and cooled in a vacuum furnace after being machined, and it has a $267MNm^{-2}$ (38.7KSI) yield stress.

Fatigue tests were conducted under sinusoidal tension-tension load control, the stress ratio R being 0.5 and 0.05, respectively. When the fatigue crack had grown approximately 20 mm long from machined starter notch, the fatigue test was stopped and the specimen was separated at the fracture surface in liquid N_2.

A Cr-Kα X-ray beam collimated by a 200 μm diam. pin-hole was set normal to a constant position from the notch root on the fracture surface, and a Debye-Scherrer pattern from the (211) crystal plane was photographed. Several patterns were obtained at stated intervals in the direction of crack propagation, and a thin layer of fracture surface was electropolished away. This process was repeatedly performed in order to measure the plastic zone size. The diameter of radiated area was about 400 μm on the specimen surface.

EXPERIMENTAL RESULTS AND DISCUSSION

Monotonic Plastic Zone

Fig. 2 shows microbeam X-ray diffraction patterns photographed at several depths fromthe fatigue fracture surface. A continuous

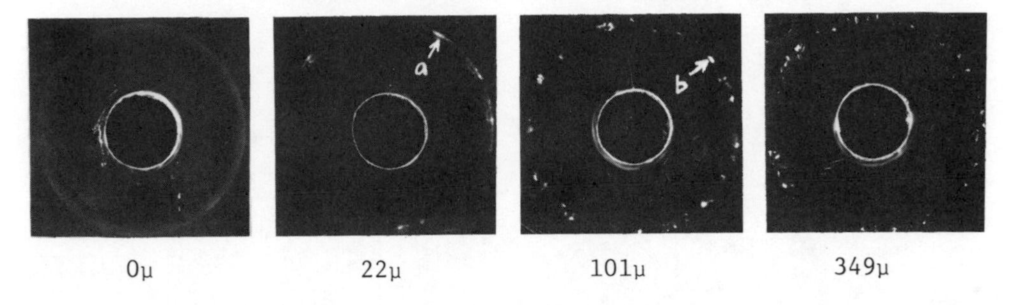

Fig. 2. Microbeam X-ray diffraction patterns photographed at several depths from the fracture surface.

Debye-Scherrer ring was observed on the surface layer; however, it gradually changes from the arcs to spots, and the size of spots finally coincides with that which is diffracted from non-fatigued specimen at a depth W. We used the tangential length, St, of such diffraction arcs as the parameter indicating the amount of plastic deformation. Fig. 3, which shows a logarithmic plot of St against depth from the fracture surface d, shows almost a straight line. The plastic zone size Wm was determined by the depth where this line intersects the St_0 line, St_0 being arc length for the non-fatigued specimen. Fig. 4 shows the relation between the depth of plastic zone size and the stress intensity factor Kmax. A linear relation could be expressed as $W = 1.77 \times 10^{-5}\ Kmax^{2.36}$. The power in these equations fairly agrees with the theoretical value 2 and this plastic zone size Wm measured in this study is recognized as the monotonic plastic zone size with regard to Kmax.

Cyclic Plastic Zone

Enlarged pictures of arcs a and b in Fig. 2 are shown in Fig.5. In the monotonic plastic region, the crystals are plastically bent

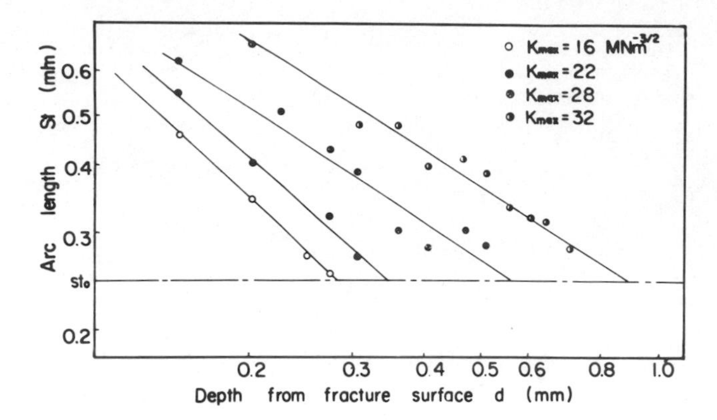

Fig. 3. Tangential arc length St against depth from the fracture surface d

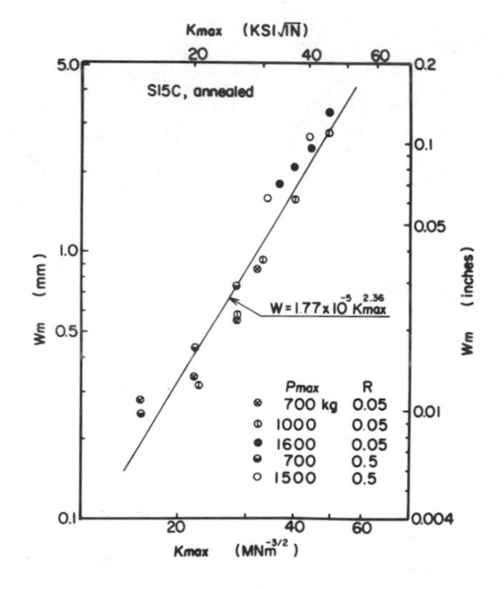

Fig. 4. Depth of plastic zone size Wm against Kmax

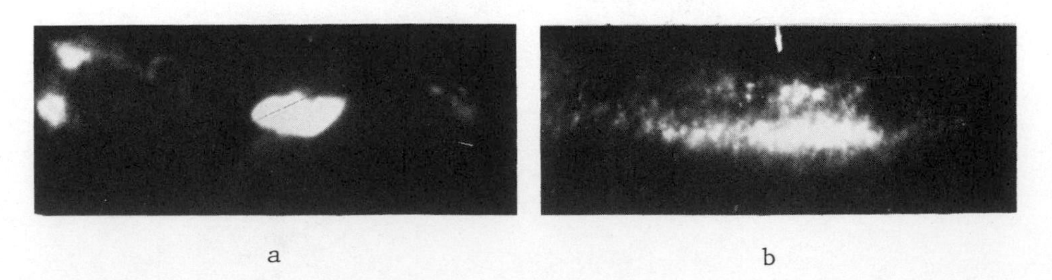

Fig. 5. Enlarged pictures of arcs a and b in Fig. 2

by a loading and the X-ray diffraction spots grow tangentially and radially as shown in Fig. 5b. However, crystals near the fracture surface would be deformed polygonally by the cyclic loading, that is, many subgrains would occur and the X-ray diffraction pattern will show the arcs which collect many small spots, as shown in Fig. 5a.

The number of such arcs decreases as the depth from the fracture surface increases. The cyclic plastic zone size Wc can be calculated theoretically from the value of monotonic plastic zone size Wm. Observing the pattern photographed at depth Wc, the number of such arcs was approximately a half of all arcs in a ring. Fig. 6 shows the relation between plastic zone size Wc and stress intensity factor range ΔK. The effect of stress ratio is seen. Thus

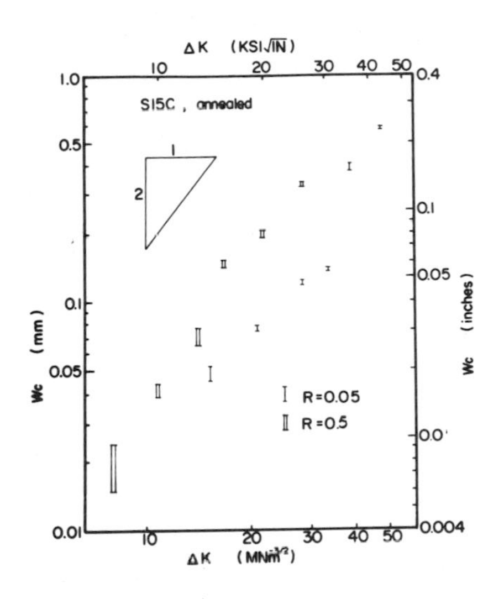

Fig. 6. Plastic zone size, Wc, against the stress intensity factor range, ΔK

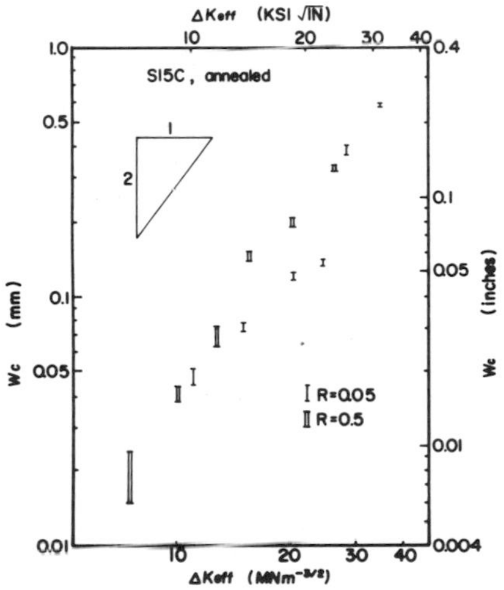

Fig. 7. Plastic zone size, Wc, against the stress intensity factor, ΔKeff

it may be considered that crack closure occurs in the interior of the specimen, and it is necessary to introduce the effective stress intensity factor ΔKeff for eliminating the effect of the stress ratio, R. In this study, ΔKeff is determined as a function of R, and Wc was plotted against this ΔKeff. The result is shown in Fig. 7. A linear relation is obtained regardless of stress ratio R, and this inclination coincides with the theoretical value of 2. This plastic zone size Wc measured in this manner shows the cyclic plastic zone with regard to the effective stress intensity factor range. The presumption of maximum stress intensity factor Kmax and stress intensity factor range ΔK is made possible by observing the microbeam X-ray diffraction patterns.

REFERENCES

1. J. R. Rice, "Mechanics of Crack Tip Deformation and Extension by Fatigue," Fatigue Crack Propagation, ASTM STP 415, Am. Soc. Testing Mats., 1967, pp. 247-311.

2. W. Elber, "The Significance of Fatigue Crack Closure," Damage Tolerance in Aircraft Structures, ASTM STP 486, Am. Soc. Testing Mats., 1971, pp. 230-242.

RESIDUAL STRESS AT FATIGUE FRACTURE SURFACE OF HEAT TREATED HIGH STRENGTH STEELS

Atsutomo Komine, Hideo Ueda, Eisuke Nakanishi,
Shotaro Araki and Kazuo Taguchi

Technical Research Center, Komatsu Ltd.

Hiratsuka, Japan

INTRODUCTION

Electron fractography has been widely applied to analyze fracture processes and service failures. However, the results obtained by the fractographic method are restricted to information in a topograph of the fracture surface. Therefore, other methods are needed to analyze the fracture. One of the hopeful advanced techniques is thought to be X-ray diffraction, because another type of structure sensitive information would be obtained.

In the present paper, the residual stresses developed near the fracture surface of heat treated, high strength steels were measured by an X-ray stress analyzer. The plastic deformation takes place at the tip of the propagating fatigue crack. The residual stress at the fracture surface is due to this plastic deformation. The relation between the applied stress intensity factor and the plastic zone size has been obtained. Therefore, it is expected that the surface residual stress is related to the applied stress intensity factor.

MATERIALS AND EXPERIMENTAL PROCEDURE

The materials tested were heat treated 4340 and 4161 steels. The mechanical properties are given in Table I. Compact specimens (W=51 mm) were used for the fatigue crack propagation tests.

The residual stresses at the fracture surface were measured by the X-ray diffraction method. The X-ray conditions are given in

Table I. Mechanical Properties

	0.2% yield strength (kg/mm^2)	Ultimate tensile strength (kg/mm^2)	HRC	Half value breadth (degrees)
4340 200°C Tempered	163.5	183.3	51.6	5.46
600°C Tempered	88.3	95.6	29.8	1.99
4161 200°C Tempered	187.2	191.5	55.8	6.41

Table II. X-Ray Conditions

Characteristic X-Ray	CrKα
Diffraction plane	(211)
Filter	V
Tube voltage	30 kV
Tube current	9 mA
Scanning speed	8 deg/min
Irradiated area	1 x 10 mm^2

Table II. The X-ray irradiated area was 1 x 10 mm^2 at the mid-thickness of the fracture surface. The residual stresses were measured in the direction of the crack propagation. The surface layer was removed by electropolishing and the distribution of the residual stresses below the fracture surface was also measured.

RESULTS AND DISCUSSION

Validity of X-Ray Residual Stress Measurements at the Fracture Surface

In many cases, the fracture surface shows high roughness in comparison with machined surfaces. The validity of X-ray residual stress measurements at such a rough surface must be confirmed. Fig. 1 shows the relation between $\mathrm{Sin}^2\psi$ (ψ; incident angle) and 2θ (2θ; diffraction angle) at the surface roughness of about 20 μm. The lineality of the $\mathrm{Sin}^2\psi$-2θ diagram is a premise in the X-ray residual stress measurements and it is confirmed in Fig. 1.

In Fig. 2, open symbols show a representative distribution of the residual stress at the fracture surface. The roughness of the fracture surface increases with crack length; roughness is about

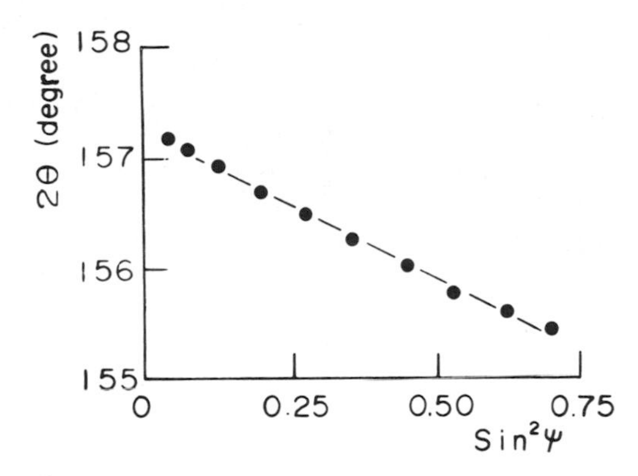

Figure 1. $\sin^2\psi - 2\theta$ Diagram. (4340 Steel Tempered at 200°C)

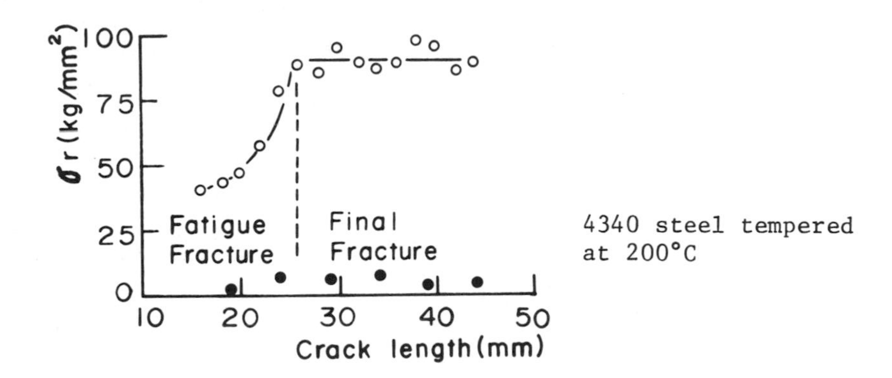

Figure 2. Relation Between Crack Length and Residual Stress

10 μm at a crack length of 15 mm, and 20 μm or more at 30 mm. Solid symbols show the result for the stress relieved specimen. It is clear that the residual stresses at the fracture surface were reduced to nearly 0 regardless of the surface roughness.

These experimental results show that the X-ray residual stress measurements at the fracture surface with surface roughness below 20 μm are valid.

Fracture Analysis by Electron Fractography and X-Ray Residual Stress Measurements

First, fractographic examinations were performed on each specimen. The fractographs of fatigue fractured 4340 steels tempered at 200°C show transgranular facets with structure sensitive appearance regardless of the stress intensity factor range, ΔK. On the other hand, the fatigue fracture surfaces of 4340 steels tempered at 600°C were characterized by striations. The fatigue fracture surfaces of 4161 steels were characterized by the intergranular facets. In the final fracture surface of each specimen, elongated and equiaxed dimples were observed. It is difficult to establish some relations between transgranular or intergranular fracture surface in the process of fatigue crack propagation and applied stress condition.

At the fatigue fracture surface, the residual stresses increase monotonically with increasing crack length and remain nearly constant at the final fracture for 4340 tempered at 200°C. Similar tendency is obtained for 4340 steel tempered at 600°C and 4161 steel tempered at 200°C. Fig. 3 shows that the residual stresses, σ_r, depend on applied maximum stress intensity factor Kmax with different stress ratios R, ($R = \sigma_{max}/\sigma_{min}$). It is emphasized that the residual stress measurements of the fracture surface, in intergranular or transgranular fracture, is effective in analyzing the fracture mode and the applied stress condition.

ANALYSIS OF PLASTIC ZONE SIZE BY X-RAY RESIDUAL STRESS MEASUREMENT

4340 and 4161 steels were thoroughly hardened by oil quenching. The initial value of the residual stresses due to quenching in each specimen were negligibly small. The residual stresses near the fracture surface are due to the plastic deformation at the tip of the propagating fatigue crack.

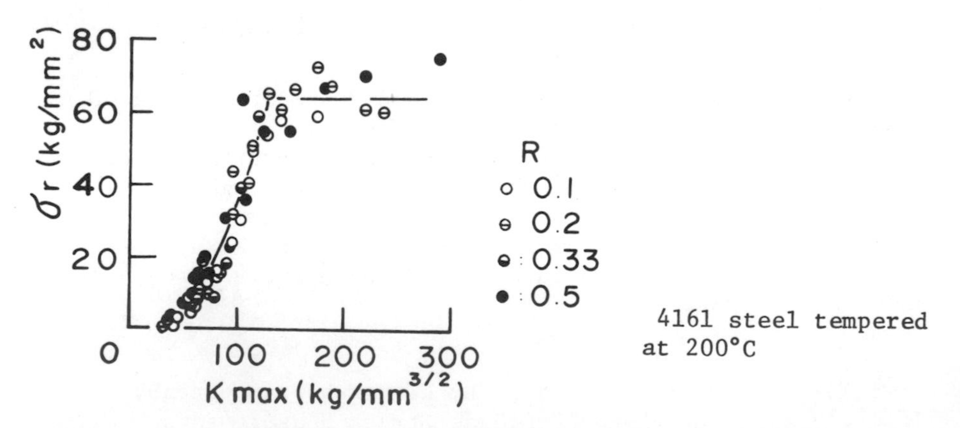

Figure 3. Relation Between Residual Stress and Maximum Stress Intensity Factor, Kmax

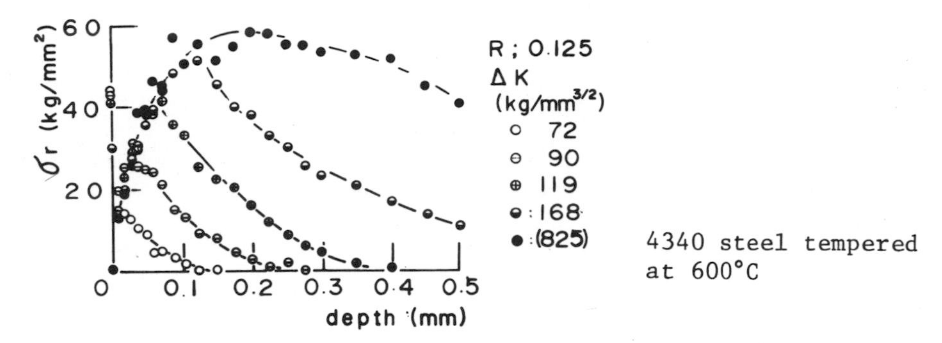

Figure 4. The Distribution of Residual Stress

Fig. 4 shows the subsurface distribution of the residual stresses at various ΔK conditions. There are two types of plastic zones at the crack tip, which are the cyclic plastic zone and the monotonic one. In Fig. 4, it is thought that an intersection point of two kinds of curves indicates the cyclic plastic zone size Ry*. The monotonic plastic zone size Ry is taken as the point where the decreasing curve with depth from the surface approximately reaches zero. Fig. 5 shows the relation between Ry* and ΔK/σy and between Ry and Kmax/σy. Ry* and Ry are expressed as follows:

$$Ry^* = 0.035\ (\Delta K/\sigma y)^2$$

$$Ry = 0.18\ (Kmax/\sigma y)^2$$

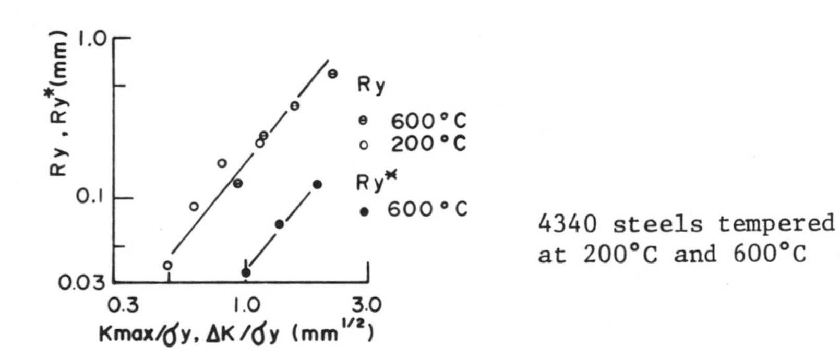

Figure 5. The Relation Between Ry* and ΔK/σy and Between Ry and Kmax/σy

CONCLUSIONS

1. It is possible to measure the residual stress at the fracture surface by X-ray if surface roughness is under 20 μm.

2. The residual stress at the fatigue fracture surface is related to Kmax and can be expressed as a function of Kmax.

3. The residual stress at the final fracture surface is nearly constant independently of Kmax.

4. Cyclic and monotonic plastic zone sizes can be determined experimentally by X-ray residual stress measurements.

5. In fracture analysis, residual stress measurement is an effective method to learn the fracture mode and the applied stress condition for high strength steels.

STUDY ON X-RAY STRESS ANALYSIS USING A NEW POSITION-SENSITIVE PROPORTIONAL COUNTER

Yasuo Yoshioka
Musashi Institute of Technology
1, Tamazutsumi, Setagaya, Tokyo 158, Japan

Ken-ichi Hasegawa and Koh-ichi Mochiki
Faculty of Engineering, University of Tokyo
7, Hongoh, Bunkyo, Tokyo 113, Japan

ABSTRACT

A position-sensitive proportional counter suitable for the X-ray stress measurement has been developed and residual stresses were measured with an apparatus that uses this PSD system. The counter was designed to have a good angular resolution over the counter length for the diffracted X-ray beam and high counting rates. The mean angular resolution measured was about 0.2° in 2θ (FWHM) at 200 mm, and the maximum allowable counting rate reached about 40,000 cps.

The time required for the data accumulation was shown to be 1/10 to 1/30 of the time required with a standard diffractometer.

INTRODUCTION

Position-sensitive proportional counters have been developed in the fields of X-ray diffraction and nuclear physics. When the PSPC system is employed for the X-ray stress analysis, the time required for the data accumulation is shortened. Moreover, this system will be capable of the residual stress analysis on small areas in which the application of the conventional counter method is impossible because of weak diffraction intensities. For example, the behaviour of residual stress around fatigue cracks will be easily analyzed with high precision in comparison with the film method, and thus the contribution of residual stress to the crack growth will be clear.

The progressive work on the stress analysis by use of PSD system was performed by James and Cohen (1); however, this system should be investigated further with the development of a detector suitable for the stress analysis.

The authors commenced a study of this subject about two years before. A description of the detector suitable for the X-ray stress analysis is reported and its properties are presented. The precision of the stress is also discussed.

DETECTOR DESIGN

We have developed a one-dimensional position-sensitive proportional counter which has high count rate and uniform angular resolution. Although there are many types of PSPC, a resistive wire cathode type which has an angular span of about 20° in 2θ at a radius of 200 mm was constructed. As shown in Fig. 1, the detector consists of two anode wires, a resistive wire cathode wound in a modified zigzag fashion for the two-dimensional detector (2) and a copper plate cathode. The depth of PSPC is 10 mm. The entrance window has the dimensions 10 x 100 mm^2 and it is covered by a beryllium foil of 0.3 mm thickness. The chamber of the detector is filled with PR gas (Ar 90%, CH_4 10%) of 1 atm. It is continuously exchanged.

The construction of cathode is schematically shown in Fig. 2. A Ni-Cr wire of 27 μm diam. is wound in a fan-shaped design on an acrylic resin plate with 140 mm width, 40 mm depth and 2 mm thickness, the number of turns being 200. The coil pitch at the window side is 0.5 mm and that at opposite side is 0.6 mm, the center of the fan shape is positioned at 200 mm from the window side, and this position is to be coincident with the position of the X-ray beam on the specimen surface. The cathode forms one long resistive readout wire with a total resistance of approximately 45 KΩ. The use of a single, long resistive wire for readout simplifies the construction procedures used in this chamber, and the relatively low resistance does not adversely affect the position resolution of the detector.

The anode consists of 20 μm diam. gold-plated tungsten wires with a spacing between wires of 5 mm. In Fig. 3a, an anode wire is stretched at the center of the PSPC, and electrons induced by X-rays distributed in the region (d sinα) can be collected on the anode, that is, the angular resolution grows worse with increase of path angle α. However, if the two wires are stretched in parallel as the anode as shown in Fig. 3b), the deterioration in angular resolution with increase of α is a half in comparison with the former case.

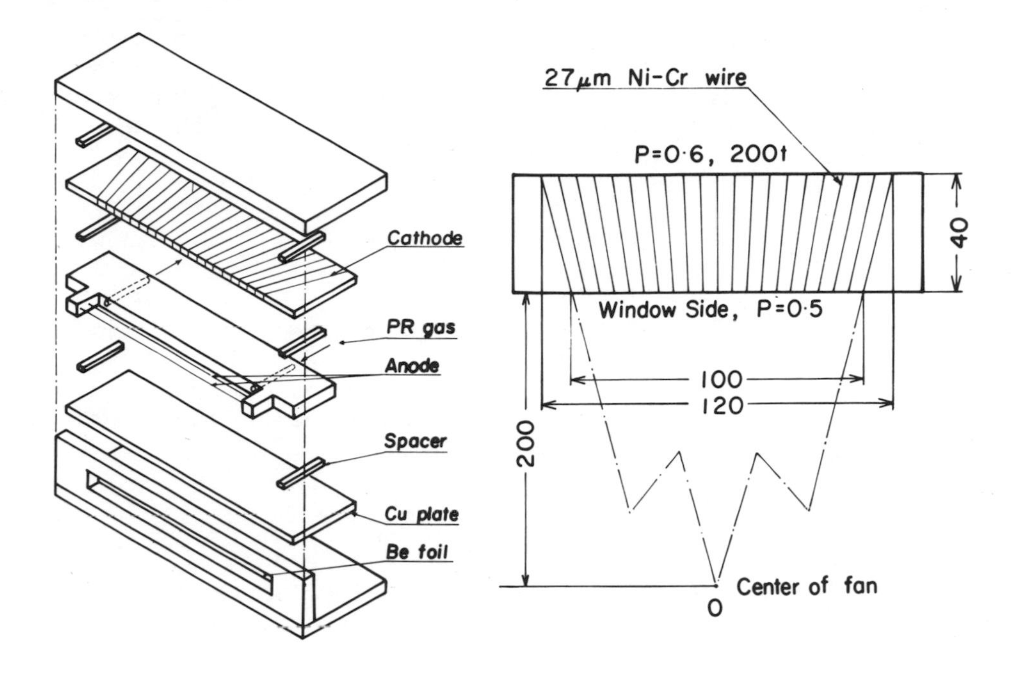

Fig.1 Schematic construction of detector

Fig.2 Detail of fan-shaped catode

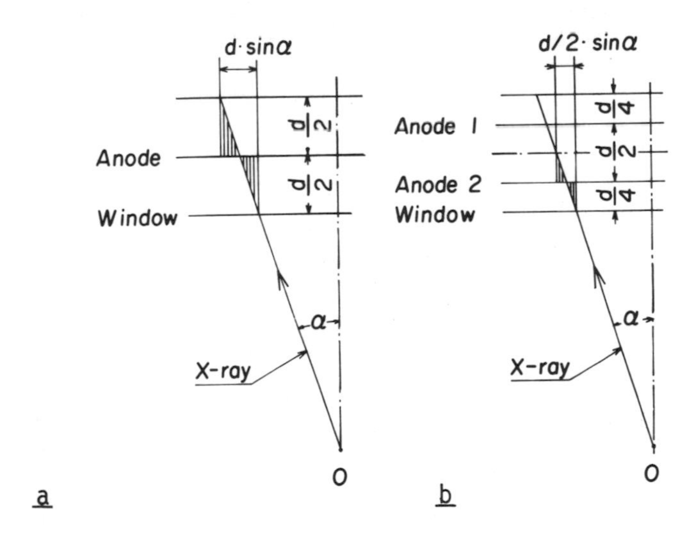

Fig.3 Spreading of angular resolution caused by arrangement of anode a) An anode wire b) Two anode wires

Each end of cathode wire is connected to a charge sensitive preamplifier, and we have adopted the charge division method with a divider as the basis for position readout. The output pulses from divider is processed by the 512 channel pulse height analyzer (PHA).

DETECTOR PERFORMANCE

Maximum Allowable Counting Rate

We first measured the maximum allowable counting rate, and it was measured to be about 40,000 counts per second. This counting rate is sufficient for stress measurement.

Angular Resolution

An X-ray beam collimated by a 50 μm wide slit was placed incident upon the PSPC with path angle of $\alpha°$ from the focus point which agrees with the center of the fan shape shown in Fig. 2. As shown in Fig. 4, the measured angular resolution was about 0.18° in 2θ (FWHM) at $\alpha = 0°$ and, although it gradually increased with increase of path angle α, it was at the most 0.3° in 2θ at $\alpha = 10°$. Since the angular resolution on a conventional parallel beam type stress analyzer is more than 0.5° in 2θ (FWHM), sharper diffraction profiles will be obtained with this PSPC system.

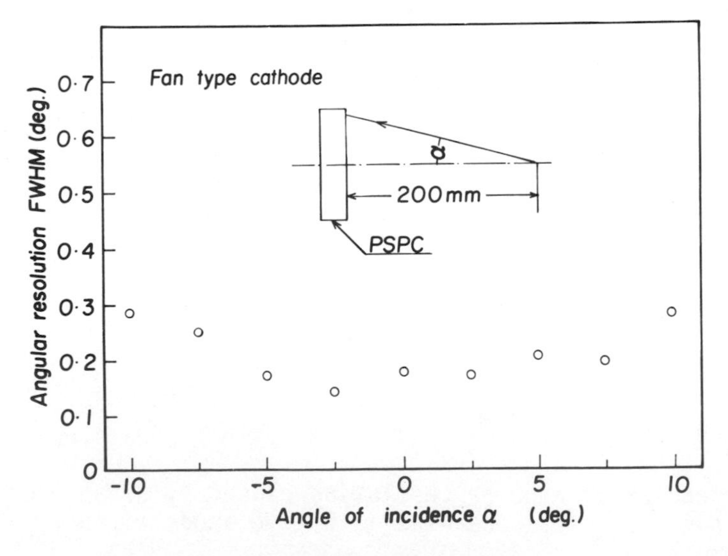

Fig. 4. Angular resolution in 2θ

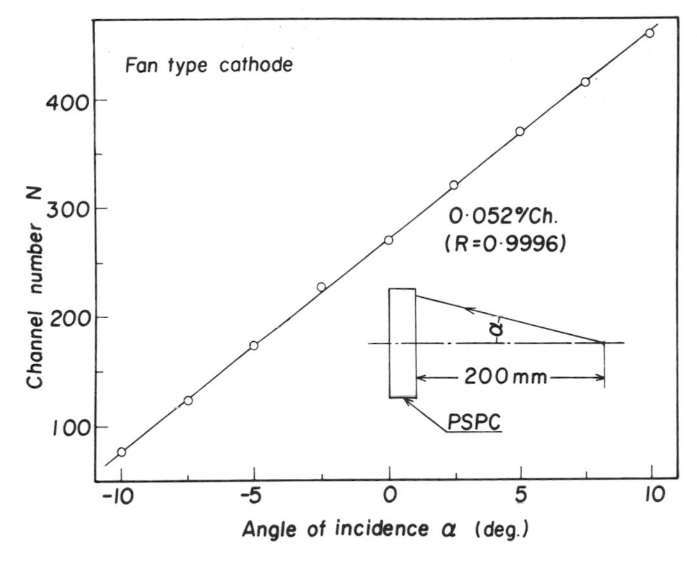

Fig. 5. Detector Linearity in 2θ

Linearity

The linear response with respect to the path angle is shown in Fig. 5. A plot is obtained that indicates the very good linearity available. The line shown in this figure was drawn by a least-squares fit and the correlation coefficient R was measured to be 0.9996, which can be approximated to a straight line.

STRESS ANALYSIS CONFIGURATION

The basic configuration of X-ray source, PSPC and specimen is shown in Fig. 6. For the diffraction angle from αFe(211) plane (Fig. 6a) of 156° in 2θ, the angle between the incident X-ray beam and the bisector of detector length was set to be 24°, and the intersection coincides with the specimen surface. For ferritic steel specimens, both the iso-inclination method and the side inclination method can be employed for the stress determination. On the other hand, for austenitic stainless steel specimens, since the diffraction angle from γFe (220) plane is about 128.8° in 2θ, the inclination angle was set to be 51°, and the plane normal to the specimen was also coincident with the bisector of inclination angle, because the side inclination method has to be used in this case.

The $\sin^2\psi$ method was employed for the determination of stress value, and the peak position of profiles is determined by the half value breadth method (3).

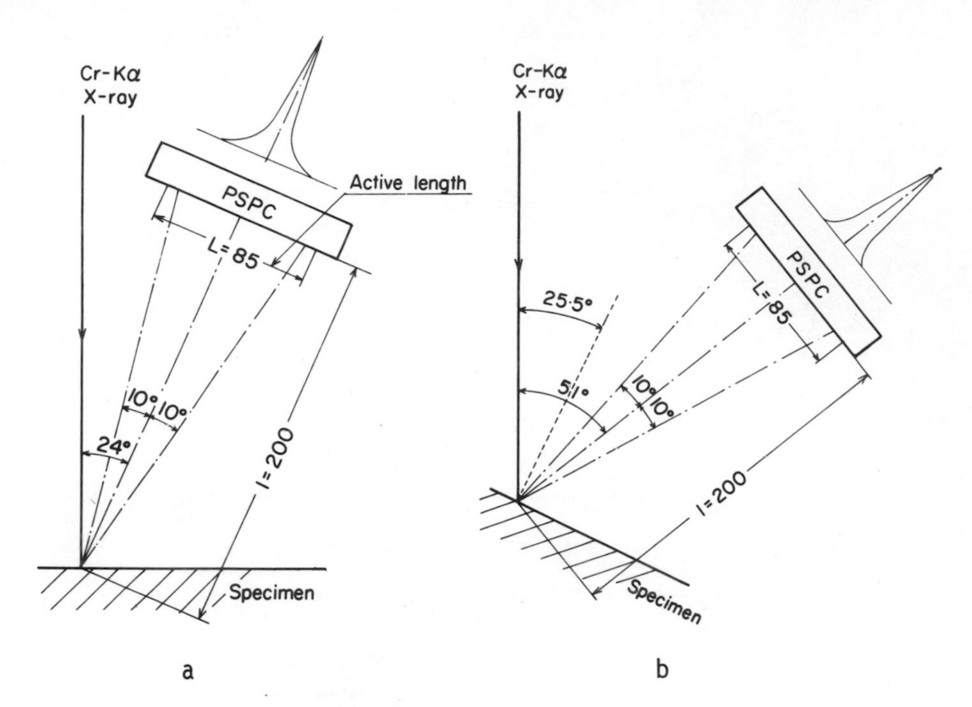

Fig.6 Schematic arrangement of incident X-ray beam, PSPC and specimen for stress analysis
a) For specimen of ferritic steel
b) For specimen of austenite stainless steel

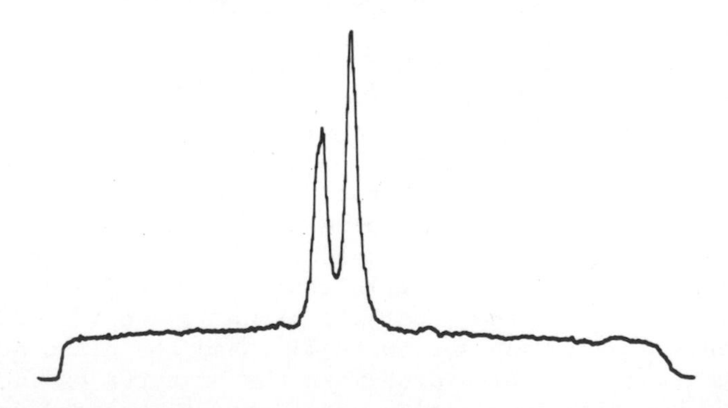

Fig.7 Diffraction profile from an annealed low carbon steel Cr-Kα radiation

EXPERIMENTAL RESULTS

Fig. 7 shows an example of profile on the X-Y recorder of the PHA from an annealed low carbon steel specimen. PC means preset peak counts. The time required was 32 sec by use of the X-ray source operating at 30KV-10 mA and 1° soller slit. The counting rate was calculated to be about 10,000 cps. As the maximum allowable counting rate is about 40,000 cps, if more intense X-ray sources can be used it would reduce the time by a quarter.

For discussing the precision of peak position of profile, replicate measurements were made under several preset count conditions. Two kinds of specimen, one with a half value breadth of 1.2° and the other with 4.2°, were prepared. The standard deviation from the mean peak position in 2θ for the seven replicate measurements is calculated. Fig. 8 shows the result. Standard deviation on the sharp profile obviously decreases with increase of peak counts and it almost agrees with that obtained by the conventional counter method when the peak count is more than 2048. The time is only 8 seconds for the peak counts of 2048. On the other hand, it takes

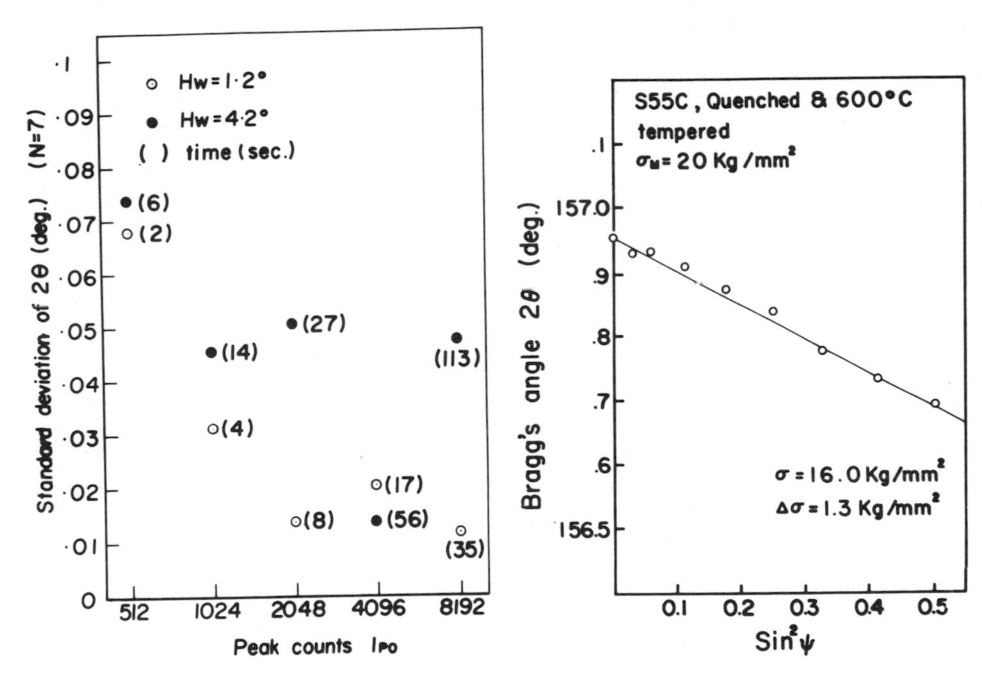

Fig.8 Standard deviation of profile peak in 2θ vs. peak counts

Fig.9 $\sin^2\psi$ diagram for a heat treated 0.55%C carbon steel specimen mechanically stressed

more than 4 minutes for the measurement by the conventional method. Thus the time is reduced by 1/30.

As the breadth of profile increases, the standard deviation becomes poor in comparison with the sharp profile, as shown in this figure. However, it is expected that the peak position could be determined with good precision by increasing the peak counts.

In general, the precision of profile peak location is proportional to the square root of peak intensity and inverse proportional to the breadth of profile. With the goniometer method, the peak intensity gradually decreases with increase of the breadth, that is, the precision worsens. Fixed count or fixed time methods are good for high sensitivity goniometer measurement; however, the time required is too long. As the PSPC method acquires the data simultaneously across the entire profile, the data accumulation time is saved and high sensitivity measurement can be expected.

Fig. 9 shows an example of $2\theta-\sin^2\psi$ diagram. $\Delta\sigma$ shows the 95% confidence limits of stress determined by the $\sin^2\psi$ method. A good linear relation is found to hold.

CONCLUSION

It is concluded based on these results that the X-ray stress analysis by use of PSD system will be a powerful means in place of the goniometer method and seems to provide a new field of the stress analysis.

ACKNOWLEDGEMENT

We wish to thank H. Asano for his experimental assistance.

REFERENCES

1. M. R. James and J. B. Cohen, "The Applications of a Position Sensitive Detector to the Measurement of Residual Stresses," Advances in X-Ray Analysis, Vol. 19, p. 695 (1975).

2. J. Alberi, J. Fischer, V. Radeka, L. C. Rogers and B. Schoenborn, "A Two-Dimensional Position-Sensitive Detector for Thermal Neutrons," Nuclear Instruments and Methods, Vol. 127, p. 507 (1975).

3. Y. Yoshioka, "A Method of Data Acquisition for X-Ray Stress Measurement," J. JSMS, Vol. 23, p. 15 (1974).

COMPARISON OF STRESS MEASUREMENTS BY X-RAYS WITH THREE DIFFERENT DETECTORS AND A STRONGLY FLUORESCING SPECIMEN

M. R. James° and J. B. Cohen

Department of Materials Science and Engineering
The Technological Institute
Northwestern University
Evanston, Illinois 60201

ABSTRACT

Measurements on the heat affected zone of a weldment are presented using a gas filled position sensitive detector and a normal diffractometer equipped with a scintillation detector and a solid state detector. The sample, a surface ground titanium alloy, provided a difficult application for the X-ray technique from which a test of the real usefulness of the position sensitive detector could be made. The diffraction profile from the Ti alloy is very broad and the fluorescence produces a high background. The fluorescence is easily rejected using a solid state detector; however, the time of analysis is very long. With the position sensitive detector, the combination of increased energy discrimination over the scintillation detector and the simultaneous measurement of many data points over the broad peak enabled the measurements to be made for the same accuracy in much shorter times than for either the solid state detector or the scintillation detector.

INTRODUCTION

The residual stress distribution near a fillet weld joining two titanium alloy plates was examined by means of X-rays. Titanium is a difficult material in which to measure the residual stress by the X-ray technique because a large fluorescent background makes accurate peak location difficult. This sample

°Presently Member, Technical Staff, Science Center/Rockwell International, 1049 Camino Dos Rios, P. O. Box 1085, Thousand Oaks, California 91360.

thereby allows a comparison to be made between a scintillation detector, a solid state detector (SSD) and a gas filled position sensitive detector (PSD) in the measurement of residual stress in a practical situation. The SSD has the advantage of excellent energy resolution which enables the fluorescence to be removed to improve the peak to background ratio. It is not clear what detector is best, and this choice is the purpose of this report.

EXPERIMENTAL DETAILS

Specimen

An alpha-beta titanium alloy having been surface ground and fillet welded was obtained from Dr. W. H. Lucke of the Naval Shipyard Research and Development Center, Annapolis, Maryland. It was stated to have a yield strength of 695 MPa (100,000 psi) and an ultimate tensile strength of 834 MPa (120,000 psi). Two such blocks 12.7 x 10.2 x 2.5 cm were fillet welded, then cut apart such that the built up material was on the measured sample.

The first welding pass causes the edges of the material next to the weld to "sink in" and form the familiar welding groove. Three subsequent weld passes were made to decrease the severity of the groove.

The X-ray method has been amply described (1). It essentially involves measuring the "d" spacing at various ψ tilts of the specimen from the parafocusing condition, and the slope of "d" vs $\sin^2\psi$ (and elastic constants) yields the stress in a direction on the surface which is the intersection of the circle of tilt. The studies conducted in this investigation were made using a stress constant of 335 MPa/$^o2\theta$ (48000 psi/$^o2\theta$) for the 114 planes and CoK_α radiation, determined from measured X-ray elastic constants (2).

Data Collection

It has been shown that the customary scanning of a diffraction profile can be eliminated using a PSD (3) thereby saving time in the data collection process. In principle, this is also possible with the SSD. This energy dispersive detector is capable of determining energy versus intensity for a series of diffraction lines without moving the detector provided the continuous spectrum of the X-ray tube is used (4). The energy and interplanar spacing are related:

$$Ud \sin \theta = h\, c/2 = 6.195 \text{ (keV-Å)} \quad (1)$$

where U is in keV, d in Å and h and c are Planck's constant and the speed of light, respectively. A shift in energy will occur in a similar manner as the shift in θ from a change in d spacing. The relation for the surface stress is then given by:

$$\sigma_{\psi} = \left(\frac{E}{1+\nu}\right)\left(\frac{1}{\sin^2\psi}\right)\left(\frac{U_o - U_{\phi,\psi}}{U_{\phi,\psi}}\right) \quad (2)$$

Unfortunately, as good as the energy resolution of the SSD is (typically 170 eV FWHM at 5.94 keV), to obtain the necessary precision in determining the energy shift, peaks in the range of 20 keV must be used (5,6). For reasonable counting times this means using high intensity peaks at low angles (~ 20°-40° 2θ). This increases sensitivity to sample positioning, reduces the range of possible Ψ tilts, and makes surface preparation more important. In view of these problems, it has been estimated that at best, an accuracy of ± 70 MPa (± 10000 psi) can be obtained in this manner (5).

The SSD was therefore used in the traditional wavelength dispersive mode via normal step scanning. Its advantage lies in the ability to reject fluorescent radiation so as to better characterize the peak shape.

Complete automation of the residual stress measurement was accomplished using the computer package described in ref. 7. The $\sin^2\psi$ method was used for measurements with the scintillation detector and the SSD using inclinations of 0°, 26.57°, 39.23° and 50.72° and five data points to locate the peak position. Background subtraction was used in determining the region of fit for the scintillation detector because the peak to background ratio was only 1.2 due to the poor energy discrimination. The correlation coefficient for d vs. $\sin^2\psi$ was always greater than 0.97 indicating a linear relationship to a 99 pct confidence level.

The two-tilt technique could be used with the PSD because d vs. $\sin^2\psi$ was linear. The collection of many data points on the peak minimizes the random errors as shown in ref. 7 so the use of the two-tilt technique with the PSD is as precise in the same time as the $\sin^2\psi$ technique using either of the other two detectors. Calibration of the PSD was accomplished using three low order peaks, 010, 002 and 011 recorded simultaneously. This gave a calibration constant of .0405°2θ/channel.

A Co tube operated at 50 kV and 6 mA was used on the Picker diffractometer with a divergent slit of 2°. For the scintillation detector and for the SSD a 0.2° receiving slit was used. In all cases the pulse height analyzer was set for 90% acceptance of the CoK_{α} radiation.

TIMING COMPARISON

The stress, σ_x, was measured at point A in the center of the sample as shown in Fig. 1 with the three detectors. A statistical counting error of ±14 MPa (±2030 psi) was specified and the resultant times to achieve this precision are given in Table 1.

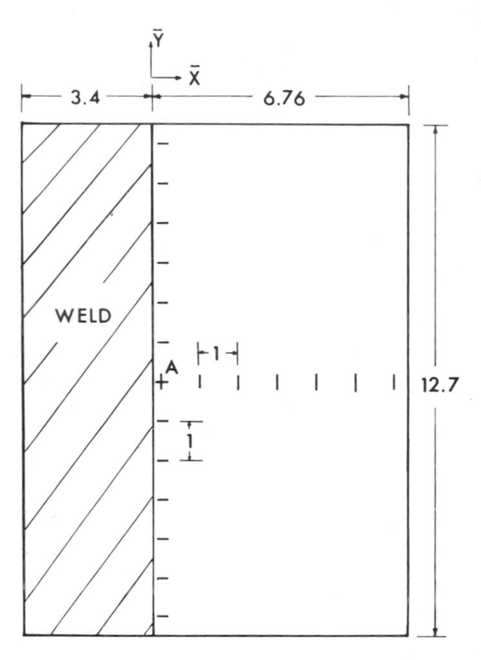

Figure 1. Location of stress measurements along the line parallel to the weld (giving σ_x) and measurements transverse to the weld (giving σ_y). Distances in cm.

Table 1

Detector Comparison: Stress, σ_x, Measured at Point A in Fig. 2

Detector	Stress MPa (psi)	Error MPa (psi)	Time Minutes
PSD	-351.2 (-50931)	± 16.3 (±2360)	2
SCINTILLATION	-363.4 (-52710)	± 17.3 (±2515)	25
SSD	-333.6 (-48379)	± 14.7 (±2140)	30

PSD - two tilt method
SCINTILLATION - $\sin^2\psi$ method with background subtraction
SSD - $\sin^2\psi$ method

The profile was better characterized with the SSD, in that the peak was sharper and had a better peak to background ratio than for the other detectors, because the Ti fluorescence was eliminated. However, the intensity was considerably decreased so that more time was taken in the measurement. The PSD provides the fastest method. In Ref. 7 it was shown that the $\sin^2\psi$ method and the two-tilt technique could be accomplished in the same time with the same error. From Table 1, it should be realized that the $\sin^2\psi$ measurement with the PSD could have been carried out in under 4 minutes, still 6 times faster than with the normal step scanning procedures and with a lower error. For the scintillation detector, the automated program had great difficulty finding the peak because of the poor peak to background ratio. This was not the case with the PSD since the energy resolution is 20% compared to 50% for the scintillation detector, so that some of the fluorescence could be eliminated.

STRESS DISTRIBUTION IN THE HEAT AFFECTED ZONE

With the coordinate system in Fig. 1, the residual stresses parallel to the weld bead, σ_y, were measured with the PSD along a line perpendicular to the weld and the stress, σ_x, was measured along a line parallel to the bead. The results are plotted in Figs. 2 and 3.

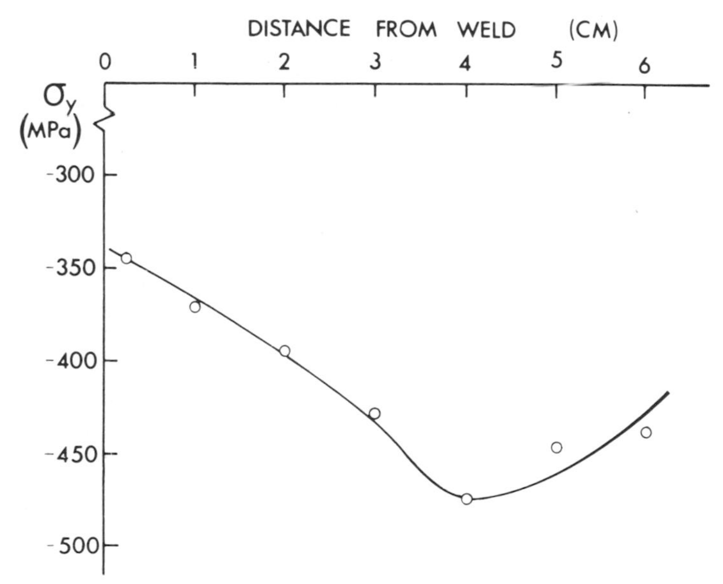

Figure 2. Stress distribution for σ_y transverse to the weld bead.

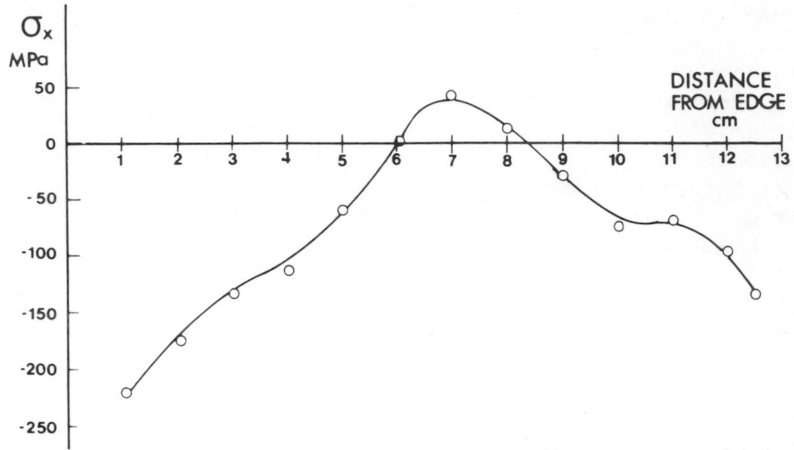

Figure 3. Stress distribution for σ_x parallel to weld bead.

Papazoglou and Masubuchi (8) predicted a typical distribution of stresses in a butt weld, taking into account thermal contraction and the phase transformation at different times at the surface and interior, and at different locations relative to the weld. There is reasonable agreement with Figs. 2 and 3.

This research was supported by ONR.

REFERENCES

1. M. E. Hilley, J. A. Larson, C. F. Jatczak and E. E. Ricklefs, eds., Residual Stress Measurement by X-ray Diffraction, SAE Information Report J-784 SAE, Inc. (1971).
2. A. L. Esquivel, "X-ray Study of the Effects of Uniaxial Plastic Deformation on Residual Stress Measurements," in Adv. in X-ray Analysis, Vol. 12, p. 260 (1969).
3. M. R. James and J. B. Cohen, "The Application of a Position Sensitive X-ray Detector to the Measurement of Residual Stress," Adv. in X-ray Analysis, Vol. 19, p. 695 (1976).
4. W. C. Giessen and G. C. Gordon, "X-ray Diffraction: New High Speed Technique Based on X-ray Spectrography," Science 159, 973 (1968).
5. L. Leonard, "The Application of Solid-State X-ray Detectors in Diffraction Techniques," ONR Technical Report F-C3454 (1973).
6. M. R. James, Ph.D. Thesis, Northwestern Univ., Evanston, IL (1977).
7. M. R. James and J. B. Cohen, "Study of the Precision of X-ray Stress Analysis," Adv. in X-ray Analysis, Vol. 20, p. 291 (1977).
8. V. Papazoglou and K. Masubuchi, ONR Contract No. N00014-75-C-0469, NR 031 -773, Sept. 1976.

USE OF CR K-BETA X-RAYS AND POSITION SENSITIVE DETECTOR FOR RESIDUAL STRESS MEASUREMENT IN STAINLESS STEEL PIPE

C. O. Ruud and C. S. Barrett

University of Denver Research Institute

Denver, Colorado 80208

Residual stresses on the inner surface of stainless steel pipe used in nuclear reactors are of exceptional importance. Apparatus for measuring these *in situ* in welded lengths of 10-inch diameter austenitic (304) stainless pipe has been developed at the University of Denver Research Institute under the sponsorship of the Electric Power Research Institute.

The dimensions of the detector system and of an X-ray diffraction tube of the usual size permit both longitudinal stresses and hoop stresses to be measured inside 10-inch Schedule 80 pipe. The detector is a scintillation position-sensitive type involving coherent fiber optic bundles that transmit fluorescent light excited by the diffracted beams to a proximity focussed image intensifier and to a solid state scanner (1). The intensified image of the diffracted beam profile is interfaced with a PDP8 minicomputer that provides curve fitting of diffraction peaks and corrections such as those for background and for the electron gain in each of the 1024 individual elements of the scanner. Laboratory tests led to the selection of Cr Kβ radiation filtered by Cr foil and diffracted at $2\theta = 147.8°$ from (311) planes of the austenite for this task. Both the single exposure technique and the $\sin^2\theta$ technique were used, with the distance from the irradiated spot to the detector surface (43 mm) and the 2θ angles being calibrated by diffraction from copper powder. Calibration of peak shifts by hoop stresses of known magnitude was provided by wire strain gauge readings on a section of the pipe 3 inches long, C-shaped, which was stressed by a bolt across the opening of the C.

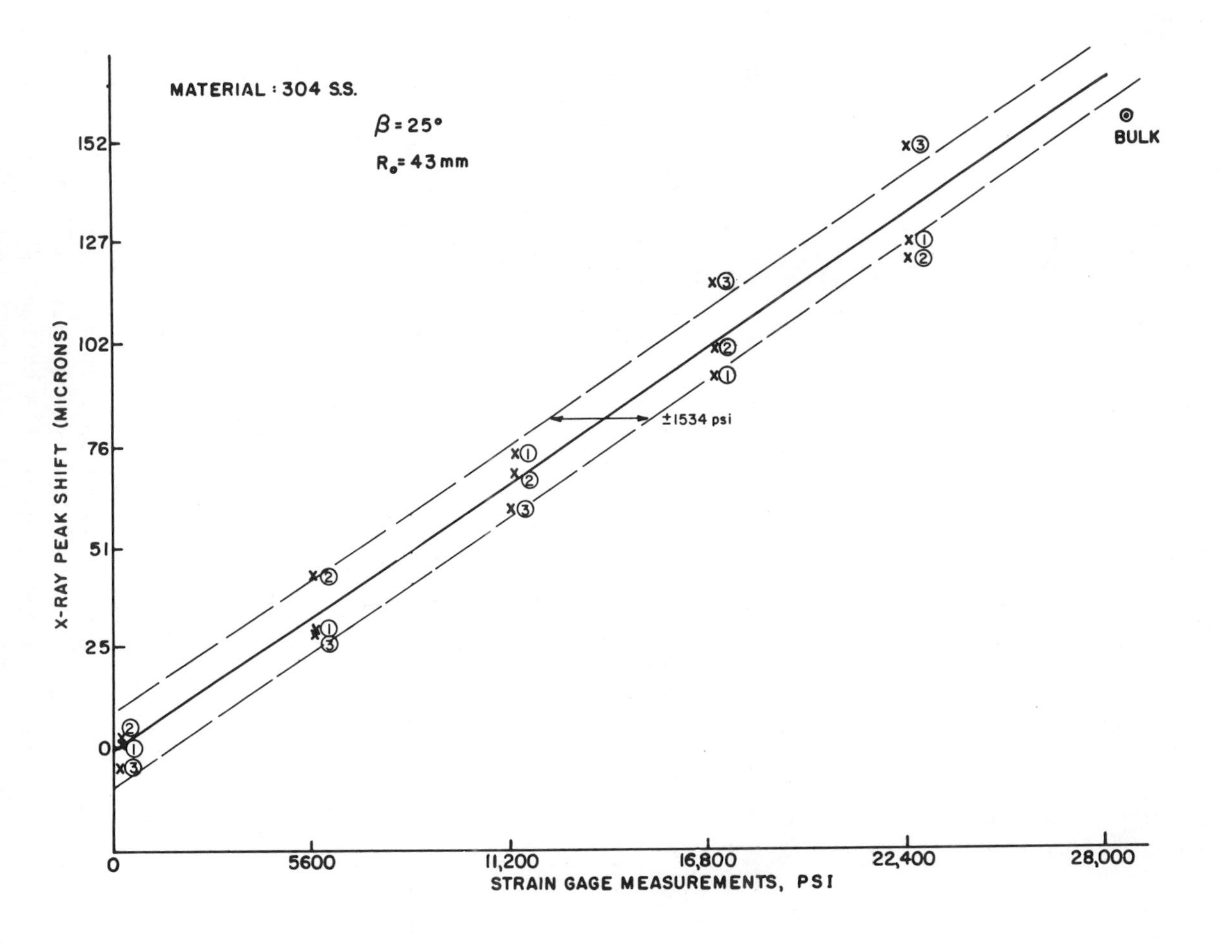

Fig. 1. Calibration curve showing CrKβ 311 peak shifts *vs* stress from 4-point bending of #304 stainless steel strip

Hoop stresses and longitudinal stresses have been determined on the machined inner surface. Hoop stresses at points in the heat affected zone 5mm from the weld were determined with the primary beam at an angle $\beta = 25°$ to the radial direction, using the single exposure technique and agreed within 1,900 psi (13 M Pa) with the stresses determined by the $\sin^2\psi$ technique (ψ = 8.9, 16.2, 41.1). At this distance from the weld the stresses varied from place to place around the circumference, but were always compressive (60,000 to 75,000 psi). Longitudinal stresses, also compressive, were in the same range but accuracies were lower with these. A series of stainless steel (#304) sheet specimens equipped with electrical resistance strain gauges and stressed by bending indicated that a 5-minute exposure with single exposure technique yielded a precision (standard deviation) of ± 1534 psi. Fig. 1 shows the calibration curve with dashed lines indicating one standard deviation from the (least squares) solid line. After these tests were finished it was discovered that the gain in the image intensifier was incorrectly adjusted so that a considerable loss of counts had occurred, these counts being submerged into the background noise. This maladjustment has now been corrected and no longer occurs. Without this loss, exposure times are now shorter than before.

REFERENCE

1. D. A. Steffen and C. O. Ruud, "A Versatile Position Sensitive X-Ray Detector," *Advances in X-Ray Analysis*, Vol. 21, 309-315, 1978.

A DIFFRACTOMETER BASED ENERGY DISPERSIVE ELEMENTAL ANALYSER

Michael F. Ciccarelli and Raymond P. Goehner

General Electric Company

Schenectady, New York 12301

Instrumentation has been developed which employs standard x-ray diffraction equipment and an energy dispersive analyzer for elemental and phase analysis. A General Electric XRD-6 diffractometer, a sample chamber, a Nuclear Semiconductor energy dispersive analyzer (EDA) with 155 eV resolution, and a Tracor Northern multi-channel analyzer are its main components.

This system was designed for the typical x-ray diffraction laboratory that does not have elemental analysis capability. Our concept gives these laboratories the means of acquiring qualitative information which makes phase identification of an unknown faster and more reliable without the increased cost of a commercial unit.

The unit is designed to allow for the utmost flexibility. A wide range of sample sizes can be analyzed using a collimator with adjustable apertures to change the size of the incident x-ray beam. A special chamber, shown in Figure 1, was designed and constructed for use under vacuum to increase sensitivity for the light elements. This attachment is easily converted to a helium path for the analysis of liquid samples. The sample chamber consists of three concentric chambers. The inner cylinder contains the sample holder. The middle and outer cylinders are designed to allow for sample and detector rotation, respectively. 'O' ring seals are used to maintain a rough pump vacuum in the innermost cylinder. Using this chamber allows the sample to rotate at half the rate of the detector in the normal θ-2θ scanning mode. This arrangement provides two important advantages. First, it provides quick identification of elemental peaks as opposed to diffraction peaks which shift when the detector is rotated, while the characteristic peaks remain

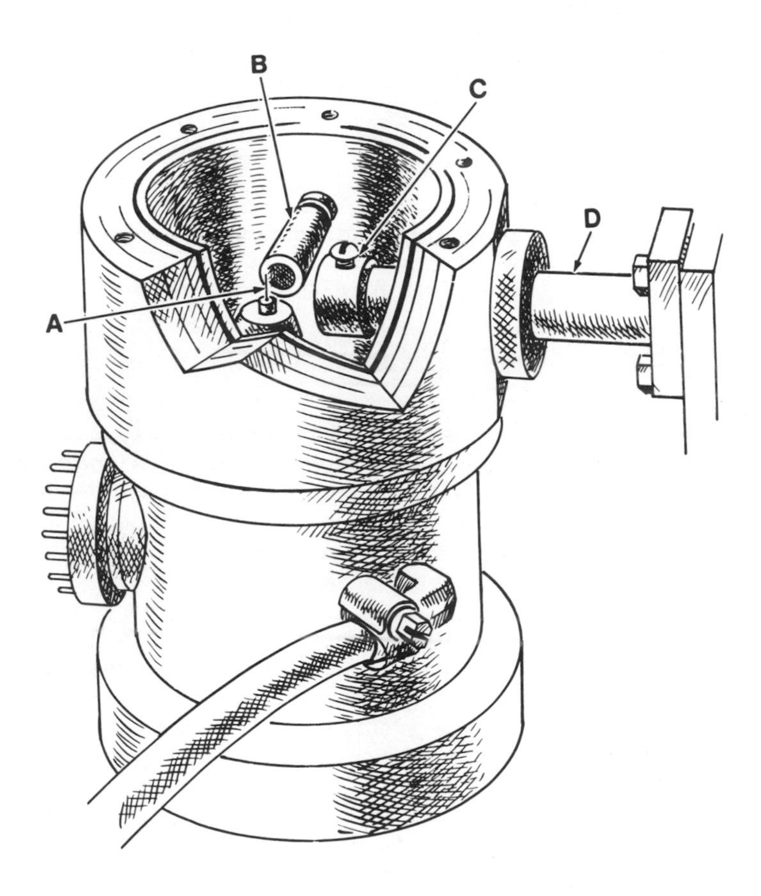

Figure 1. Sample Chamber

A. Sample on nylon fiber
B. Tube collimator
C. Analyzer aperture
D. Retractable snout

stationary. Second, the sample chamber can be used as a vacuum attachment on a diffractometer. The energy dispersive detector has a retractable snout, and a modification has been made to the sample chamber which allows for placing the detector close to the sample to increase the solid angle subtended by the detector. In the case of high count rates, where dead time would become a serious problem, an aperture can be placed in front of the beryllium window of the EDA.

The unit is presently being used to analyze elementally a wide range of material, both in solid and liquid form. It has been extremely useful for handling the routine qualitative analysis normally run on an x-ray fluorescence wavelength dispersive unit. Elements less than atomic number nine cannot be detected using this technique because there is a fixed beryllium window in front of the detector. However, we find that in the case of many different matrices we are able to detect elements in the part per million range. A number of linear calibration curves have been collected and used for quantitative analysis.

Both large bulk and small particle samples have been analyzed using this instrumentation. For example, a particle on the order of ten microns in size was mounted on a nylon fiber and an elemental analysis was made using a pin hole collimator. This particle was then mounted in a Debye-Scherrer camera, and a phase analysis was obtained.

A sample stage was designed with X and Y translation capability for use with the sample chamber. The stage is driven by dual stepping motors which can be controlled manually or automatically. Three important functions are obtained using the feature. First, we are able to check samples for homogeneity using a small beam while moving the sample in a raster fashion. Second, concentration profiles across the surface of the sample can be measured. Third, by using this stage, the possibility also exists of producing x-ray maps.

A NEW METHOD FOR FAST XRPD USING A POSITION SENSITIVE DETECTOR

Herbert E. Göbel

Forschungslaboratorien der SIEMENS AG

D 8000 München 83, West Germany

ABSTRACT

The scanning speed of powder diffractometers can be increased by about two orders of magnitude by replacing the normal detection system (receiving slit/scintillation counter) by a linear position sensitive detector (PSD) which collects diffracted x-rays of a larger part of the diffractometer measuring arc simultaneously. The velocity increase is proportional to the section of the arc 2θ in which the diffracted x-rays are detected. With respect to the Bragg-Brentano focussing geometry, a range of several degrees (typically 5°) can be tolerated without affecting the resolution remarkably. Diffractograms over extended angular regions of 2θ are received by slewing the PSD continuously along the arc while the diffracted x-rays are accumulated with their accurate 2θ angle address on a digital base.

The method has been named the continuous position sensitive detection technique (CPSD). With this technique, diffraction patterns with similar quality (resolution, linearity, peak/background ratio) as conventional, scintillation counter recorded diagrams can be received at slewing rates of several hundred degrees per minute. Additionally, the CPSD technique augments the number of crystallites contributing to the diffraction intensity by the same amount, thus producing much better crystallite statistics for quantitative measurements.

INTRODUCTION

For fast structure identifications and dynamical studies of phase transformations or chemical reactions in solids, methods for fast or "*in-situ*" x-ray powder diffraction (XRPD) have been developed in recent years. The most important ones are film techniques with high power or flash x-ray sources and the energy-dispersive XRPD using solid state detectors with high energy resolution and polychromatic x-ray sources. In the latest years, however, proportional counters with high position resolution, so-called position sensitive detectors (PSD), have advanced into the fields of x-ray scattering, where so far films or scanning devices of conventional x-ray detectors like scintillation counters have been applied. These detectors combine the position resolution of a photographic film with the high sensitivity dynamics, energy selectivity and quantitative performance of a single quantum registration technique. The main PSD applications have been x-ray small angle scattering and stress analysis, where a linear PSD is used to analyse only a small portion of the diffracted pattern. For analytical powder diffraction, R. A. Sparks *et al.* (1) presented a Debye-Scherrer camera using a full angle curved PSD instead of the x-ray film.

The method presented here relates to the Bragg-Brentano geometry, which is verified in commercial powder diffractometers. A PSD collects the diffracted x-rays over a section of a few degrees of the arc 2θ simultaneously and assigns their position of incidence with a resolution of about 0.02 degrees. A full angle diffractogram is recorded by a continuous scan along the arc in a coupled $\theta/2\theta$ mode. The technique therefore was called the continuous position sensitive detection (CPSD) method.

PHYSICAL PRINCIPLES

The Bragg-Brentano focussing geometry has found a widespread application for the construction of commercial focussing x-ray diffractometers. As detectors, x-ray counters are used which are slewed around the arc of the diffraction angle 2θ with double the angular velocity of the flat polycrystalline specimen. The angular resolution must be produced by a tiny receiving slit of a few hundredths of a degree, thus utilizing only a disappearingly small portion of the diffracted beam. The original Bragg-Brentano goniometers with photographical recording, however, were designed for larger openings of the receiving window. Thus, the normal counter diffractometers cover only a small part of the potentiality of the Bragg-Brentano geometry. The full electronic verification of this idea is the CPSD technique. The use of large receiving windows, however, is accompanied by a lack of definition of the peaks out of focus. J. C. M. Brentano (2) estimated these line broadening effects due to angular displacements of the sample plane out of the correct $\theta/2\theta$ focussing condition.

It can be shown that for normal applications, an angular range of up to 5° (or 10° at higher angles) of 2θ can be utilized without broadening the lines seriously. The effect of defocussing in the CPSD-method is exaggeratedly illustrated in Figure 1.

A reflection is recorded during all the time when the angular range of the PSD is passing it while rotating along the diffractometer arc. It is exactly focussed only if it hits the PSD in the center. Before and after the center of the PSD passes the reflection, the diffraction line is defocussed due to the change of the focussing circle caused by the rotation of the sample in the primary beam. However, the integral intensity and the center of gravity as the most important quantities for chemical analyses remain nearly unchanged.

ELECTRONIC DESIGN

The increase of the velocity for the CPSD recording compared to conventional techniques is proportional to the ratio of the receiving apertures. Since this ratio is at about 100, the slewing rate reaches angular velocities of several hundred degrees per minute, which can be performed only by the fast gear of modern step-motor driven diffractometers. The CPSD method was tested on a Siemens Type-F diffractometer equipped with a step motor providing a fast gear of 200°/min with 200 steps per degree and on a standard Siemens D-500 diffractometer providing 400°/min with 500 steps per degree.

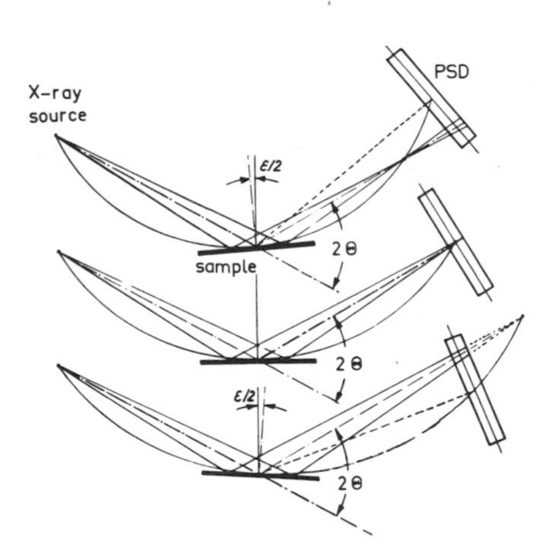

Fig. 1. Defocussing of a CPSD recorded diffraction peak due to the mismatch of the $\theta/2\theta$ condition

The PSD was a commercial high pressure argon-methane (P10) proportional counter (3) with a linear high resistive counting wire and chamber dimensions of 5 cm x 1 cm x 0.3 cm (length x height x depth). The position of incidence is determined by the pulse shape technique after Kopp and Borkowski (4) with a resolution of 50 μm over the length of 5 cm. The other specifications of the detector are:

Time difference for positioning:	about 2 μs full scale
Position linearity:	better than 1%
Quantum efficiency (for Cu-Kα radiation):	about 50%
Maximum counting rate:	2-3 x 10^5 counts/sec
Energy resolution (for Cu-Kα radiation):	about 25%
Working pressure:	11 bar
Beryllium entrance window:	0.5 mm

The PSD detects the event of an incoming x-ray quantum and assigns to its position of incidence a digital address. Since the time and the position of the x-rays are statistical, every event has to be saved in a random access memory, for example in a multichannel analyzer (MC). For the CPSD method the address from the PSD has to be coupled with the angular position of the PSD on the diffractometer. The aim is to store a diffracted x-ray quantum under its diffraction angle address (2θ) regardless of its incidence on the PSD and the instantaneous diffractometer step, so that 2θ becomes independent of the diffractometer motion. This means that x-rays contributing to a diffraction line are added into the same memory channels during the PSD passes over the line. For this purpose the angle 2θ has to be imagined as a digital sum of the angular position of the PSD on the diffractometer (A) and the incidence position (a) of the x-ray in the PSD (Figure 2). The motion of the detector can be represented by an infinitesimal shift ε (Pos. 2). If the increase of A is compensated by the decrease of a on the same digital scale, ε is eliminated in the second equation in Figure 2. For a constant slewing rate, a digitized diffraction pattern is collected similar to the familiar strip chart recordings or step scanned diagrams.

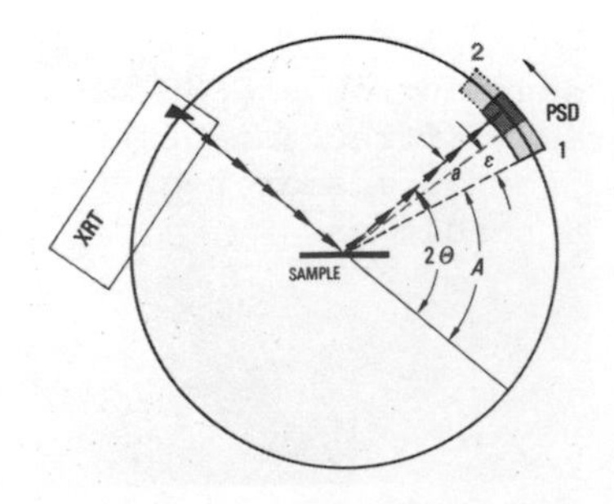

Fig. 2. Additive principle of the CPSD addressing; Position 1: $2\theta = A + a$ Position 2: $2\theta = (A+\varepsilon) + (a-\varepsilon)$

The electronic problem that has to be solved is to tune the digital scales of A and a very accurately. Since A can be tuned only by increments of the step motor drive, the fine adjustment has to be achieved with the digitization of a. There are two possible ways for a vernier tuning of the digitization of a:

1) A combination of a time-to-amplitude converter with a continuously adjustable output amplifier (5) and an external analogue-to-digital converter. This technique, however, is too slow to handle the possible incoming pulse rates of more than 10^5 counts per second without great dead time losses. It was therefore replaced by

2) a fast time-to-digital converter (TDC) with a tunable clock frequency. Since the available time difference of 2 μs is marginally short to be digitized into maximal 1000 steps, the clock frequency has to be chosen at about 500 MHz to utilize the full detector resolution. The TDC used here is tunable around 500 MHz. By dividers the digitization scale can be selected in steps of 0.01°, 0.02°, 0.05°, 0.1° and 0.2° of 2 theta. The same subdivisions are selectable at the motor-step counter/divider (MSC). Both the TDC and the MSC are 1/12 NIM modules.

In Figure 3 a block diagram of the electronic equipment is shown. In addition to the TDC described above, the module AMP/TSCA has to be pointed out. It is a combined dual pulse shaping amplifier and timing single channel analyzer which is now a single 1/12 NIM module. It has been developed especially for the PSD and replaces as many as eight standard NIM modules previously necessary. As the only adjustment the optimum gate for the quantum energy has to be set. For best operation of the pulse shaping circuitry, this pulse height has to be adjusted by the detector bias at the high voltage supply (HVS). The energy discrimination can be controlled via an added (1+2) amplifier output at the AMP/TSCA module.

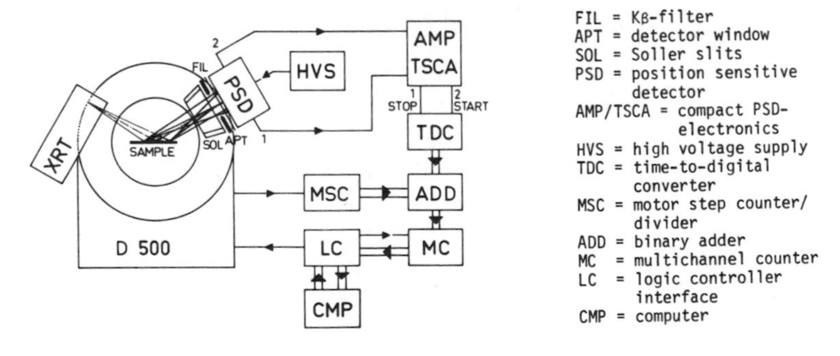

Figure 3. Block diagram of the electronic equipment

The address 2θ of a diffracted x-ray is formed in a 12 bit-adder (ADD, 1/12 NIM module), where the binary addresses from the TDC (a) and the MSC (A) are added. An addition is started and the sum transferred to the multichannel counter (MC) by a data-enable signal originated by an event at the PSD and transferred via the TDC. The MSC address can be looked upon as quasi static changing its value stepwise with frequencies in the 100 Hz region. As MC, a 4k x 20 bit binary addressable memory of a Canberra 8100 MCA system was used. A minicomputer (CMP) in combination with a logic controller interface (LC) take charge of the diffractometer and specimen exchange control, the data transfer from the MC and the evaluation of the measured data.

The electronic concept of the CPSD technique can also be favorably applied to comparative measurements of single line profiles, where the detector is not moved. In this case a binary address applied to the MSC input of the adder can be used as digital offset for subsequent profile measurements. This is a very comfortable way to record time-resolved profile changes.

Finally, the electronics can also be used for data collection in conventional diffractometry. In this case, the signal from a scintillation counter is used as the data-enable signal at the adder which then transfers the MSC address as 2θ to the MC.

PERFORMANCE AND EXAMPLES

The data collection in the CPSD diffraction technique is shown in Figure 4. Photographs taken at time differences of about 1 second during a 200°/minute scan are shown. The detector window

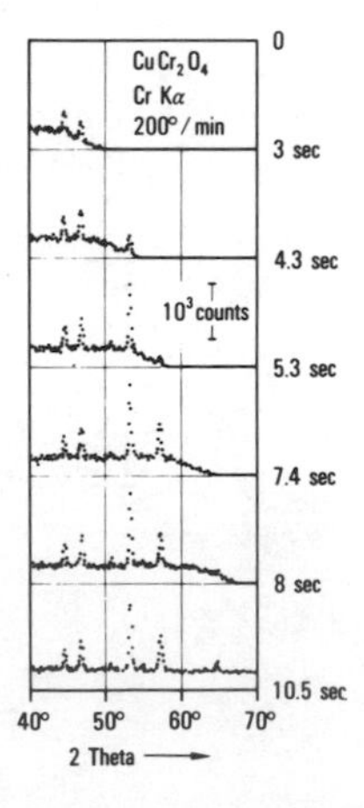

Figure 4. Sequence of Photographs of the CPSD Recording

is situated at the slope at the right end of the diffraction pattern. Counts are collected into each channel as long as the detector passes across their angular position. For an opening area of about 5° and a slewing rate of 200°/min, data are accumulated for 1.5 sec into each channel. A channel width of 0.1° was chosen. Similar conditions have been chosen for the examples in Figures 5 and 6. To compare with conventional diagrams recorded with the

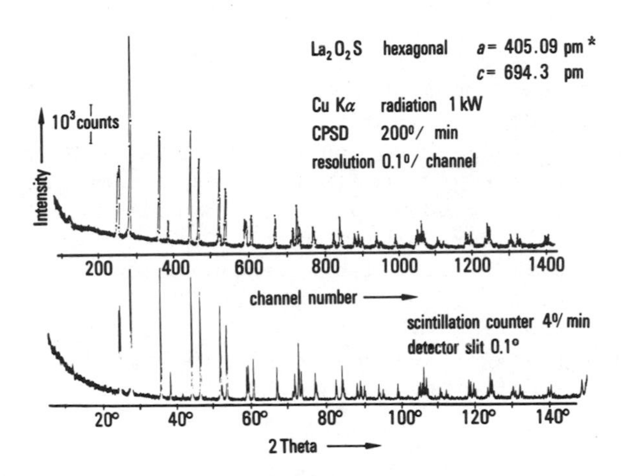

Figure 5. Diffraction pattern by Cu-Kα-radiation

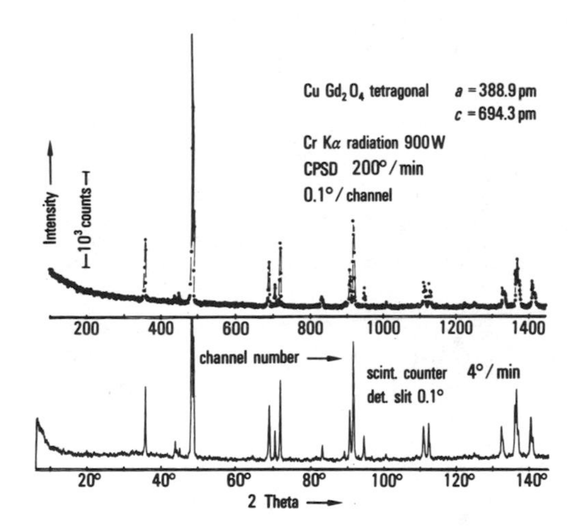

Figure 6. Diffraction pattern by Cr-Kα-radiation

*100 pm - 1 angstrom unit

same resolution (= receiving slit), the same samples have been measured under equal conditions with 4°/min slewing rate. For both examples samples with very closely adjacent diffraction lines were chosen to demonstrate the efficient resolution of this fast technique.

It can be seen that the method works for CrKα-radiation as well as for CuKα-radiation thus covering the most important wavelength range for diffraction work. It also is evident that the information content of the compared diagrams is equivalent.

In a larger scale the resolution is demonstrated in Figure 7 with the aid of the famous three $CuK\alpha_1/\alpha_2$-doublet diffraction peaks of α-quartz. Instead of six lines only five should be distinguished since the (203)-α_2-reflection and the (301)-α_1-reflection overlap almost perfectly. The vertical scale is the same for the three sections compared. The upper part is a step scan with an increment of 0.01° and a similar receiving slit. The same resolution is received if the PSD is fixed and the lines reside close to its center. In the lower part, the resolution of the CPSD technique is shown with an open area of 5°. The "effective" counting time (= shown angular detail/slewing rate) was about 0.3 seconds. But even here the resolution is still appreciable.

DISCUSSION

The CPSD technique produces obviously very good results in a minimum of time. The question remains as to whether these results are optimum for all applications in X-ray powder diffraction.

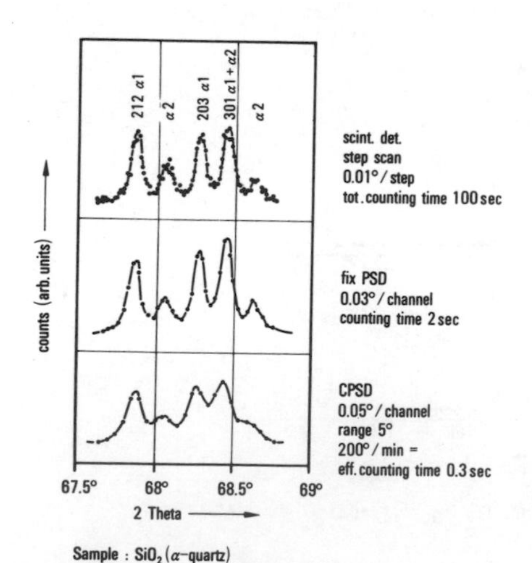

Fig. 7. Resolution test of the CPSD method

The time for the data accumulation is a matter of counting statistics. The lines in diffraction diagrams are determined as distributions around maxima above a continuous background. The minimum measuring time to receive the best results therefore must not be produced by the fastest data collection technique. Here the criteria of linewidth and low background are at least as important. For reliable diffraction data also very high demands are set for the linearity and absolute accuracy of the diffracted beam angular positions.

The usefulness of the CPSD method should therefore be discussed also under the following points of view:

1) Limitations of data collection velocity

In a single quantum registration technique (all counter methods with serial pulse registration) the maximum velocity for the data collection is determined by the maximum pulse rate that can be processed by the electronic system. A physical limitation is set by the dead time of the X-ray detector. The proportional counter used here has a maximum capacity of about $2 \cdot 10^5$ counts per second. To evaluate the incidence position by the pulse shape technique (4) only the rise time is consumed. So no further dead times are produced in the analogue signal part. Also, the digitizer part with a very fast time-to-digital converter produces no additional dead time.

The fact that the CPSD technique accumulates incident X-rays over an area of a few degrees simultaneously has two important advantages: First the position has to be digitized only into a few steps (address a, figure 2) compared to the full channel address 2 theta. So a direct time-to-digital conversion with sufficient resolution is possible. Secondly, the total counting rate is averaged between peaks and background. So the total intensity fluctuations are not extremely high along the arc.

The average intensity can also be influenced by a variable entrance window (APT in Figure 3). The opening of this window can be coupled with the theta/2 theta motion of the diffractometer and so the higher intensities in the low angle region which perhaps overload the detector can be suppressed. With a special device the opening can be varied to produce LP-factor free diagrams. A reduction of the entrance window at low angles is also very suitable because of the line broadening effects discussed above which are more serious at low angles.

2) Peak-to-background ratio

The peak-to-background ratio is determined by the resolution (peak height $\cong$ integral intensity/half width) and the background reduction. The CPSD method produces similar resolution as the normal diffractometry. For background reduction, however, no diffracted beam monochromators can be utilized. The only way to reduce the background of the diffracted beam is by absorption filters in front of the detector. Except for strongly fluorescing samples, the background can also be efficiently reduced by primary monochromators. If the fluorescence radiation cannot be avoided by using a suitable target material, the CPSD technique may in some cases be inferior to normal diffractometry equipped with a secondary monochromator to detect small peaks in a minimum of time [see for example the arrangement, W. Parrish *et al.* (6)].

3) Linearity and angular accuracy

In the CPSD technique all nonlinearities for the positioning and changes of the quantum efficiency along the counter wire are averaged by the continuous mode of operation. Nonlinearities of the PSD result in line broadenings and false tuning of the digitizer frequency (address a) in an additional zero shift. If the zero point is fixed by a calibration specimen, the angular accuracy and linearity of a diffraction pattern is as good as the mechanical precision of the used diffractometer.

4) Crystallite statistics

In powder diffractometers only lattice planes nearly parallel to the specimen surface contribute to the diffracted intensity. Normally the angular spread of these crystallite orientations is less than 0.1° of theta. In the CPSD technique the sample follows the theta rotation in the primary beam while the detector passes a diffraction peak. The rotation angle is half the value of the open area of the PSD, which is several degrees (see Figure 1). Consequently the number of crystallites contributing to the diffraction intensity is increased proportional to the detector window, which is the same amount as the data accumulation is accelerated. Therefore, for quantitative analyses, the intensities recorded by the CPSD technique are as reliable as intensities with oscillating and spinning samples in normal diffractometers.

5) Expenses and complication of the system

The CPSD detection system has been simplified to a high extent. There are now only few additional NIM modules necessary to adapt the system to commercial step motor driven diffractometers. The most expensive parts of the system, which of course are not necessarily confined in their application to this system are the multichannel analyzer and the interfacing system with the host computer. In automated diffractometers the additional expenses should be low compared with the high gain of measuring speed and soon compensated according to the low time consumption of analyses.

6) Applications

The CPSD technique is fundamentally applicable to all powder diffractometer tasks. The short time for perfoming analyses makes it useful for frequent process controls in preparative chemistry of metallurgy to look for mainly contained phases. However, there are also applications where a fast data collection is crucial. Those are investigations where the specimen changes its properties by decomposition or radiation damage, for example. The CPSD method is especially useful for dynamical studies of phase transformations and solid state reactions.

REFERENCES

(1) R. A. Sparks, *et al.*, U.S. Patent 4,076,981, Feb. 28, 1978; and Advances in X-Ray Analysis, Vol. 20.

(2) J. C. M. Brentano, Proc. Phys. Soc. 49, 61-77 (1936).

(3) Manufactured by M. Braun Co., Munich. The PSPC and the associated electronics are part of the SIEMENS analytical equipment programme.

(4) C. J. Borkowski, M. K. Kopp, Rev. Sci. Instrum. 38, 1515-1522, (1968).

(5) ORTEC 457 biased time to pulse height converter.

(6) W. Parrish, T. C. Huang, G. L. Ayers, Paper 144-13, XI. Int. Congress of Crystallography, Warszawa, Poland, Aug. 3-12 (1978).

LASER-PRODUCED PLASMAS AS AN ALTERNATIVE X-RAY SOURCE FOR SYNCHROTRON RADIATION RESEARCH AND FOR MICRORADIOGRAPHY

P. J. Mallozzi, H. M. Epstein, and R. E. Schwerzel

Battelle Memorial Institute

505 King Avenue, Columbus, Ohio 43201

ABSTRACT

The radiation from plasmas produced by the interaction of a pulsed laser and a solid target can be made to fall in the soft x-ray regime. The x-rays can serve as an alternative to the increasingly important synchrotron radiation facilities for a variety of techniques such as Extended X-ray Absorption Fine-Structure Spectroscopy and X-ray Lithography. In addition, the x-rays are of special interest for general microradiography of thin samples.

I. An Alternative X-Ray Source For Synchrotron Radiation Research

In recent years, synchrotron radiation has evolved from a laboratory curiosity to the foundation of a technique widely exploited by physicists, chemists, and biologists. (1) Among the research areas explored with synchrotron radiation are the investigation of chemical structure by means of EXAFS (Extended X-ray Absorption Fine Structure) and ESCA (Electron Spectroscopy for Chemical Analysis) techniques, and the replication of sub-micrometer linewidth patterns by an x-ray lithographic technique. (1) The purpose of this paper is to discuss how these and other applications of synchrotron radiation might be performed with a laser-plasma x-ray source. Because the laser system is far smaller and less expensive than a synchrotron, it is felt that laser-produced x-rays could make the x-ray techniques that are currently employed only at synchrotron facilities accessible to a large number of organizations which might otherwise be precluded from the use of the limited number of synchrotron facilities available. In addition, laser-produced x-rays would allow the performance of novel x-ray experiments which are inherently beyond the capabilities of synchrotron radiation sources, particularly those which make use of the short, intense pulse structure of laser-produced x-rays for fast-kinetic studies.

X-Ray Lithography. It is well established that x-ray lithography is an effective means for replicating sub-micrometer line-width patterns. (2) Besides replicating test patterns, the technique has been used to fabricate surface acoustic wave devices, bubble domain devices, pn diodes, bipolar transitors, and MOS transitors. The basic concept of x-ray lithography is to use the smaller wavelength of an x-ray source instead of the long wavelength of an ultraviolet source. This essentially eliminates the diffraction limitation of the ultraviolet source. With this eliminated, x-ray lithography is capable of producing line patterns with an "error" or "replication accuracy", δ, of 1000 Å. Patterns this small are near the theoretical limit for microcircuit fabrication.

Many difficult problems must be solved before small geometry circuit manufacture by x-ray lithography becomes a commercial reality. The two main obstacles appear to be (a) alignment of the mask with respect to the wafer, and (b) development of an adequate x-ray source.

Recent work by Austin, Smith, and Flanders suggests that the problem of mask alignment is well on its way to solution. (3) These authors have developed a new laser interferometric technique that in principle should be capable of aligning masks relative to wafers with a superposition precision of 100 Å. The technique is compatible with x-ray lithography, and is adaptable to automation.

It should be noted that the second problem, development of an adequate x-ray source, is also solvable, primarily because of work performed at Battelle. (4,5) Experiments have been performed at Battelle in which a COP photoresist was exposed a distance of 20 centimeters from a laser plasma x-ray source. (5) The basic experimental configuration is shown in Figure 1. The x-rays passed through a mask consisting of a relatively transparent layer of silicon, 3 or 4 microns thick, on which a relatively opaque layer of gold, shaped in a circuit pattern, was arrayed. The mask to wafer separation was 30 microns. Four laser shots were required to expose the photoresist at 20 centimeters. At 10 centimeters, a single laser shot sufficed. Pattern replication was perfectly faithful within the limits of optical microscopy, as shown in Figure 2.

The technique used for generating the x-rays involves vaporizing and ionizing material at the surface of a solid target with an approximately 1-joule, approximately 10-nanosecond prepulse, and laser

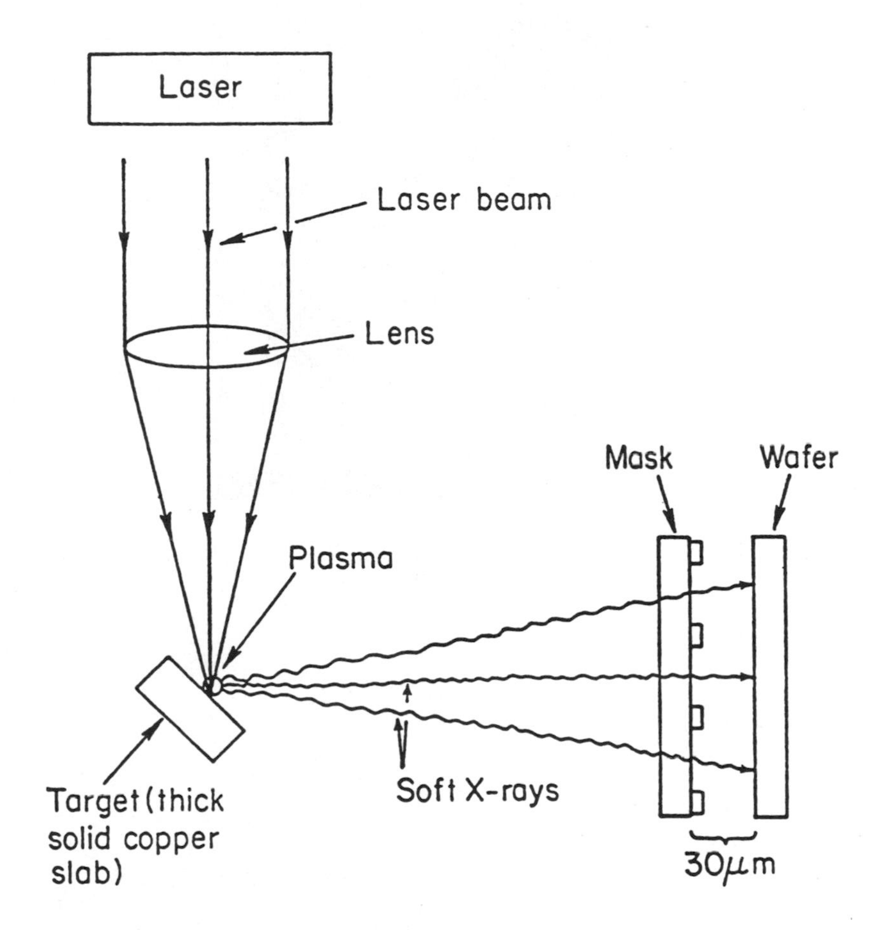

FIGURE 1. BASIC EXPERIMENTAL CONFIGURATION USED FOR X-RAY LITHOGRAPHY APPLICATION. LASER BEAM IS FOCUSED BY A LENS ONTO A SOLID COPPER TARGET AND PRODUCES A PLASMA. X-RAYS GENERATED BY THE PLASMA PASS THROUGH A MASK AND EXPOSE A PHOTORESIST-COATED WAFER.

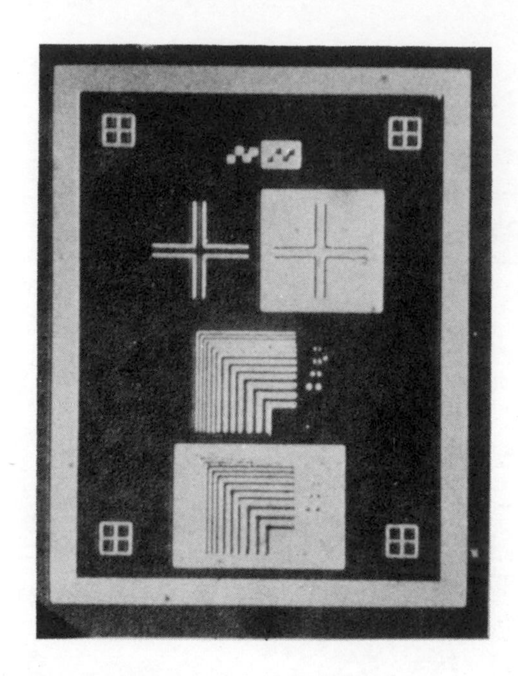

FIGURE 2. TYPICAL LITHOGRAPHIC PATTERN MADE WITH LASER-PRODUCED X-RAYS.

heating the resulting low-temperature plasma to the multi-kilovolt regime with a 10 to 100-joule, approximately 1 nanosecond main pulse via the inverse bremsstrahlung absorption process. The prepulse strikes a 100- to 200-micron diameter focal spot at an incident intensity of about 10^{11} W/cm^2, whereas the main pulse strikes it at about 10^{14} W/cm^2. The x-rays are produced in the plasma by bremsstrahlung, recombination radiation, and line radiation. More than 20 joules of x-rays, with energies between 0.3 and 1.5 keV, have been produced in this way in a single beam, multinanosecond laser shot.(4)

<u>Chemical Structure</u>. Two powerful new structural analysis techniques, both involving the use of "soft" x-rays (with energies of roughly 0.1 to 10 keV), have recently been developed. Because these techniques can provide direct structural information on complex materials which lack long-range order, they promise to revolutionize the analysis of chemical structure.

One of these techniques is known as EXAFS. The essential features of EXAFS spectroscopy are schematically depicted in Figure 3. In EXAFS, the local configuration in the vicinity of a given atom embedded in a solid, liquid, or gas molecule is indicated by the tiny "wiggles" on the x-ray absorption curve of that atom. Because this fine structure is caused by the scattering of photo-electrons from the neighboring atoms, the mathematical analysis of the fine structure provides direct information about the positions of these atoms. The other technique, which is schematically depicted in Figure 4, is a refinement of ESCA. In the ESCA approach, a highly monochromatic beam of x-rays strikes the surface of the sample, and the energy spectrum of the photoelectrons ejected from the surface is measured. The configuration in the vicinity of a given atom is indicated by the energy shifts in the photoelectron spectrum relative to the energy of an electron ejected from the same type of atom in free space.

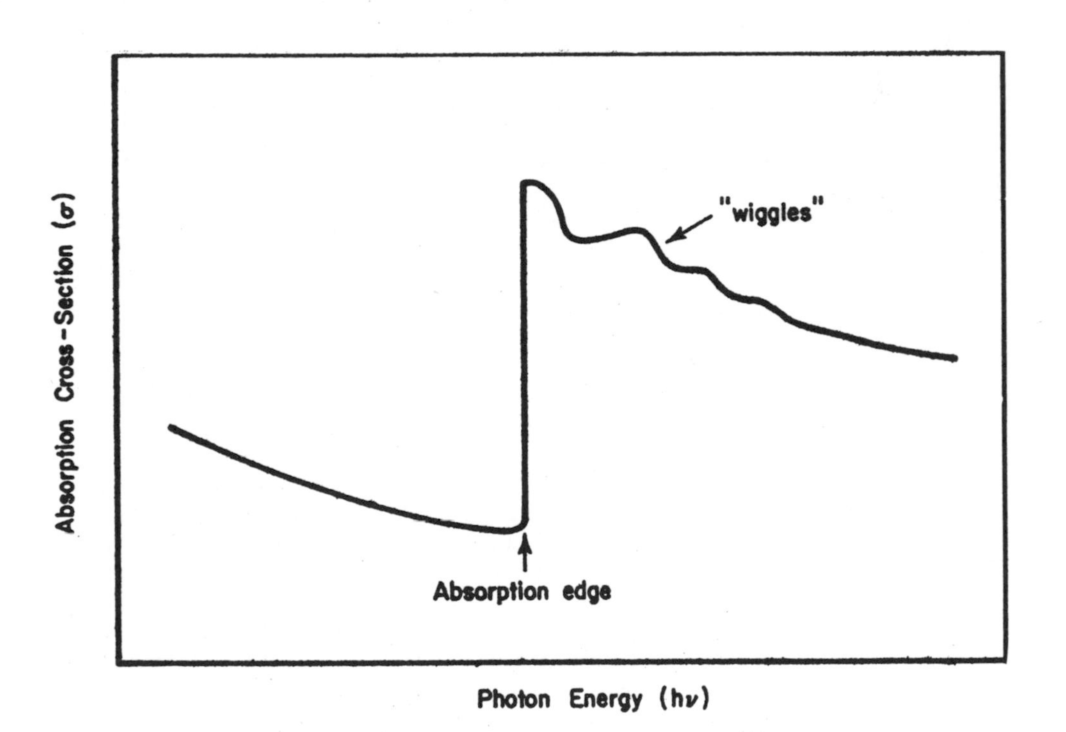

FIGURE 3. SCHEMATIC ILLUSTRATION OF AN EXTENDED X-RAY ABSORPTION FINE STRUCTURE (EXAFS) SPECTRUM.

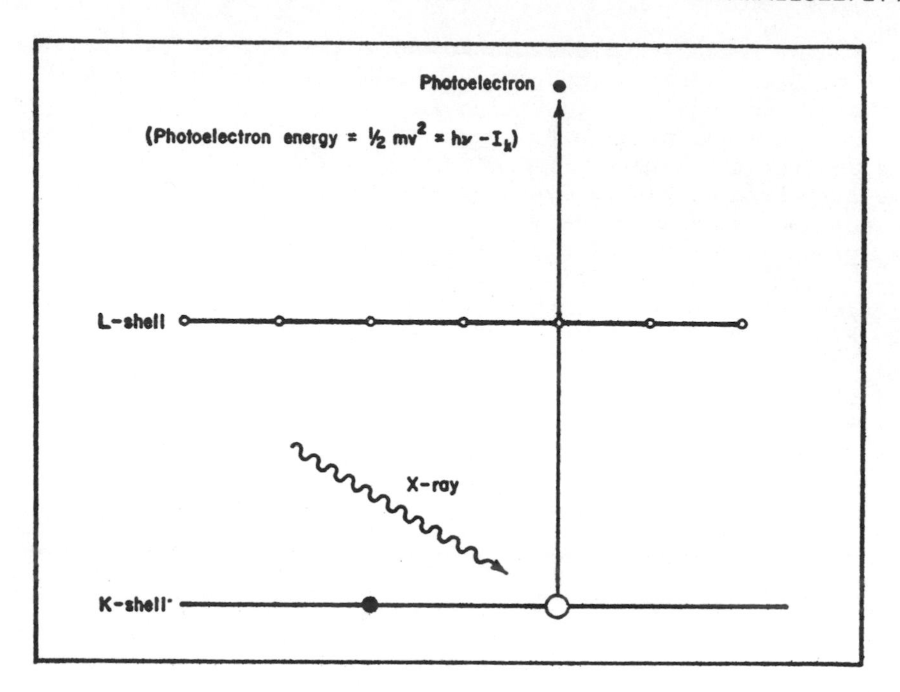

FIGURE 4. SCHEMATIC ILLUSTRATION OF ELECTRON SPECTROSCOPY FOR CHEMICAL ANALYSIS (ESCA).

It would be useful to compare laser x-rays with synchrotron x-rays, particularly from the point of view of performing EXAFS and ESCA experiments.

Synchrotron facilities harnessed for x-ray research typically deliver a time-average, broad-band x-ray power of several watts in a well collimated 1 mm diameter beam. They can therefore deliver 10 joules of more-or-less CW x-rays to a sample in a time interval of several seconds. This is about the same number of joules that are radiated into 2π steradians in a few nanoseconds by an approximately 100 μm diameter laser plasma x-ray source. It should be noted that the laser-produced x-rays are concentrated in a much narrower spectral band than the synchrotron x-rays.

A major treatise could be written concerning the relative utility of laser-produced x-rays and synchrotron x-rays for various research purposes. In general, the two sources are rather different, and their relative utility depends on the particular application one has in mind. There is little question, however, that the laser x-ray source can play an absolutely unique role in fast time-resolved EXAFS studies. This is most easily seen on contemplation of laser-produced x-ray spectra taken with a bent crystal spectrometer. These spectra can be taken with a single multinanosecond x-ray exposure and the optical density of the film used in these experiments indicates that the number of photons in the intense spectral lines is in excess of 10^8 per eV. This suggests that a signal-to-noise ratio of $\sqrt{10^8} = 10^4$ may be ultimately realized in fast kinetic EXAFS studies. Resolution such as this on a nanosecond time scale is inherently beyond the capability of synchrotron beams.

II. Microradiography With a Laser Plasma X-Ray Source

X-rays emitted from laser generated plasmas are a unique source for microradiographic analysis of thin samples. Among their advantages over conventional sources are: (1) typical pulse widths of nanoseconds or less can stop almost any motion; (2) strong emission in the 100's of eV to the few keV range provide excellent contrast; (3) emissions of 10's of joules per pulse produce flash exposures intense enough for sub-micron resolution in typical contact radiograph configurations; (4) the small x-ray source size, ~100 micrometers, reduces penmubra effects; and (5) the approximate exponential decay of the output in the range of 1 to several keV is useful in the quantitative analysis of radiographs.

A laser plasma x-ray source emits both spectral lines and bremsstrahlung radiation, (6,7,8) with a pronounced grouping of L lines in the vicinity of $h\nu \approx 1$ keV for targets in the atomic number range of ~25-30. However, samples thicker than ~10 micrometers H_2O severely attenuate the lines, and the plasma bremsstrahlung spectral shape,

$$\phi(h\nu) \propto \exp(-h\nu/KT) \quad ,$$

is dominant. ($h\nu$ is the photon energy and KT is the plasma temperature.) For specimens thicker than ~4 mm H_2O, the effective plasma temperature begins to increase because of a small component of higher energy radiation. The energy density, E_f, absorbed in a photographic emulsion after transmission through a specimen of thickness, x, and absorption cross section, α, is

$$E_f \approx c_1/2\pi r^2 \int_0^\infty S(h\nu)\exp[-\alpha x - h\nu/KT]\,dh\nu \quad j/cm^2 \quad ,$$

where $S(h\nu)$ is the film spectral sensitivity and r is the source-film distance. The Battelle laser can produce about 10 j of x-rays in a 2π solid angle at an equivalent temperature of ~1 keV. For a 100 j laser pulse, $c_1 = 10$.

For low atomic number material in an energy range where attenuation is dominated by the photoelectric absorption,

$$\alpha \approx c_2 \ (h\nu)^{-3}$$

Since the x-ray emission from the plasma source falls off exponentially with increasing energy while the x-ray transmissivity of the specimen falls off rapidly with decreasing energy, the resultant x-rays absorbed in the film, R(hν), form a very narrow energy band as seen in Figure 5. (If an absorption edge lies within the R(hν) peak a differential absorption analysis would be applicable instead of the technique described here.) These x-rays can be considered as approximately monoenergetic with an equivalent energy $(h\nu)_m$ corresponding to the peak of R(hν), which depends only on the thickness and composition of the specimen for a given film and source.

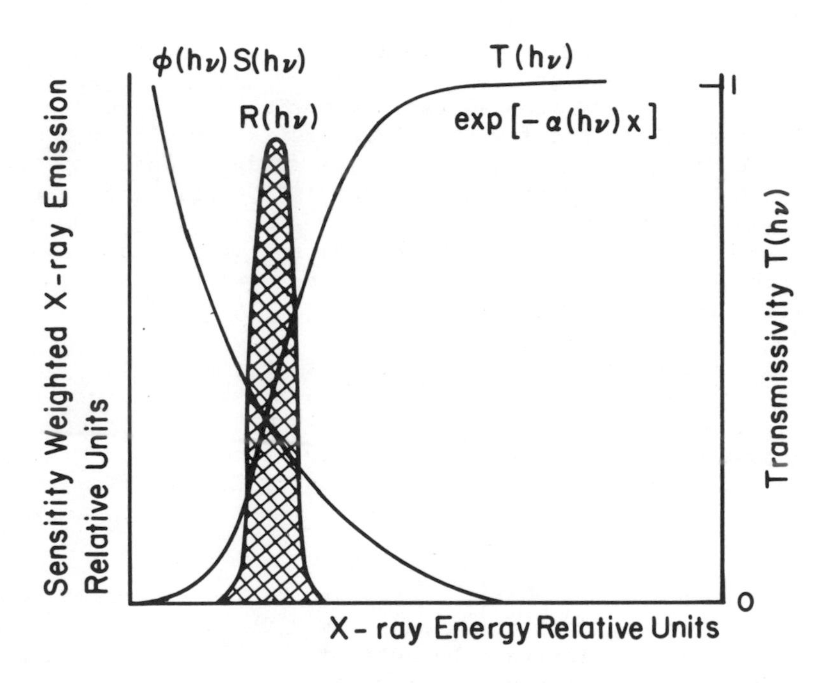

FIGURE 5. EFFECT OF X-RAY SPECTRUM φ, DETECTOR SENSITIVITY S, AND FOIL TRANSMISSIVITY ON DETECTOR RESPONSE R.

Any quantitative analysis requires information on the sensivity function S(hν) of the film. (9) The thick emulsion films absorb essentially all of the incident x-rays between ~1 and 5 to 10 keV. Below ~1 keV, the coating over the emulsion is sufficiently absorbing to reduce the sensitivity, and above ~5 keV transmission becomes important. However, the peak of R(hν) lies in the ~1-5 keV range for a wide range of thin sample radiography applications, and this case is worth analyzing carefully.

For constant $S(h\nu)$, the equivalent energy is given by

$$(h\nu)_m = [3c_2x(KT)]^{1/4} \quad .$$

Because of the sharply peaked integrand, E_f can be evaluated by the saddle point method. For KT = 1 keV,

$$E_f \approx 2.280\ r^{-2}\ (c_2x)^{1/8}\ \exp\ [-1.76\ (c_2x)^{1/4}] \qquad j/cm^2 \quad .$$

Any change in absorption cross section, density, or thickness will cause a change in E_f, and the resultant energy contrast is given by

$$C = \left| dE_f/d(c_2x))E_f^{-1}\Delta(c_2x)\right| \approx 0.44\ (c_2x)^{-3/4}\ \Delta(c_2x) \quad .$$

Commonly, we wish to determine the absorption cross section of a small inclusion of thickness, t, to evaluate its composition. In the linear range of the film,

$$D \approx \gamma \log E_f$$

and

$$\Delta D/\gamma \approx 0.19\ (c_2x)^{-3/4} t\Delta c_2 \quad .$$

By taking the difference in optical density, D, between the inclusion and the immediately adjacent region, the change in c_2 can be determined. This method has been used to find the salt concentration of small ducts in thin biological specimens, for example.

Resolution. The quality of a contact microradiograph is determined by several factors(10): (1) geometric resolution, (2) blurring due to motion, (3) diffraction limitations, (4) mottling due to statistical fluctuations, and (5) grain size and noise limitation of the film. The effect of (2) and (3) can generally be neglected for most laser-plasma x-ray applications. In a "contact" microradiograph, the resolvable diameter based only on geometry is given by(11)

$$R_s = D_s \ell / r \quad ,$$

where ℓ is the distance between film and specimen (limited by the specimen thickness), and D_s and r the source diameter and distance.

Statistical fluctuations impose a limit on the resolvable diameter given by

$$R_s = 2.5 / C \sqrt{\pi n} \quad ,$$

where n is the fluence in photons/cm^2 on the film. Considering both geometry and statistics, we can chose a distance to yield an optimum resolution

$$R_{opt} = 4.0 \times 10^{-3} D_s^{1/3} c_2^{0.27} x^{0.60} \exp[0.29(c_2x)^{1/4}](\Delta c_2)^{-1/3}$$

as seen in Figure 6.

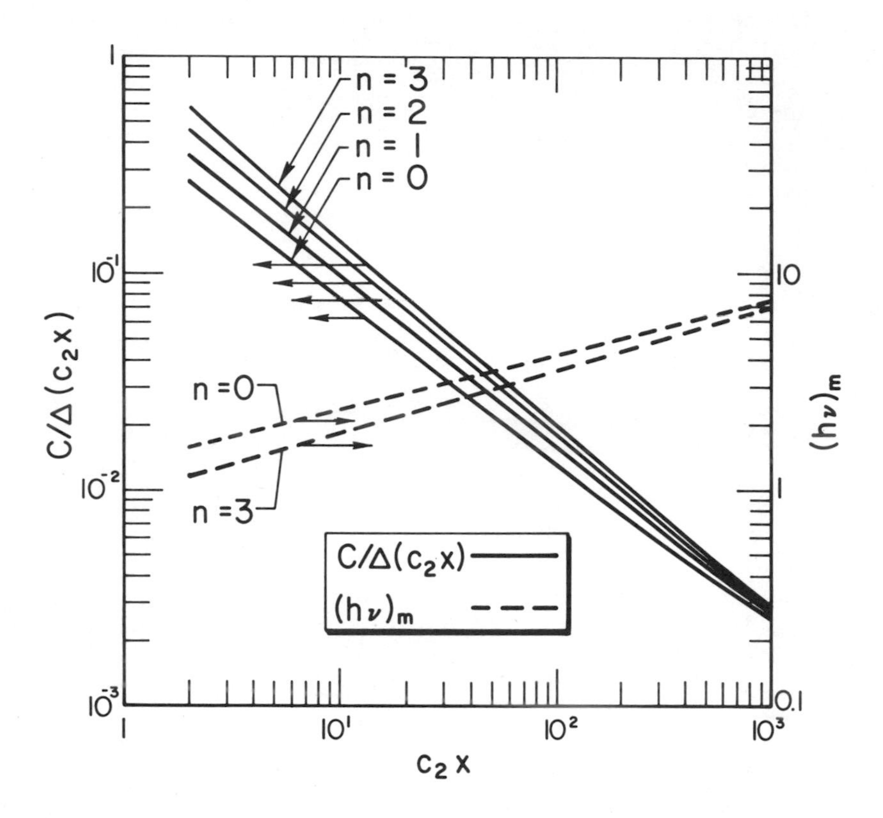

FIGURE 6. CONTRAST AND EFFECTIVE X-RAY ENERGY AS A FUNCTION OF THE ABSORPTION THICKNESS PARAMETER c_2x.

In most x-ray film, both the sensitivity and resolution are determined by silver halide grain size. An increase in grain size produces an increase in sensitivity and a decrease in resolution. The sensitivity is roughly proportional to the area of the grain while the resolving power is inversely proportional to the grain diameter. The limit on resolution imposed by the film grain is of the same order of magnitude as that imposed by statistical fluctuations. The resolution for maximum contrast, C = 1, and exposures to an optical density of ~1 are shown for several films in Table 1.

When the graininess of the film is taken into account, R_{opt} is increased by about an order of magnitude. Contrasts below about .01 are difficult to resolve because of the signal to noise ratios of typical films, but statistical averaging can help reduce this limitation.

TABLE 1. RESOLUTION AND SENSITIVITY FOR VARIOUS KODAK FILMS

Film Type	Resolution C = 1 (lines/mm)	Exposure D = 1 at 2 keV ($ergs/cm^2$)
NS 2T	25	0.1
SO 424	1250	100
RAR 2490 RAR 2497	160	2
649	4000	500
Type M	50	0.2

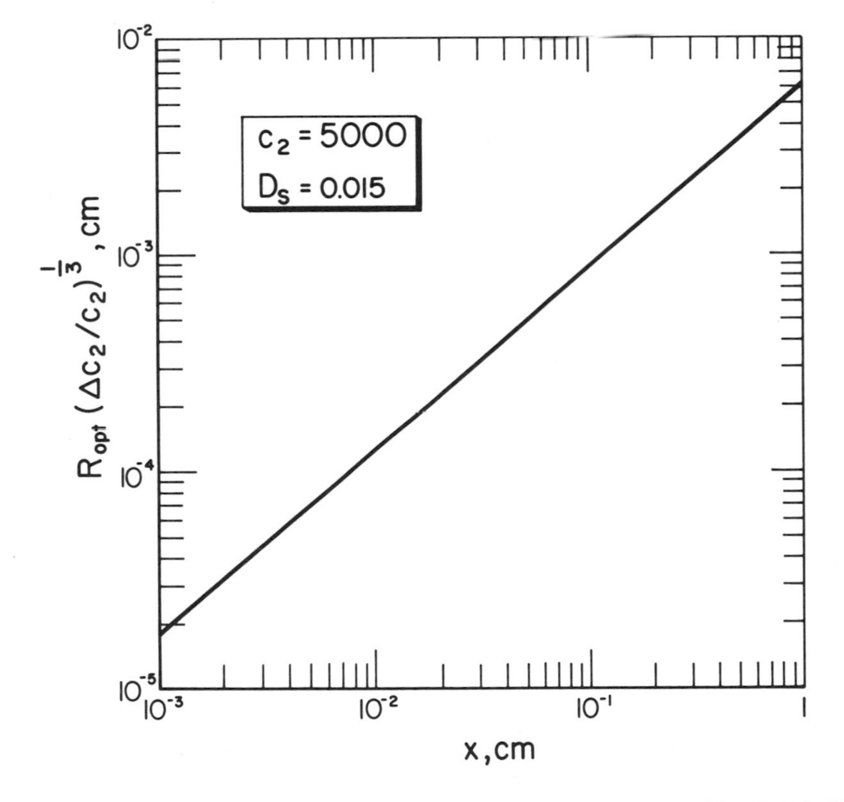

FIGURE 7. MINIMUM RESOLVABLE DIAMETER VS. THICKNESS OF WATER WITH PERFECT RECORDING MEDIA.

For x-ray energies at which the sensitive emulsion of a film becomes significantly transmissive, the films exhibit a strong spectral sensitivity. Thin emulsions will have a sensitivity given by

$$S(h\nu) \alpha (h\nu)^{-3} \quad ,$$

if no silver absorption edges appear in the $R(h\nu)$ peak. In case silver absorption edges are present or the film is in the transition region between thick and thin, $S(h\nu)$ can usually be adequately approximated by

$$S(h\nu) \alpha (h\nu)^{-n} \quad .$$

It can be seen from Figure 7 that $(h\nu)_m$ is not a strong function of the choice of n, and that the effect of n on contrast is not large. Thus, precise characterizations of the films is unnecessary.

ACKNOWLEDGMENTS

This paper was supported in part by AFOSR Grant No. AFOSR-78-3575, and by NSF Grant No. DAR-76-15224A01.

REFERENCES

1. K. O. Hodgson, H. Winick, and G. Chu, Eds. "Synchrotron Radiation Research", SSRP 76/100, Stanford University (1976).
2. H. I. Smith, and D. C. Flanders, Japanese J. Appl. Phys. 16, Suppl. 16-1, 61-65 (1977).
3. Stewart Austin, "Alignment of X-ray Lithography Masks Using a New Interferometric Technique-Experimental Results", to be published in the Journal of Vacuum Science and Technology as part of the proceedings of the 14th Symposium on Electron Ion and Photon Beam Technology.
4. P. J. Mallozzi, H. M. Epstein, R. G. Jung, D. C. Applebaum, B. P. Fairand, and W. J. Gallagher, J. Appl. Phys. 45, 1891 (1975).
5. D. J. Hamman, D. C. Applebaum, O. A. Ullrich, and P. J. Mallozzi, "Pulsed X-ray Lithography for High-Density Integrated Circuits", Battelle Columbus Laboratories Report (1977).
6. P. J. Mallozzi, and H. M. Epstein, "Diagnostic Techniques in Laser Fusion Research", R&D, 27:2-30-42 (1976).
7. P. J. Mallozzi, H. M. Epstein. R. G. Jung, D. C. Applebaum, B. P. Fairand, W. J. Gallagher, R. L. Uecker, and M. C. Muckerheide, "Laser-Generated Plasma as a Source of X-Rays For Medical Applications", J. Appl. Phys. 45:4, 1891 (1974).
8. P. J. Mallozzi, H. M. Epstein, R. G. Jung, D. C. Applebaum, B. P. Fairand, and W. J. Gallagher, "X-Ray Emission from Laser-Generated Plasmas", in Fundamental and Applied Laser Physics:

Proceedings of the Esfahan Symposium (edit. by M. S. Feld, A. Javan, and N. A. Kurnit) John Wiley and Sons, New York (1973).
9. R. F. Benjamin, P. B. Lyon, and R. H. Day, LA-UR-76-1502 (1976).
10. M. M. Ter-Pogassian, The Physical Aspects of Diagnostic Radiology, Harper and Row, New York, 195 (1967).
11. V. E. Cosslett, and W. C. Nixon, X-Ray Microscopy, Cambridge University Press, 19 (1960).

A REVIEW OF EMPIRICAL INFLUENCE COEFFICIENT METHODS IN X-RAY SPECTROMETRY.

Ron Jenkins

Philips Electronic Instruments

Mahwah, New Jersey 07430

INTRODUCTION and HISTORICAL DEVELOPMENT of EMPIRICAL CORRECTION MODELS.

In the X-ray fluorescence analysis of homogeneous specimens, the correlation between the characteristic line intensity of an analyte element and the concentration of that element, is typically non-linear over wide concentration ranges, due to interelement effects between the analyte element and other elements making up the specimen matrix. Although in many cases limiting the analyte concentration range may allow the use of linear calibration curves based on type-standardisation, it is usually more desirable to work with general purpose calibration schemes which are applicable to a variety of matrix types over wide concentration ranges.

Since the mid 1950's, empirical correction procedures have been used for the minimisation of these interelement interactions allowing the use of flexible algorithms for the determination of elemental concentrations. In principle, an empirical correction procedure can be described as the correction of an analyte element intensity for the influence of an interfering element, using the product of the interfering element intensity and a constant factor, as the correction term. This constant factor is typically referred to as an influence coefficient and is assumed to represent the influence of the interfering element on the analyte. In practice, several such terms may be used to correct the analyte intensity in multi-component matrices and the total correction is made by summing the individual terms. Commonly employed influence coefficient methods may use either the intensity or the concentration of the interfering element in the correction term. These methods are referred to as intensity correction methods or concentration correction methods.

$$W_a = R_a [W_a + K_a^b . W_b + K_a^c . W_c]$$

$$W_b = R_b [W_b + K_b^c . W_c + K_b^a . W_a]$$

$$W_c = R_c [W_c + K_c^a . W_a + K_c^b . W_b]$$

Figure 1 Early models for the correlation of X-ray intensity (Beattie & Brissey; Sherman and Marti)

Fig. 1 shows three equations for the determination of elements a, b and c in a ternary mixture. These equations are typical of the early empirical correction models (1) in which the R terms are characteristic line intensities and the influence coefficients are given by K_a^b, K_a^c, etc. For calibration purposes one uses a number of standards of known composition to determine the constant factors. There were several problems associated with the practical application of such models, not the least of which was the fact that concentration terms appear on both sides of the equations and calculation of the concentration values requires solution of a series of simultaneous equations. It will be remembered that these models were being proposed some time before the advent of the minicomputer hence the computation involved was not an insignificant problem.

In 1961 a more practical approach was proposed by Lucas-Tooth and Price (2), and this involved a series of linear equations of the form

$$W_{nm} = \alpha_n + I_{nm}\left(K_o + \sum_x K_{nm} I_{xm}\right)$$

In this expression, an intensity term Ixm is used for correction rather than concentration. With certain minor modifications, this intensity correction model is still used today.

In the succeeding 20 years or so, there have been many published attempts to refine or improve these earlier models and the advent of the minicomputer in the mid 1960's has done much to enhance the popularity of this method of quantitative analysis. There are a wide variety of empirical correction models employed today (3) and it is the purpose of this paper to review the current state of the art and pinpoint the advantages and disadvantages of the empirical correction approach.

Linear model:

$$W_i/R_i = K_i$$

Lachance-Traill:

$$W_i/R_i = K_i + \sum \alpha_{ij} \cdot W_j$$

Rasberry-Heinrich:

$$W_i/R_i = K_i + \sum \alpha_{ij} \cdot W_j + \sum \beta_{ik} \cdot W_k / 1 + W_i$$

Claisse-Quintin:

$$W_i/R_i = K_i + \sum \alpha_{ij} \cdot W_j + \sum \gamma_{ij} \cdot W_j^2$$

Figure 2 Forms of the "Alpha" correction models

Fig. 2 shows several of the more commonly employed empirical models. All of these models are concentration correction models, in other words, the product of the influence coefficient and the concentration of the interfering element is used to correct the slope Ki of the calibration curve for the analyte. In the Rasberry-Heinrich model, absorbing and enhancing elements are separated as α and β terms respectively, since these authors have shown that the enhancing effect cannot be adequately described by the same hyperbolic function as the absorbing effect. In the Claisse-Quintin model, higher order terms are added to allow for so-called "crossed effects", which would include enhancement and third element effects. In all of these models, one or more of three basic approaches is used to determine the influence coefficients. Intensity measurements are first made using a suitable series of well analysed standards. Multiple regression analysis techniques can then be used to give the best fit for slope, background and influence coefficient constants. Alternatively, the same data set can be used to graphically determine the individual influence coefficients. Thirdly, calculation of the influence coefficients can be performed using a fundamental type algorithm based on physical constants. In this latter approach, the only standards required are those needed to establish the sensitivity of the spectrometer for the various analyte elements.

Application of Empirical Coefficient Methods

In evaluating the applicability of these models, a useful starting point is to realise that although the influence coefficients are referred to as empirical constants, they do in fact have a

For a binary mixture ab: $W_a = R_a[W_a + K_a^b.W_b]$

since: $W_a + W_b = 1$

substitution gives: $W_a = R_a[1 + \alpha_{ab}.W_b]$

where: $\alpha_{ab} = K_a^b - 1$

Note also that $$\alpha_{ab} = \left[\frac{\mu_b(\lambda) + A.\mu_b(\lambda_a)}{\mu_a(\lambda) + A.\mu_a(\lambda_a)}\right] - 1$$

Figure 3

theoretical basis (4) and are calculable. As a simple example, Fig. 3 shows the case of absorption in a binary mixture ab. Here it will be seen that the α_{ab} term contains only primary and secondary absorption coefficients and a geometric constant A. In this example, λ_a is the analyte wavelength and λ the so-called "equivalent" wavelength of the exciting continuum (5). Although use of the full spectral distribution of the primary continuum, plus inclusion of possible enhancement terms, certainly complicates the expression, the α terms are still calculable and indeed programs are available (6) to perform such calculations.

If then these empirical correction terms are so well theoretically based, why is it that they sometimes fail to provide the correct solution in certain instances? There are, in fact, many reasons why such a situation can occur and following are six of the more commonly encountered problem areas:

1) Failure to adequately separate instrumental and matrix dependent effects.
2) Poor judgement on the part of the analyst as to whether or not a correction term should really be included.
3) Poor technique on the part of the analyst in the determination of the coefficients.
4) Poor quality and/or range of calibration standards.
5) Inadequacy of the regression analysis program used in the determination of the coefficients.

6) Application of the technique in cases where the specimens under analysis are insufficiently homogeneous.

It is the intention to now explore these six items in turn and to consider whether or not we are able to eliminate or, at least, minimize each of them.

1) Failure to adequately separate instrumental and matrix dependent effects

The intensity of the analyte line is subject not only to the influence of other matrix elements but also to random and systematic errors due to the spectrometer and the counting procedure employed. Provided that a sufficient number of counts is taken and provided that the spectrometer is reasonably well designed and calibrated, random errors from these sources are generally insignificant relative to other errors. Systematic errors of this type, however, are by no means insignificant and effects such as dead time, background, spectral line overlap, and so on, can all contribute to the total error in the measured characteristic line intensity. Unless these instrument dependent terms are completely separated from the matrix dependent terms, the instrumental effects will tend to become associated with the influence coefficients. This is obviously undesirable since completely different physical laws apply to the various cases. As an example, Fig. 4 shows some of the basic shapes of calibration curves subject to absorption and enhancement. The deviation from linearity in the lower curve is due to dead time effects and certainly does not have the same form as either of the other two. It is vital, therefore, to correct the analyte intensity for any instrumental artifact before attempts are made to evaluate or apply empirical correction procedures.

Fig. 5 shows an example of an empirical correction model in which matrix and instrumental dependent terms have been separated. In this instance, the sensitivity of the spectrometer for the analyte element is defined by the slope and background terms D and E. These constants are derived using intensities free of instrumental artifacts. Two additional practical benefits accrue by use of this procedure. First, since the influence coefficients are instrument independent, one has the capability to transport a set of coefficients from one spectrometer to another of similar type. This fact alone saves a tremendous duplication of effort. A second advantage is that curve updating can be readily applied to concentration correction models. Such an updating may be called for, for example, where an X-ray tube has to be replaced. In this instance, only the D and E constants need to be updated. One should also remember, however, that intensity correction models are at a disadvantage in this context since their influence coefficients are dependent upon spectrometer sensitivity.

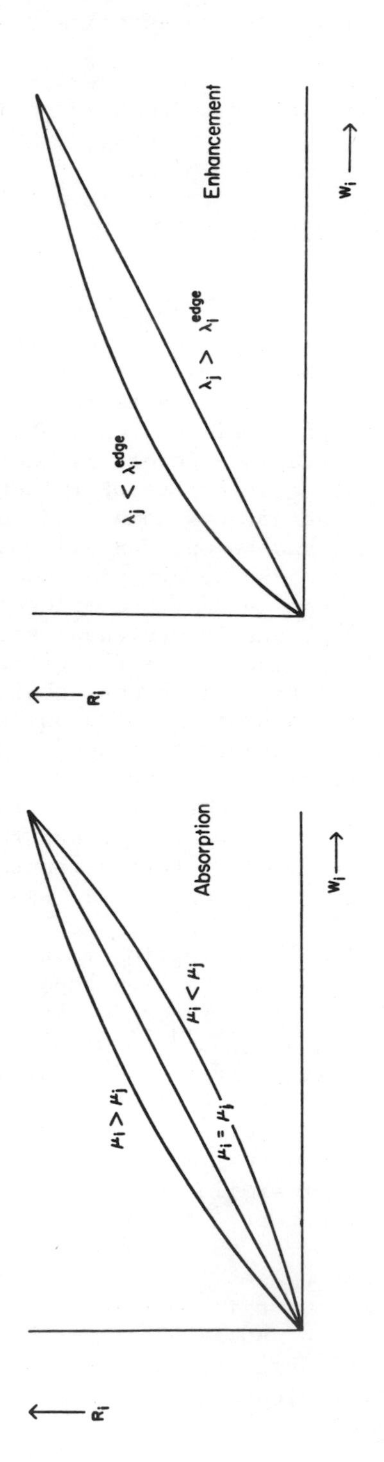

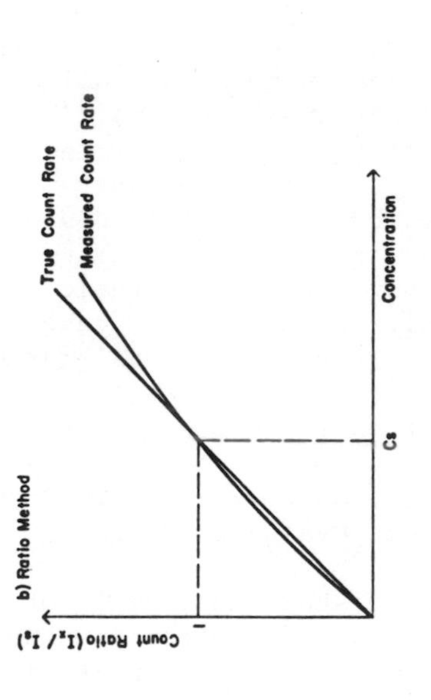

Figure 4. Basic Shapes of Intensity/Concentration Curves

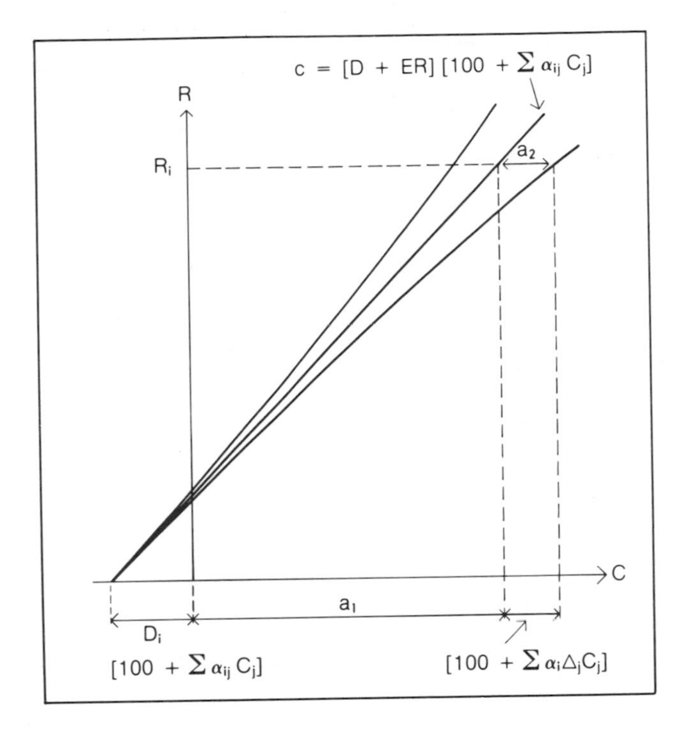

Figure 5

2) Poor judgement as to whether a correction term is really required.

One of the biggest complications in the application of empirical correction models is the difficulty in assessing whether or not a correction term should really be applied. This may be particularly troublesome for the less experienced analyst, particularly since the inclusion of more correction terms in a regression analysis scheme invariably improves the standard deviation of the fitted data. In multicomponent mixtures, one can be misled sometimes about the effect of one element upon another. For example, in the measurement of CuKα in Cu/Zn/Sn/Pb alloys if the tin and lead concentrations vary at the same rate in the same direction, a search for the influence of tin on CuKα may show an effect which is really due to lead.

We have found that graphical techniques are particularly useful in aiding the analyst to judge the potential effect of one element upon another. As an example, Figure 6 shows a graphical plot of the concentration to intensity ratio for iron against the concentration of chromium, in other words, the slope of the calibration curve of the analyte against the concentration of the interfering element. Here, there is a clear effect of chromium on

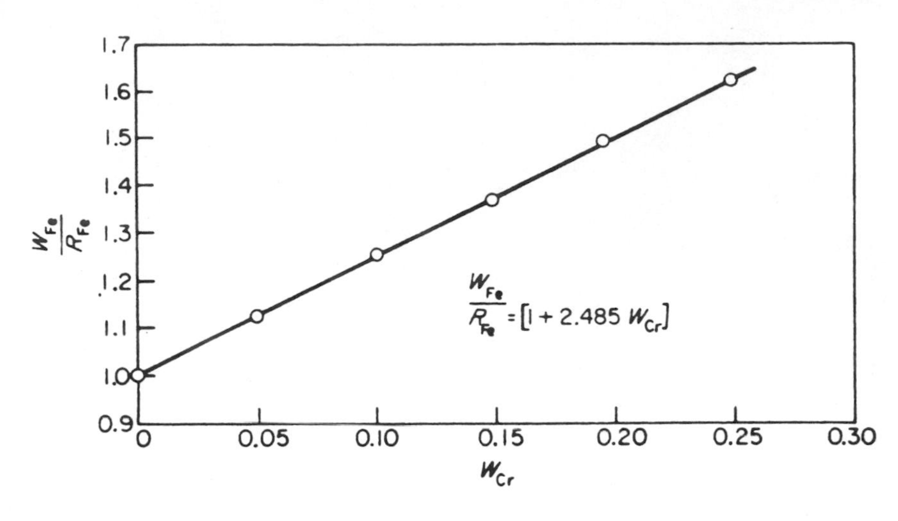

Figure 6

the iron and a correction term should be applied. In this special binary case, the slope of the line is exactly the influence coefficient and even in multicomponent systems, the slope is a good first approximation of the influence coefficient.

Further checks can be made after a set of influence coefficients have been determined. For example, if a matrix effect was due predominantly to absorption, there should be some correlation between the influence coefficient and the total absorption of the specimen.

As an example, Fig. 7 shows such a correlation for data obtained on a series of lubricating oils containing Ba, Zn, Ca, Cl, S and P. The correlation curves for barium, zinc and calcium are shown, there being five points on each curve representing the five interfering elements. An obvious correlation is seen indicating that the influence coefficients truly reflect the absorption situation.

3) Poor technique in the determination of the influence coefficients.

Even though the analyst is able to predict a potential interference, poor technique in the determination of the influence coefficient can lead to incorrect values. In our own laboratory, we use three basic rules in applying regression techniques to determine influence coefficients:

a) Is there a sufficient number of degrees of freedom in our data set, in other words, do we really have enough standards to give credence to our data?

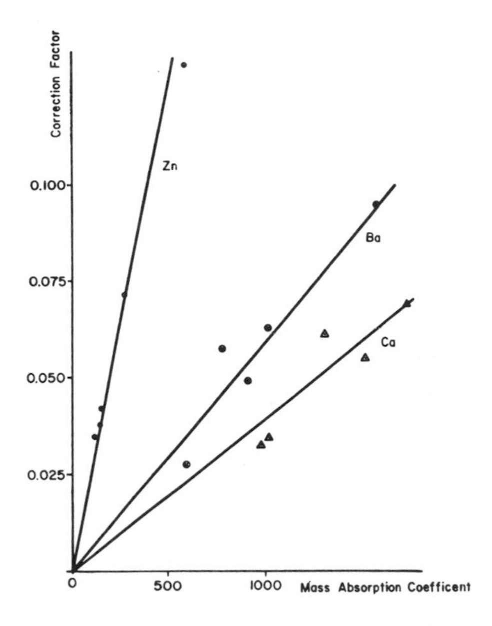

Figure 7

b) Did the application of the correction term really do something significant to the fit? In other words, is the standard deviation of the corrected data significantly better than the standard deviation of the uncorrected data?

c) Is the quality factor K of our data within reasonable limits? ie, in

$$\sigma = K\sqrt{\bar{c} + 0.1}$$

where $\bar{c}$ is the average concentration level is $0.03 > K > 0.005$. If the answer to any one of these three questions is no, then there is almost certainly something wrong and a re-evaluation should be made.

4) <u>Poor quality and/or range of standards</u>.

Part of the technique of the analyst involves the correct selection of a range of standards. In the example already cited, the influence of tin on CuKα was masked by the effect of lead. This problem is unlikely to have arisen if the range of standards truly reflected the range of element concentrations actually covered. The analyst himself must make the judgement as to which standards are to be employed. He must also judge whether the quality of the standards is sufficient for the analytical accuracy sought. Inaccurate standards lead not only to a poor correlation being obtained between corrected X-ray data and assumed composition, but what is worse, can lead to incorrect judgement as to whether a matrix effect is genuine or not. What might appear as a significant influence of one element on another could simply be due to a systematic bias in the reported concentration of the analyte relative to the interferer. Concentration levels are almost by definition interdependent and coupling of concentration levels can confuse the issue.

Another common source of error is to attempt to apply the correction model beyond the range for which it was calibrated. The data are usually fit to an average line defined by the regression analysis programs as best fitting the given data set. Extrapolation of the applied concentration range beyond the calibrated limits may well require a new average line.

5) Inadequacies of the regression analysis program.

Many of the regression analysis programs used with the earlier computer controlled spectrometers were rather restricted in their scope, mainly because of the limited computer core space available. Add to this the fact that the analysts using these programs were not particularly mathematically oriented and the result was that too much credence was often given to rather suspect data. More recently, the advent of sophisticated low-cost computer systems has provided the analyst with better programs (7) which generally include some statistical analysis routines to aid in the evaluation of data.

Our experience indicates that reasonably up-to-date regression analysis programs do a good job of data fiting in most cases, with perhaps two exceptions. Firstly, the case of a binary or psuedo binary systems where a self absorption correction is being made for the analyte element. In this instance, the problems may be mathematically overdefined in that three equations result from a binary AB: an equation for A in terms of B, an equation for B in terms of A, plus the fact that A + B = 100%. This may result in several solutions to the fitting of the data set and although the fits may be mathematically correct, they may not obey the laws of X-ray physics (8).

A second and similar problem area can occur where all matrix elements are being determined. Again, in this instance, the matrix may be overdefined, leading to several minima in the fitting algorithm. These problems are more important in determining the influence coefficients than in applying them. When fitting standard concentrations and intensity data to determine influence coefficients, the concentration terms on left and right hand side of the equations are not exactly the same. The "true" concentrations are those on the left hand side, whereas those on the right hand side are derived during the fitting process. The regression analysis programs generally work line by line, in other words, determining the constants first for A, then for B, and so on, iterating this process until some predefined fitting criterion is met. In applying the influence coefficients, the concentration terms on left and right hand sides are identically equal.

6) Application of the correction models to non-homogeneous specimens

As has been indicated earlier, the influence coefficients do have a theoretical significance based on well-defined physical processes. The mathematical expressions which define these processes break down, however, where the specimen is not homogeneous to the depth of penetration of the characteristic analyte photons. In such a case, fundamental type correction procedures will tend to break down and empirical correction procedures will become more empirical and less reliable. The problem is most pronounced in the analysis of powders, particularly when the absorption range is large and low atomic number elements are being determined. Our experience indicates that accuracies obtainable for low Z elements in pressed powder specimens are 1.5 to 2 times worse than accuracies obtained for similar elements in fused beads. One must be careful, therefore, not to expect _too_ much in terms of quality of corrected data on powder specimens. Quality factors quoted earlier of 0.005 to 0.03 may deteriorate to perhaps 0.01 to 0.05 for pressed powders.

Although there have been attempts to provide empirical correction procedures for heterogeneous specimens (9,10), these methods are limited in their range of application and my personal opinion is that the only solution lies in adequate specimen preparation.

Conclusions

The major advantage to be gained from use of a well-defined empirical correction procedure is that wide concentration range analysis of multi-component mixtures can be applied with a relatively small computer. The major disadvantage is that a large number of standards may be required for the initial determination of the influence coefficients. This disadvantage can, however, be offset to a certain extent by use of a more sophisticated program to calculate certain influence coefficients, allowing a combination of calculated and experimental coefficients to be used (11). Adequate separation of instrumental and matrix dependent terms not only allows the calculation of influence coefficients, but also allows use of a set of coefficients determined on one spectrometer, to be used on a different spectrometer of similar type.

The major drawbacks in the application of the empirical correction procedures invariably stem from poor judgement on the part of the analyst as to whether a given correction term is required or not, and to poor selection of standards from which the influence coefficients are to be determined. The solution here is to aid the decision making process by provision of easily understood statistical data analysis.

The empirical correction technique has served the X-ray spectroscopist well for the past 20 years or so, even though we have frequently abused the method by poor technique. In my opinion, these methods will continue to be useful for many years to come, particularly where more sophisticated data evaluation techniques are employed.

REFERENCES

1. Beattie, M. J. and Brissey, R. M., Anal. Chem., 26 (1954) 980.
2. Lucas-Tooth, J. and Price, R. J., Metallurgica, 54 (1961) 149.
3. Jenkins, R., Adv. in X-Ray Anal., 20 (1976) 1.
4. Jenkins, R., "An Introduction to X-Ray Spectrometry" 2nd Ed., Heyden:London (1974) p. 132.
5. Kalman, Z. M. and Heller, L., Anal. Chem. 34 (1962) 946.
6. de Jongh, W. K., X-Ray Spectrom. 2 (1973) 151.
7. Schreiner, W. and Jenkins, R., Adv. in X-Ray Anal. 22 (1978).
8. Jenkins, R. and Schreiner, W., X-Ray Spectrom., In Press.
9. Rhodes, J. R. and Hunter, C. B., X-Ray Spectrom., 1 (1972) 113.
10. Criss, J., Anal. Chem., 48 (1976) 179.
11. Jenkins, R. *et al.*, Adv. in X-Ray Anal. 18 (1974) 372.

QUASI-FUNDAMENTAL CORRECTION METHODS USING BROADBAND X-RAY EXCITATION

A. P. Quinn

Corning Glass Works

Corning, New York 14830

ABSTRACT

A critical discussion of the equivalent wavelength representation of polychromatic primary radiation as applied to the fundamental parameters method is given. This representation raises problems with appropriate selection of equivalent wavelengths and with accurate calculation of secondary fluorescence. Methods for reducing these difficulties are discussed and have been incorporated into a mini-computer program which achieves reasonable accuracy for alloy test cases.

INTRODUCTION

In quantitative x-ray spectrometry, fundamental parameters methods have been employed to compensate for matrix effects because very few standards are required and unknown specimens of widely different composition may be analyzed accurately. The first fundamental parameters methods required not only the appropriate "fundamental parameters," such as absorption coefficients and fluorescent quantum yields, but also a quantitative description of the primary exciting radiation, commonly polychromatic or broadband radiation from x-ray tube (1,2). Stephenson (3) applied the concept of an equivalent wavelength (λ_e) to the fundamental parameters approach, in which monochromatic primary excitation at a given wavelength was postulated which would yield elemental excitation equivalent to that actually caused by the polychromatic x-ray tube output. This assumption became popular because it obviated the need for measurement of tube spectra and, secondarily, reduced the complexity of calculation such that this method has been implemented with relatively small computers (4,5). On the other hand, it will be shown that the representation of polychromatic excitation by an equivalent wavelength may degrade analytical accuracy if the equivalent wavelengths are inappropriately chosen or if secondary fluorescence, i.e. enhancement, is a large effect. Methods are presented to minimize these two difficulties, however. Analyses of alloys are given as illustration, since both difficulties arise therein.

APPROPRIATE SELECTION OF EQUIVALENT WAVELENGTHS

As first postulated, the selection of an equivalent wavelength sought to describe the primary excitation source as a monochromatic source which, at some intensity, would have produced atomic excitation of an element in the specimen 'equivalent' to that actually caused by the polychromatic output of an x-ray tube. From the above conception, one would expect a unique elemental equivalent wavelength for a given x-ray tube, and two rules of thumb have been proposed for its selection. An element's equivalent wavelength was to be taken as an energy just above (0.3KeV) the appropriate absorption edge or, alternatively, two-thirds of the absorption edge wavelength. When primary excitation via an x-ray tube characteristic line predomininated over excitation due to Bremsstrahlung, that tube line might be selected as the equivalent wavelength. These three selections of equivalent wavelengths were tested for an Fe-Ni binary alloy using Stephenson's quasi-fundamental parameters method, but with rather poor resultant accuracies (Table I).

The above selections of equivalent wavelengths failed because they ignored attenuation of the primary radiation by the specimen. For certain specimen compositions, attenuation by elements other than the analyte may come to dominate absorption of the analyte's

Table 1. Calculated Compositions of 30.36 wt.% Ni/69.64 wt.% Fe Alloy Using Various Equivalent Wavelengths

	Edge +0.3KeV	2/3 λEdge	Rh K-Alpha	Computed
Fe λ_e(Å)	1.673	1.162	0.614	0.756
Ni λ_e(Å)	1.436	0.992	0.614	0.756
Fe wt.%	80.02	86.04	63.92	69.75
Ni wt.%	33.22	31.78	29.14	30.02
Total	113.24	117.82	93.06	99.78

Rh tube, 57KV, STD: Fe = 73.61, Ni = 8.14, Cr = 18.25

Table II. Iteration of Equivalent Wavelengths for 10.05 wt.% Ni/ 89.95 wt.% Fe Alloy. Computed Compositions for Selected Equivalent Wavelengths. Pure Element Standards.

Iteration	1	3	5	7
Fe λ_e (Å)	0.614	0.680	0.725	0.750
Ni λ_e (Å)	0.614	0.680	0.725	0.750
Fe wt.%	90.52	90.50	90.48	90.48
Ni wt.%	12.10	10.92	10.19	9.81
Total	102.62	101.41	100.67	100.29

equivalent wavelength. In such cases, a different analyte equivalent wavelength should be selected such that analyte excitation predominates over matrix attenuation. Thus appropriate selection of equivalent wavelengths is a function of specimen composition. Since x-ray analytical methods are ratio methods, in general, both the composition of the unknown and of the standard should be considered, such that an equivalent wavelength be chosen which produces analyte excitation in the unknown which is 'equivalent' to that produced in the standard (1,6). Since rigorous selection of equivalent wavelengths would require the composition of the unknown, among other parameters, requiring a 100% analysis total provides a more reasonable criterion for selection. For this work, initial equivalent wavelengths were chosen from 1) the Bremsstrahlung intensity maximum, 2) two-thirds the analyte edge (Å) or 3) an x-ray tube characteristic line, selecting that closest to the analyte edge. A composition for the unknown was then computed and the analysis total found. If that total was greater than 100.3% or less than 99.7%, the equivalent wavelengths were altered, with the step size (in Å) being a function of the difference in absorption for the equivalent wavelength between the standard and the unknown. This process of iterating equivalent wavelengths is illustrated in Table II for an Fe-Ni alloy where seven passes through the computation were required to achieve an acceptable analysis total. This approach gave reasonable accuracy when applied to the aforementioned Fe-Ni alloy in Table I. Results for a series of Fe-Ni alloys illustrate that the nickel equivalent wavelength required for this method changes with the composition of the unknown (Table III).

SECONDARY EXCITATION (ENHANCEMENT)

One may compute the analyte intensities expected for a given composition employing the following equation:

Table III. Optimal Nickel Equivalent Wavelengths Using Pure Element Standards

Known Ni wt.%	Calc. Ni wt.%	Optimal Ni λ_e(Å)
10.05	9.81	0.749
15.16	14.89	0.712
20.09	19.29	0.729
24.64	24.10	0.710
30.36	29.80	0.708

$$I_i^{calc} = A_i I_o^{ie} + \sum_j B_{ji}\ I_o^{je} \qquad (1)*$$

This formulation is similar to that of Stephenson, where the intensity of each element's equivalent wavelength, I_o^{ie}, is determined using measured analyte intensities for standards. Equation (1), however, is often unrealistic since it predicts intensities at the sample surface, rather than at the detector. Since instrumental detection efficiencies may vary widely for the analyzed elemental characteristic lines, especially in wavelength dispersive spectrometers, one must include an instrument sensitivity factor, g_i, which yields equation (2).

$$I_i^{calc} = g_i(A_i I_o^{ie} + \sum_j B_{ji}\ I_o^{je}) \qquad (2)$$

There are now twice the number of unknowns, i.e. either too many unknowns or too few equations for solution. There are at least three approaches which allow rigorous solution of equation (2). First, one may measure the instrument sensitivity, g_i, experimentally. This is essentially the approach of Harmon, et al. (7), who measured 'calibration constants' for their energy dispersive spectrometer, though determination of such constants for wavelength dispersive spectrometers is more complicated than for energy dispersive spectrometers due to differences in crystal reflectivities, secondary collimator transmissions, etc. Second, one can employ two standards per element, or third, measure two lines for each analyte edge (such as K-alpha and K-beta) to double the number of equations.

For analyses where the above requirements are too confining, it is proposed that the primary intensity exciting the fluorescing element (I_o^{je}) may be nearly equal to the primary intensity exciting the fluoresced element (I_o^{ie}), since the absorption edge of the

*See appendix for definition of terms.

enhancing element (j) will not be of greatly higher energy than that of the analyte element (i). Thus, the intensities of the primary radiation at these two equivalent wavelengths may be considered to be equal for computation of secondary fluorescence. The assumption that I_o^{je} equals I_o^{ie} for enhancement yields

$$I_i^{calc} = g_i \ (A_i I_o^{ie} + \sum_j B_{ji} I_o^{ie}) \qquad (3)$$

which may be rearranged to give

$$I_i^{calc} = A_i \ (g_i I_o^{ie}) + \sum_j B_{ji} (g_i I_o^{ie}) \qquad (4)$$

This form allows the product $g_i I_o^{ie}$ to be treated as a single variable, and therefore there are as many variables as equations.

It is noteworthy that for specimens devoid of secondary fluorescence, no difficulty exists with equation (1), since then $g_i I_o^{ie}$ is treated as a single variable. The results presented in Table IV are for Fe-Ni binaries with appreciable secondary fluorescence and illustrate the preceding contentions. First, the assumption of equal instrument sensitivities, i.e. $g_j = g_i = 1$ or equation (1), yields accurate results when the same secondary collimator is used for both Ni and Fe measurement (column 1), yet fails when a different collimator is employed for Ni than for Fe measurement (column 2). Second, the assumption that the primary excitation intensity at the fluorescing element's equivalent wavelength is equal to that at the fluoresced element's equivalent wavelength, i.e. $I_o^{je} = I_o^{ie}$ or equation (4), provides reasonable accuracies independent of detection sensitivities, e.g. choice of secondary collimators.

Table IV. Analysis of Fe-Ni Binaries Using Pure Elements as Standards

Mode	$g_j = g_i = 1$		$I_o^{je} = I_o^{ie}$	
Fe coll. Ni coll.	0.075° 0.075°	0.4° 0.075°	0.075° 0.075°	0.4° 0.075°
Ni (wt.%) Known	Ni (wt.%) Calc.	Ni (wt.%) Calc.	Ni (wt.%) Calc.	Ni (wt.%) Calc.
10.05	9.81	16.86	11.17	11.17
15.16	14.89	22.73	16.18	16.18
20.09	19.29	28.22	20.93	20.93
24.64	24.14	32.61	26.23	26.23
30.36	29.83	37.57	32.39	32.39

COMPUTER PROGRAM

A FORTRAN computer program, CORRAL, was written to explore the equivalent wavelength concept as applied to the fundamental parameters method, wherein the analyst may select either the $g_j = g_i = 1$ or the $I_o^{je} = I_o^{ie}$ assumption for computation of enhancement, the former mode being essentially that of Stephenson's CORSET. Also, the analyst may enable or disable iteration of equivalent wavelengths. The program requires about 13K words of memory in a Digital Equipment Corp. PDP-11 computer under the RT-11 F/B monitor.

RESULTS AND DISCUSSION

Suitable alloy test cases have been published. Analyses of Fe-Cr binaries using pure element standards are given in Table V. employing the method of iterating equivalent wavelengths. No estimate of detection sensitivities can be inferred from the data published by Rasberry and Heinrich (8), and it is not surprising that results obtained with the $I_o^{je} = I_o^{ie}$ assumption for computation of enhancement are more accurate (av. deviation 14% rel.) than those results obtained with the $g_j = g_i = 1$ assumption (av. deviation 29% rel.).

The alloy analyses published by Criss, et al. (9), were processed with CORRAL. Results obtained (Table VI) were virtually as accurate as those obtained with NRLXRF, a full fundamental parameters method which treats the polychromatic primary excitation explicitly. It is expected that there are inappropriate intensities in the data set, especially for silicon, and the accuracy of either NRLXRF or CORRAL was limited thereby.

Table V. Analysis of Fe-Cr Binaries Using Pure Element Standards (Ref. 8)

Alloy	Known Cr(wt.%)	$I_o^{je} = I_o^{ie}$ CORRAL Cr(wt.%)	$g_j = g_i = 1$ CORSET Cr(wt.%)
4061	3.53	4.26	5.00
4062	6.08	7.14	8.36
4065	12.14	13.88	15.96
4173	19.00	20.80	23.61
4181	25.03	28.05	31.26
4183	31.94	35.54	38.63
4184	36.58	40.27	43.45

Table VI. Analysis of NBS Alloys (wt.%), Ref. 9

	1154	1155			1159			1160		
	Reference	NBS	NRLXRF	CORRAL	NBS	NRLXRF	CORRAL	NBS	NRLXRF	CORRAL
Si	1.09	0.50	0.83	0.84	0.32	1.59	1.56	0.37	0.92	0.90
P	0.038	0.020	0.038	0.037	0.003	0.003	0.003	0.003	0.036	0.035
S	0.033	0.018	0.041	0.040	0.003	0.032	0.031	0.001	0.016	0.015
Cr	19.58	18.45	18.81	18.66	0.06	0.43	0.45	0.05	0.43	0.48
Mn	1.74	1.63	1.64	1.62	0.30	0.27	0.31	0.55	0.52	0.58
Fe	65.09	64.45	64.21	63.96	51.0	49.6	50.6	14.3	13.4	15.0
Ni	10.25	12.18	12.32	12.13	48.2	47.9	47.2	80.3	80.0	78.9
Cu	0.56	0.17	0.17	0.17	0.038	0.096	0.087	0.021	0.093	0.072
Mo	0.46	2.38	2.38	2.36	0.01	0.009	0.009	4.35	4.40	4.15
Total	98.84	99.91	99.93	99.82	99.9	99.8	100.2	99.9	99.9	100.2

CONCLUSIONS

Two difficulties arising from the equivalent wavelength representation of the polychromatic primary excitation of an x-ray tube are identified as 1) existing methods for selecting an energy for an elemental equivalent wavelength are often inadequate, and 2) previous calculations of secondary fluorescence are invalid. A computer program (CORRAL) is presented which calculated appropriate equivalent wavelengths assuming only that the analysis total should be 100%, and which offers improvement in the calculation of secondary fluorescence by assuming that the intensity of the equivalent wavelength exciting the fluorescing element is equal to the intensity of the equivalent wavelength exciting the fluoresced element. Results for alloy test cases using the CORRAL quasi-fundamental parameters method gave accuracies similar to that of the NRLXRF fundamental parameters method which considers the full polychromatic primary excitation.

REFERENCES

1. T. Shiraiwa and N. Fujino, "Theoretical Calculation of Fluorescent X-Ray Intensities in Fluorescent X-Ray Spectrochemical Analysis," Jap. J. Appl. Phys. 5, 886 (1966).

2. J. W. Criss and L. S. Birks, "Calculation Methods for Fluorescent X-Ray Spectrometry, Empirical Coefficients vs. Fundamental Parameters," Anal. Chem. 40, 1080 (1968).

3. D. A. Stephenson, "Theoretical Analysis of Quantitative X-Ray Emission Data: Glasses, Rocks, and Metals," Anal. Chem. 43, 1761 (1971).

4. R. B. Shen and J. C. Russ, "A Simplified Fundamental Parameters Method for Quantitative Energy-dispersive X-Ray Analysis," X-Ray Spectrom. 6, 56 (1977).

5. M. F. Ciccarelli, "QUAN - A Computer Program for Quantitative X-Ray Fluorescence Analysis," Anal. Chem. 49, 345 (1977).

6. R. Tertian, "Quantitative X-Ray Fluorescence Analysis Using Solid Solution Specimens - A Theoretical Study of the Influence of the Quality of Primary Radiation," Spectrochim. Acta 26B, 71 (1971).

7. J. C. Harmon, G. E. Wyld, T. C. Yao, and J. W. Otvos, "X-Ray Fluorescence Analysis of Stainless Steels and Low Alloy Steels Using Secondary Targets and the EXACT Program," presented at the Denver Conference on Applications of X-Ray Analysis, Denver, Colo., Aug. 4, 1978.

8. S. D. Rasberry and K. F. J. Heinrich, "Calibration for Inter-element Effects in X-Ray Fluorescence Analysis," Anal. Chem. 46, 81 (1974).

9. J. W. Criss, L. S. Birks and J. V. Gilfrich, "A Versatile X-Ray Analysis Program Combining Fundamental Parameters and Empirical Coefficients," Anal. Chem. 50, 33 (1978).

APPENDIX

Definition of Terms for Equations (1) through (4)

$$A_i = \frac{1}{\sec\theta_1 \mu_s^{ie} + \sec\theta_2 \mu_s^{ia}} \cdot \sec\theta_1 \mu_i^{ie} W_i \frac{K_i - 1}{K_i} \omega_i R_i$$

$$B_{ji} = \frac{1}{2} \frac{1}{\sec\theta_1 \mu_s^{ie} + \sec\theta_2 \mu_s^{ia}} \cdot \sum_{\ell} D_{j\ell} \sec\theta_1 \mu_j^{je} W_j \frac{K_j - 1}{K_j} \omega_j R_j \cdot$$

$$\mu_i^{j\ell} W_i \frac{K_i - 1}{K_i} \omega_i R_i \cdot \left[\frac{1}{\sec\theta_1 \mu_s^{je}} \ln\left(1 + \frac{\sec\theta_1 \mu_s^{je}}{\mu_s^{j\ell}}\right) + \right.$$

$$\left. \frac{1}{\sec\theta_2 \mu_s^{ia}} \ln\left(1 + \frac{\sec\theta_2 \mu_s^{ia}}{\mu_s^{j\ell}}\right) \right]$$

I_i^{calc} is the calculated x-ray intensity of the ith element,

I_o^{ie} is the intensity of the equivalent wavelength exciting element i,

θ_1 is the angle of incidence of the excitation energy, measured with respect to the sample normal,

θ_2 is the angle at which characteristic x-rays are observed, measured with respect to the sample normal,

μ_i^{ie} is the mass absorption coefficient of the ith element for radiation at i's equivalent wavelength,

μ_i^{ia} is the mass absorption coefficient of the ith element for the analyzed characteristic line of element i,

μ_i^{jl} is the mass absorption coefficient of the ith element for the lth line of the jth element,

μ_s^{jl} is the mass-fraction average absorption coefficient of the elements in the specimen for the lth line of the jth element, that is:

$$\mu_s^{jl} = \sum_{i=1}^{n} W_i \, \mu_i^{jl},$$

W_i is the weight fraction of element i,

K_i is the absorption edge jump ratio of element i,

ω_i is the fluorescent yield of element i,

R_i is the fraction of the ith element's analyzed characteristic x-rays in the analyzed line's series, e.g.,

$$K_\alpha \;/\; (K_\alpha + K_\beta),$$

D_{jl} **is unity when the lth line of the jth element is sufficiently energetic to excite element i to yield analyzed characteristic x-rays for element i, and zero in all other cases.**

PROGRESS IN X-RAY FLUORESCENCE CORRECTION METHODS USING SCATTERED RADIATION*

K. K. Nielson

Battelle Pacific Northwest Laboratory

Richland, Washington 99352

INTRODUCTION

This paper reviews x-ray fluorescence (XRF) matrix correction methods which use scattered radiation and discusses their typical analytical applications. It also discusses some of the underlying assumptions of these methods and thus provides a basis for estimating their merits and limitations in quantitative analytical applications.

The use of scattered radiation in XRF analysis provides an alternative to the common problem of matching samples to standards of similar composition. The backscatter peaks are sometimes treated as fluorescent peaks from internal standards since they suffer the same matrix absorption as fluorescent peaks and behave similarly with instrumental variations. They are better described as total matrix peaks, however, because the scattering originates from all elements in the sample matrix rather than from a single element. They thus provide the only direct spectral measure of the total or average matrix of biological, geological or other materials containing significant quantities of carbon, oxygen, or other light elements not observed by fluorescent peaks.

XRF correction methods using scatter peaks have been used under a wide variety of experimental conditions, including excitation by both discrete and continuum sources and detection by both wavelength- and energy-dispersive instruments. The scattered radiation is usually measured at the incoherent or coherent scatter

*This work supported by US DOE Contract EY-76-C-06-1830.

peaks from line excitation sources and at the intense high energy scatter region from continuum sources. The background continuum beneath or near the analyte peak has also been used, particularly in the $\lesssim$3 keV region. Analyses using these corrections can be broadly categorized as either relative or absolute, relying respectively on ratios of fluorescent/scatter intensities or on absolute intensity measurements.

RELATIVE INTENSITY METHODS

The use of a fluorescent/scatter intensity ratio to replace the fluorescent intensity in empirical calibration curves was first reported in 1958 by Andermann and Kemp (1). This relative intensity approach has been the most widely used correction method employing x-ray scattering, and may be subdivided according to four spectral bases for scatter measurement: 1) Combined incoherent and coherent scatter; 2) incoherent scatter; 3) coherent scatter; and 4) background continuum. The first and fourth may be used with either continuum or discrete line excitation, while the other two require discrete line sources which permit resolution of the incoherent and coherent components. Andermann and Kemp demonstrated the important features of fluorescent/scatter ratio methods, including partial compensation for varying x-ray tube voltage and current and for sample position, particle size, and matrix absorption. More recent use of these relative methods with energy dispersive detection has also obviated the need for accurate timing and dead time control since simultaneous intensity measurement eliminates the time parameter. Although independence from excitation intensity, timing, and sample positioning parameters is of great importance in many field and even laboratory applications, the extra degree of freedom afforded by carefully controlling these parameters yields potentially more information about the sample, as will be illustrated in the section on absolute intensity methods.

The use of the fluorescent/combined (incoherent plus coherent) scatter ratio was shown by Andermann and Kemp (1) to provide partial compensation for matrix absorption for vanadium in oil ($\sim$5-200 ppm V), nickel in iron ores (10-65% Fe), and lead in ores (1-40% Fe; 1-60% Zn). Similar applications by Lytle et al. (2) yielded improved matrix compensation for Mn, Fe, Co, Ni, Cu, Zn, Ga, Pb, and Mo in plant ash, but inferior compensation compared to a Ce internal standard for Sn, Sc, Ba, V, and Cr. Concentrations ranged over three orders of magnitude and relative systematic errors amounted to ±10%. Shenberg et al. (3) used the same intensity ratio for solutions (0.6-25% Cu; 0.4-23% Fe) and used solution standards to analyze solids containing up to 80% copper. However, a 3.2% decrease was noted in the Cu Kα/scatter ratio for each percent concentration of iron in solution. The same ratio was used by Clark and Baird (4) to develop an in-situ isotope-excited XRF system for geological analyses, by Nicholson and Hall (5) to analyze ∿1% potassium in submicrogram samples with 5-20% precision, and by Clark and Pyke (6) to determine uranium in powdered exploration samples with 4% relative precision. Shenberg and Amiel (7) also used the fluorescent/scatter ratio in place of analyte intensities in the Lachance-Traill method with a greater range of application.

When incoherent scatter can be explicitly measured, the fluorescent/incoherent ratio is the preferred parameter for empirical calibration curves. Boyd and Dryer (8) reported linear calibrations for sulfur (0.05-1.2%) and chlorine (0.2-1.2%) in oil using the WLα Compton peak. Burkhalter (9) obtained relative errors of 10% in analyzing silver (50-900 ppm) in a wide variety of ores using TeKα Compton scatter. The errors increased to 20% in heavy (lead) matrix ores where a slight over-compensation for the heavy matrix was observed. Shenberg and Amiel (10) reported a linear uranium (Lα) calibration over a 0.1-500 mg U/ml range of liquid concentrations, using I Kα Compton scatter for normalization. Clark and Mitchell (11) used a quadratic calibration function of fluorescent/incoherent scatter to determine copper (0.05-34%) in metallurgical samples, achieving ∿10% relative accuracies. Similar analyses for iron (1-56%) and silica (0.4-98%) were not as successful. Larson et al (12) achieved relative random and systematic errors of 1% and 2% for lead in gasoline (0.5-1300 ppm) using Pb Lα/Mo Kα (Compton) for quantitation. For samples with high lead concentrations, 90% dilution with isooctane was recommended. Hansel and Martell (13) analyzed powdered (-325 mesh) stream sediments for Ni, Cu, W, Pb, Bi, Nb, Ag, Cd, and Sn, normalizing to the Mo Kα Compton peak with 10% relative precisions.

The fluorescent/coherent scatter ratio was used by Stever et al (14) to analyze up to 1.4% Mo and W in liquids. They used the Pt Lα_1(coherent) line, and obtained relative random and systematic errors of 2.5% and 3.8%. Cullen (15) used the WLβ_1 coherent peak to normalize Cu and Ni analyte peaks (up to 0.5% Cu or Ni in refinery solutions) and reported ±1% relative accuracy. In a separate report (16), Cullen described analyses of copper alloy briquets for Zr, Te, Cr, and Ag (0.01-3%) using the WLγ_1 coherent scatter peak. Relative precisions of ±0.5-1.2% were obtained. Stulov (17) determined Zn, Pb, and Fe (<4%) in ore pulp slurries (5-30% solids) using the Re Lβ_1 coherent scatter peak and obtained relative random and systematic errors of 4% and 5.5-40%, depending on particle size effects.

The ratio of fluorescent/background intensity was used by Champion et al. (18) to determine strontium (40-4000 ppm) in widely varied sample matrices using aqueous $Sr(NO_3)_2$ standards and a "sample scattering factor" estimated from atomic and Compton scattering powers. Relative standard deviations were as low as 1% while relative accuracies were ∿5% for biological samples and 10% for rocks. Although samples as thin as 20% of infinite thickness for strontium (Kα) were successfully analyzed, such "intermediate thickness" samples are not generally analyzed quantitatively by fluorescent/background ratios when the analyte peak is further separated from the Compton peak. James (19) used a simple fluorescent/background ratio to calibrate uranium analyses (1-100 ppm) in thick, loose powdered (-200 mesh) rock samples with a relative uncertainty of 1-2%. Kalman and Heller (20), although technically relying on absolute intensities for graphical calibrations, also used the background intensity near the analyte line to adjust calibrations for Zn, Cu, Ni, Cr, and V (≤280 ppm) in various geological matrices. Relative accuracies of ∿1% were obtained, but enhancement and major absorption edges reduced accuracy in cases of high Fe_2O_3 content (>2.5%).

ABSOLUTE INTENSITY METHODS

Although absolute intensity measurements require careful control of sample geometry, excitation intensity, and other analytical parameters, they afford at least one and often two extra degrees of freedom which, if properly used, yield more information about the sample. Equations for using absolute scatter intensities are widely varied and, like the relative intensity methods, are predominantly empirical. A parametric classification of the absolute scatter intensity methods includes calculation of a) mass absorption coefficients, b) background or "true" background intensities from which mass absorption coefficients are then calculated, c) intensity correction factors, d) mass or thickness of intermediate-thickness samples, e) average atomic number, and f) representative light element concentrations for fundamental parameters matrix corrections. Clearly, most of these parameters are not fundamentally dependent on scatter intensity but may be empirically related to it under many of the conditions commonly employed in XRF analysis.

Explicit calculation of mass absorption coefficients, μ, from Compton intensities was first reported by Reynolds (21), who graphically determined μ as a linear function of the reciprocal Compton intensity. The precise (±2%) coefficients permitted determination of Ni, Rb, Sr, and Zr in various thick sample matrices using the μ-ratios at 0.9 Å,

$$C = \frac{I_f}{I_f'} \quad \frac{\mu_{0.9}}{\mu_{0.9}'} \quad C' \qquad (1)$$

where C is the analyte concentration, I_f is its fluorescent peak intensity, and the primes indicate the corresponding values for a standard. Reynolds (22) later extended the calibration for $\mu_{0.9}$ to sample matrices as heavy as CuO and adopted a curved (graphical) function for μ versus the reciprocal Compton intensity. A method to estimate $\mu_{1.9}$ (across the iron absorption edge) was also given for cases of known iron concentration. Walker (23) simplified Reynold's method of crossing absorption edges by simply plotting the μ ratio across the edge as a function of the major element peak intensity.

A recent report by Bazan and Bonner (24) shows a linear relation between μ and the incoherent/coherent intensity ratio for intermediate thickness samples. The coefficients of the calibration line varied somewhat with matrix, however. Franzini et al. (25) reported a logarithmic relation between μ and Compton scatter and achieved relative accuracies of ±2.5-3.2% for 43 coefficients in widely varied matrices. Their application (26) of the μ's, like that of Reynolds, also provided a means of crossing the iron absorption edge if the iron concentration were known and enhancement were negligible.

Feather and Willis (27) reported a simple linear relationship between $1/\mu$ and background intensity. A non-zero intercept ascribed to "residual background" was determined for several elements and was subtracted from the observed background intensity to yield the "true background." The net fluorescent/true background ratio was then related to concentration with a simple calibration line. An earlier report by Giauque et al. (28) gave a simple function to calculate background intensity (B) from coherent (I_c) and incoherent (I_i) scatter intensities,

$$B = k_0 I_c + k_1 \ (I_c + I_i), \tag{2}$$

where k_0 and k_1 define the calibration line. This equation was used for backgrounds in air filter analysis (29), but no extension to matrix corrections was reported. Giauque et al. (30) also reported using the "true background" method with appropriate absorption edge and enhancement corrections and obtained relative errors of $\leq$10% in briquetted geological samples.

Absorption corrections using scatter peak intensity ratios were reported in 1964 by Ryland (31) and evaluated by Carman (32) for Cl, P, Mn, Mo, and Te at 2-50% concentrations in polyethylene pellets. Concentrations were computed as

$$C = k_0 I_f \ (I_c/I_i), \tag{3}$$

where k_0 is a calibration constant. Relative uncertainties varied widely using this method, which only partially compensates for matrix absorption. Reed et al. (33) approached the difficult problem of solids analysis in ore slurry streams by using scatter intensity ratios. Analyte concentrations were computed as

$$C = k_o + I_f[k_1 + (k_2 + k_3\ I_c/I_{cw})^{-3}], \tag{4}$$

where I_c and I_{cw} are the respective coherent scatter intensities from the slurry and a water blank, and k_o - k_3 are empirical calibration constants. Reed et al. (34) later reported a simplified method for similar applications,

$$C = k_o + k_1\ [I_f(I_{cw}/I_c) - I_{fw}], \tag{5}$$

where I_{fw} is the analyte peak intensity in the water stream blank. Several related methods for slurry analysis were reported by Hietala and Viitanen (35), Gurvich et al. (36), and Parus et al. in these proceedings.

The calculation of sample thickness or mass as a linear function of the total (incoherent and coherent) scatter by Zietz (37) is applicable to small samples having similar bulk composition to the standards. Relative uncertainties of 2-4% were obtained for biological specimens using a separate proportional counter as the mass monitor. A similar approach by Ong et al. (38) also used a linear function of mass (2-13 mg/cm^2) versus Compton intensity. They later reported (39) relative uncertainties of several percent for a variety of biological tissues using Compton intensities and a correction factor which depended on the type of tissue. We have found in unpublished work at Battelle that this correction to sample mass (0-63 mg/cm^2) for varying sample atomic number ($\bar{Z}$)is conveniently found by plotting or fitting the mass per unit Compton intensity as a function of $\bar{Z}$ which, in turn, is easily determined from a quadratic function of the coherent/incoherent intensity ratio. Jaklevic et al. (40) also estimated mass (1-100 μg/mm) from Compton scatter in hair samples with 10% relative accuracy. Van Espen and Adams (41) estimated mass loading of air filters from a linear function of the incoherent/coherent scatter ratio, which in effect uses $\bar{Z}$ as the measure of mass loading. A related method for calculating air filter mass loadings and attenuation corrections was reported by Nielson and Garcia (42).

The variation in coherent/incoherent intensity ratios with $\bar{Z}$ has been known for over four decades (43), but has only recently been quantitatively used. Dwiggins (44) used this ratio in a simple linear function to measure the percent carbon and, by difference, hydrogen in various hydrocarbons with less than 1% relative error. Kunzendorf (45) showed that $\bar{Z}$ can be measured over a range of 4-82 using various excitation sources and a log-log plot of the incoherent/coherent ratio versus $\bar{Z}$. Fookes et al. (46) used Compton scatter from ^{241}Am (60 keV) to directly determine Fe and Pb in high-grade ores from their varying $\bar{Z}$.

The use of coherent and incoherent scatter to define the total sample matrix permits explicit fundamental parameter calculations for the various matrix effects (47). The sample is modeled as a composite of heavy elements which are observed by fluorescence, and light elements which are observed by the difference between the total scatter intensities and their heavy element contribution. Two "representative" light elements are chosen from the ratio of the light element fractions of the incoherent and coherent scatter, and their concentrations, C_1 and C_2, are then calculated from the simultaneous equations,

$$C_1\sigma_{c1} + C_2\sigma_{c2} = K_c I_c - \sum_k C_k \sigma_{ck}$$
$$C_1\sigma_{i1} + C_2\sigma_{i2} = K_i I_i - \sum_k C_k \sigma_{ik}, \qquad (6)$$

where σ_c and σ_i are the total coherent and incoherent scatter cross sections (48), k refers to the heavy elements, and the constants K_c and K_i account for excitation intensity and system geometry and are empirically determined from at least one standard of known composition. This inherently multielement method thus permits quantitative analysis of samples of varying intermediate thickness and bulk composition while requiring only a one-time calibration of the spectrometer scatter constants and thin-sample sensitivities. Relative accuracies are usually in the 2-5% range for homogeneous samples and up to $\sim$10% for briquetted particulate-cellulose mixtures (47).

DISCUSSION

Previous reviews of the application of scattering to XRF matrix corrections (49-51) deal with relative intensity methods and give three helpful guidelines for using these methods. First, incoherent scatter gives better matrix compensation than coherent scatter. Second, fluorescent/background ratios are not useful in all cases, particularly at low energies. Finally, current theory (the isolated atom model) does not suggest the existence of an optimum scatter energy for general use, although high energies (preferably the incoherent peak) are usually best in the absence of major absorption edges. The isolated atom model discussed by Taylor and Andermann (50) also produced some discrepancies with experimental results which they attributed to diffraction effects. Although diffraction may contribute to coherent scatter intensity, this has not been quantitatively reported under enough XRF analytical conditions to permit quantitative evaluation. Since the common relative intensity methods do not mathematically conform to the isolated atom model, the underlying assumptions of the relative methods could also cause the often observed "partial matrix compensation" or residual matrix effects. Table 1 lists some of the assumptions made by these methods and may indicate for which cases the relative methods and even certain absolute methods will be useful.

TABLE 1. Limiting Assumptions for Using Backscatter in XRF Matrix Corrections

	Assumption	Methods[a]
1.	The sample is either infinitely thick or of thickness equal to the standards.	F/I, F/C, F/(I + C), F/B, F/TB
2.	The ratio of μ's (sample/std) at the analyte line equals the ratio of μ/σ_s ratios (sample/std) at the scatter energy.[b]	F/I, F/C, F/(I + C), F/B
3.	Diffraction is negligible.	F/C, F/(I + C)
4.	Background at the analyte energy originated as such in the sample (i.e., no detector background).	F/B
5.	True and residual background intensities can be accurately partitioned.	F/TB

[a]Methods involve intensity ratios of analyte fluorescent line (F), incoherent scatter (I), coherent scatter (C), background (B), and true background (TB).

[b]μ = mass attenuation coefficient, σ_s = mass scattering coefficient.

Some of the absolute methods which rely on assumptions 1-3 in Table 1 include the simple methods for estimating μ and methods using scatter peak ratios. Other assumptions of the absolute methods are less obvious due to the wide variety of empirical relationships used without theoretical explanation. The resulting calculations of μ, $\bar{Z}$, and even sample mass are thus subject to the same uncertain applicability as the relative intensity methods.

Our fundamental parameters method based on the "representative light element" approximation (47) avoids the assumptions in Table 1, except number 3. This method is the first to explicitly apply the scattering equations of the isolated atom model and is thus unique in its ability to work with samples of widely varying thickness and composition. Since incoherent and coherent scatter are explicitly computed for observed analyte elements, the representative $\bar{Z}$ estimation is typically only done for the $5<Z<12$ range of light elements commonly encountered in environmental samples. Since the multielement calculation describes the entire sample matrix, corrections for enhancement and, if known, particle size, are also conveniently included.

Due to the wide variety of analytes, sample matrices, and excitation and detection methods reported with scatter-based matrix corrections, evaluation of potential precision and accuracy is difficult. The literature suggests relative uncertainties on the order of 1-10% are achievable if care is taken to ensure the suitability of the method to the analytical problem. Although the range of application of XRF standards may be greatly increased by scatter-based matrix corrections, such corrections may be limited by basic assumptions and should always be checked by analyzing suitable standard reference materials.

REFERENCES

1. G. Andermann and J. W. Kemp, "Scattered X-Rays as Internal Standards in X-Ray Emission Spectroscopy," Anal. Chem. 30, No. 8, 1306-1309 (1958).

2. F. W. Lytle, W. B. Dye and H. J. Seim, "Determination of Trace Elements in Plant Material by Fluorescent X-Ray Analysis," Advances in X-Ray Anal. 5, 433-446 (1962).

3. C. Shenberg, A. B. Haim and S. Amiel, "Accurate Determination of Copper in Mixtures and Ores by Radioisotope-Excited X-Ray Fluorescence Spectrometric Analysis Using Peak Ratios," Anal. Chem. 45, 1804-1808 (1973).

4. B. C. Clark and A. K. Baird, "Ultraminiature X-Ray Fluorescence Spectrometer for In-Situ Geochemical Analysis on Mars," Earth and Planetary Science Letters 19, 359-368 (1973).

5. W.A.P. Nicholson and T. A. Hall, "X-Ray Fluorescence Microanalysis of Thin Biological Specimens," J. of Physics 6, 1781-1784 (1973).

6. N. H. Clark and J. G. Pyke, "The Determination of Uranium in Exploration Samples by X-Ray Emission Spectrometry," Anal. Chim. Acta 58, 234-237 (1972).

7. C. Shenberg and Saadia Amiel, "Critical Evaluation of Correction Methods for Interelement Effects in X-Ray Fluorescence

Analysis Applied to Binary Mixtures," Anal. Chem. 46, No. 11, 1512-1516 (1974).

8. B. R. Boyd and H. T. Dryer, "Analysis of Nonmetallics by X-Ray Fluorescence Techniques," Develop. Appl. Spectrosc. 2, 335-349 (1963).

9. P. G. Burkhalter, "Radioisotopic X-Ray Analysis of Silver Ores Using Compton Scatter for Matrix Compensation," Anal. Chem. 43, 10-17 (1971).

10. C. Shenberg and S. Amiel, "Analytical Significance of Peaks and Peak Ratios in X-Ray Fluorescence Analysis Using a High Resolution Semiconductor Detector," Anal. Chem. 43, 1025-1030 (1971).

11. N. H. Clark and R. J. Mitchell, "Scattered Primary Radiation as an Internal Standard in X-Ray Emission Spectrometry: Use in the Analysis of Copper Metallurgical Products," X-Ray Spectrom. 2, 47-55 (1973).

12. J. A. Larson, M. A. Short, S. Bonfiglio and W. Allie, "An X-Ray Fluorescence Technique for the Analysis of Lead in Gasoline," X-Ray Spectrom. 3, 125-129 (1974).

13. J. M. Hansel and C. J. Martell, "Automated Energy-Dispersive X-Ray Determination of Trace Elements in Stream Sediments," LA-6869-MS, June 1977.

14. K. R. Stever, J. L. Johnson and H. H. Heady, "X-Ray Fluorescence Analysis of Tungsten-Molybdenum Metals and Electrolytes," Advan. X-Ray Anal. 4, 474-487 (1961).

15. T. J. Cullen, "Coherent Scattered Radiation Internal Standardization in X-ray Spectrometric Analysis of Solutions," Anal. Chem. 34, 812-814 (1962).

16. T. J. Cullen, "X-Ray Spectrometric Analysis of Alloyed Copper," Develop. Appl. Spectrosc. 3, 97-103 (1964).

17. B. A. Stulov, "Determination of the Change in the Chemical Composition of a Solid During X-Ray Spectral Fluorescence Analysis of Pulp," Ind. Lab. 36, 1184-1187 (1970).

18. K. P. Champion, J. C. Taylor and R. N. Whittem, "Rapid X-Ray Fluorescence Determination of Traces of Strontium in Samples of Biological and Geological Origin," Anal. Chem. 38, 109-112 (1966).

19. G. W. James, "Parts-per-Million Determinations of Uranium and Thorium in Geologic Samples by X-ray Spectrometry," Anal. Chem. 49, 967-969 (1977).

20. Z. H. Kalman and L. Heller, "Theoretical Study of X-Ray Fluorescent Determination of Traces of Heavy Elements in a Light

Matrix--Application to Rocks and Soils," Anal. Chem. 34, 946-951 (1962).

21. R. C. Reynolds, "Matrix Corrections in Trace Element Analysis by X-ray Fluorescence: Estimation of the Mass Absorption Coefficient by Compton Scattering," Am. Mineral. 48, 1133-1143 (1963).

22. R. C. Reynolds, "Estimation of Mass Absorption Coefficients by Compton Scattering: Improvements and Extensions of the Method," Am. Mineral. 52, 1493-1502 (1967).

23. D. Walker, "Behavior of X-ray Mass Absorption Coefficients Near Absorption Edges: Reynolds' Method Revisited," Am. Mineral. 58, 1069-1072 (1973).

24. F. Bazan and N. A. Bonner, "Absorption Corrections for X-ray Fluorescence Analysis of Environmental Samples," Advances in X-ray Anal. 19, 381-390 (1976).

25. M. Franzini, L. Leoni and M. Saitta, "Determination of the X-ray Mass Absorption Coefficient by Measurement of the Intensity of Ag Kα Compton Scattered Radiation," X-ray Spectrom. 5, 84-87 (1976).

26. L. Leoni and M. Saitta, "Matrix Effect Corrections by Ag Kα Compton Scattered Radiation in the Analysis of Rock Samples for Trace Elements," X-ray Spectrom. 6, 181-186 (1977).

27. C. E. Feather and J. P. Willis, "A Simple Method for Background and Matrix Correction of Spectral Peaks in Trace Element Determination by X-ray Fluorescence Spectrometry," X-ray Spectrom. 5, 41-48 (1976).

28. R. D. Giauque, F. S. Goulding, J. M. Jaklevic and R. H. Pehl, "Trace Element Determination with Semiconductor Detector X-ray Spectrometers," Anal. Chem. 45, 671-681 (1973).

29. P. Van Espen and F. Adams, "Evaluation of a Practical Background Calculation Method in X-ray Energy Analysis," X-ray Spectrom. 5, 123-128 (1976).

30. R. D. Giauque, R.B. Garrett and L. Y. Goda, "Energy Dispersive X-ray Fluorescence Spectrometry for Determination of Twenty-Six Trace and Two Major Elements in Geochemical Specimens," Anal. Chem. 49, 52-67 (1977).

31. A. L. Ryland, "A General Approach to the X-ray Spectroscopic Analysis of Samples of Low Atomic Number," 147th Natl. Meeting, ACS, Philadelphia, April, 1964.

32. C. J. Carman, "X-ray Fluorescent Determination of Major Constituents in Multielement Matrices by the Use of Coherent to Incoherent Scattering Ratios," Develop. Appl. Spectros. 5, 45-58 (1966).

33. D. J. Reed and A. H. Gillieson, "X-ray Fluorescence Applied to the On-Stream Analysis of Sulphide Ore Fractions," X-ray Spectrom. 1, 69-80 (1972).

34. D. J. Reed, J. L. Dalton and A. H. Gillieson, "The On-Stream Analysis of Hematite Ore Fractions Using Radioisotopes," X-ray Spectrom. 3, 15-20 (1974).

35. M. Hietala and J. Viitanen, "A Radioisotope On-Stream Analyzer for the Mining Industry," Advan. X-ray Anal. 21, 193-205 (1978).

36. Y. M. Gurvich, A. N. Mezhevich, R. I. Plotnikov, I. M. Rogachev and A. N. Smagunova, "Comparison of Calculation Methods for Ore Slurry Density in X-ray Spectrometric Analysis," Apparat. Meth. X-ray Anal. 14, 43-52 (1974).

37. L. Zeitz, "X-ray Emission Analysis in Biological Specimens", in K. M. Earle and A. J. Tousimis, Editors, X-ray and Electron Probe Analysis in Biomedical Research, p. 35-73 (1969).

38. P. S. Ong, E. L. Cheng, G. Sroka, "Use of Multiple Standards for Absorption Correction and Quantitation with Frieda," Advan. X-ray Anal. 17, 269-278 (1974).

39. H. L. Cox and P. S. Ong, "Sample Mass Determination Using Compton- and Total Scattered Excitation Radiation for Energy-Dispersive X-ray Fluorescent Analysis of Trace Elements in Soft Tissue Specimens," Med. Phys. 4, 99-108 (1977).

40. J. M. Jaklevic, W. R. French, T. W. Clarkson and M. R. Greenwood, "X-ray Fluorescence Analysis Applied to Small Samples," Advan. in X-ray Anal. 21, 171-185 (1978).

41. P. Van Espen and F. Adams, "Tube-Excited Energy-Dispersive X-ray Fluorescence Analysis--Part II. Energy-Dispersive X-ray Fluorescence Analysis of Particulate Material," Analytica Chim. Acta 75, 61-85 (1974).

42. K. K. Nielson and S. R. Garcia, "Use of X-ray Scattering in Absorption Corrections for X-ray Fluorescence Analysis of Aerosol Loaded Filters," Advan. in X-ray Anal. 20, 497-506 (1977).

43. A. H. Compton and S. K. Allison, "X-rays in Theory and Experiment," New York, Van Nostrand, Inc. (1935).

44. C. W. Dwiggins, Jr., "Quantitative Determination of Low Atomic Number Elements Using Intensity Ratio of Coherent to Incoherent Scattering of X-rays, Determination of Hydrogen and Carbon," Anal. Chem. 33, 67-70 (1961).

45. H. Kunzendorf, "Quick Determination of the Average Atomic Number $\bar{z}$ by X-ray Scattering," Nucl. Inst. Meth. 99, 611-612, (1972).

46. R. A. Fookes, V. L. Gravitis and J. S. Watt, "Determination of Iron in High-Grade Iron Ore and of Lead in Lead Concentrate by

Compton Scattering of 60 keV γ-rays from ^{241}Am," Anal. Chem. 47, 589-591 (1975).

47. K. K. Nielson, "Matrix Corrections for Energy Dispersive X-ray Fluorescence Analysis of Environmental Samples with Coherent/ Incoherent Scattered X-rays," Anal. Chem. 49, 641-648 (1977).

48. W. H. McMaster, N. K. Del Grande, J. H. Mallett and J. H. Hubbell, "Compilation of X-ray Cross Sections," UCRL-50174, Sec. II, Rev. 1 (1969).

49. D. L. Taylor and G. Andermann, "Evaluation of Soft and Hard Scattered X-rays as an Internal Standard for Light Element Analysis," Advan. X-ray Anal. 13, 80-93 (1970).

50. D. L. Taylor and G. Andermann, "Evaluation of an Isolated Atom Model in the Use of Scattered Radiation for Internal Standardization in X-ray Fluorescence Analysis," Anal. Chem. 43, 712-716 (1971).

51. D. L. Taylor and G. Andermann, "Quantitative Evaluation of Degree of Internal Standardization in X-ray Fluorescence Analysis Using Scattered X-rays," Appl. Spectros. 27, 352-355 (1973).

DEVELOPMENT OF THE DETECTOR RESPONSE FUNCTION APPROACH FOR THE LIBRARY LEAST-SQUARES ANALYSIS OF ENERGY-DISPERSIVE X-RAY FLUORESCENCE SPECTRA

L. Wielopolski and R. P. Gardner

Department of Nuclear Engineering
North Carolina State University
Raleigh, North Carolina 27650

ABSTRACT

A procedure to obtain analytical models for the elemental X-ray pulse-height distribution libraries necessary in the library least-squares analysis of energy-dispersive X-ray fluorescence spectra is outlined. This is accomplished by first obtaining the response function of Si(Li) detectors for incident photons in the energy range of interest. Subsequently this response function is used to generate the desired elemental library standards for use in the least-squares analysis of spectra, or it can be used directly within a least-squares computer program, thus eliminating the large amount of computer storage required for the standards.

INTRODUCTION

Computer techniques are being employed more and more by the XRF analyst to help automate the analysis of samples. In particular energy-dispersive XRF analyzers have been highly automated in the analysis of airborne particulates collected on filter papers. Here the major mathematical problem is in the determination of characteristic x-ray intensities. The three basic techniques used for this purpose include: (1) stripping, (2) least-squares with library spectra, and (3) least-squares with spectral models (1-4).

While the least-squares method is probably accepted as being more accurate than the stripping method (1-4), the library least-squares method requires the use of computers with more storage capability than the present minicomputers have which are supplied with most XRF analyzer systems. Therefore, it is a primary purpose of the present paper to develop spectral models so that the least-squares method can be implemented in these small computers. In addition, the spectral model approach is capable of: (1) providing better accuracy, (2) greatly reducing the experimental effort required in preparing pure elemental standards, and (3) correcting the library standards for changes in detector calibration and resolution shifts.

The approach used here to develop library spectra is to first obtain the detector response function for the detector of interest, the Si(Li) detector, in the energy range of interest. This function is defined as the pulse-height spectrum obtained for monoenergetic incident photons. Then the fundamental parameters pertinent to the X-ray spectrum for each element are used to generate the elemental library spectra of interest.

DETECTOR RESPONSE FUNCTION

In the X-ray energy region up to about 30keV we have identified five spectral features. The first is the Compton scattering resonse within the detector. Since this occurs at pulse-height energies lower than the level at which most lower level discriminators are set, this feature is not treated here. The second feature is a flat continuum from the low energy end up to the main photopeak. The third feature is a truncated exponential tail to the left of each photopeak. The fourth feature is a Gaussian silicon X-ray escape peak. The fifth and final feature is the main Gaussian photopeak.

The mathematical forms of the four features discussed are:

Flat Continuum

$$C(E',E,\sigma) = A_1(E)F(E',E,\sigma) \tag{1}$$

Truncated Exponential

$$T(E',E,\sigma) = A_2(E)\exp\left[-A_3(E)E'\right]F(E',E,\sigma) \tag{2}$$

Gaussian Escape Peak

$$G_E(E',E,\sigma) = A_4(E)G(E',E-1.74,\sigma) \tag{3}$$

Gaussian Photopeak

$$G(E',E,\sigma) = (2\pi)^{-\frac{1}{2}}\sigma^{-1}\exp\left[-(E'-E)^2/2\sigma^2\right] \tag{4}$$

where E is the incident photon energy, E' is the pulse-height energy, the A_i are experimental constants for the detector system of interest, and $F(E',E,\sigma)$ is given by

$$F(E',E,\sigma) = \int_{E'}^{\infty} G(E'',E,\sigma)dE'' \tag{5}$$

or

$$F(E',E,\sigma) = (1/2)\ \mathrm{erf}\left[(E'-E)/\sqrt{2}\sigma\right] \tag{6}$$

ELEMENTAL LIBRARY SPECTRA

An elemental library spectrum is obtained by summation of the response functions for each characteristic X-ray emitted. This is given by

$$R(E') = A_o \sum_{j=1}^{n} B_j(E_j)\Big[\{A_1(E_j) + A_2(E_j)\exp\left[-A_3(E_j)E'\right]\}F(E'E_j,\sigma_j) + A_4(E_j)G_E(E',E_j,\sigma_j) + G(E',E_j,\sigma_j)\Big] \tag{7}$$

where n is the number of characteristic X-rays in the element of interest, the B_j are the experimental relative intensities of each X-ray, and A_o is the normalizing constant for the detector system of interest. The four features of Equation 7 are shown for Gallium in Figure 1.

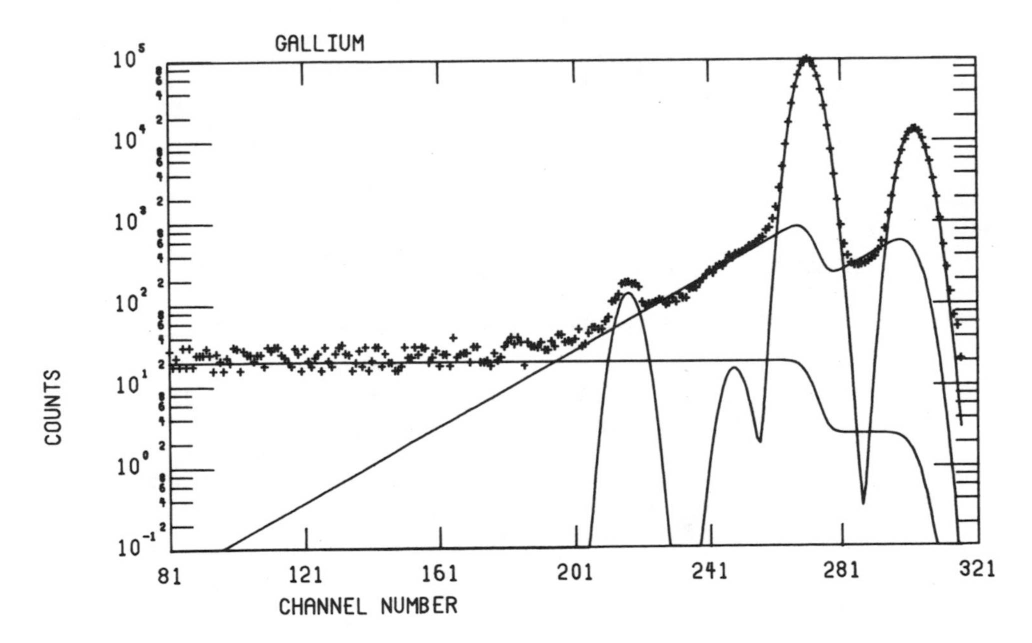

Figure 1. Illustration of the four features of a Si(Li) spectrum.

RESULTS AND DISCUSSION

The four parameters required in the response function were obtained for 13 pure element spectra for a double guard-ring Si(Li) detector system. The experimental and model pulse-height spectra for four representative elements are shown in Figures 2, 3,4, and 5. This data had to be manipulated somewhat since contaminants were present in almost all of the original spectra. In many cases unwanted photopeaks were stripped from the original data. As a result of this some of the model spectra do not correspond to experimental results as well as others. Of the four spectra given two are representative of the better model fits to experimental data (manganese and gallium) and two are representative of the worst (arsenic and nickel). This points up the fact that the experimental preparation of our elemental standards is still a difficult problem. The response function approach should be capable of eliminating this problem by providing consistent,accurate synthetic elemental library spectra.

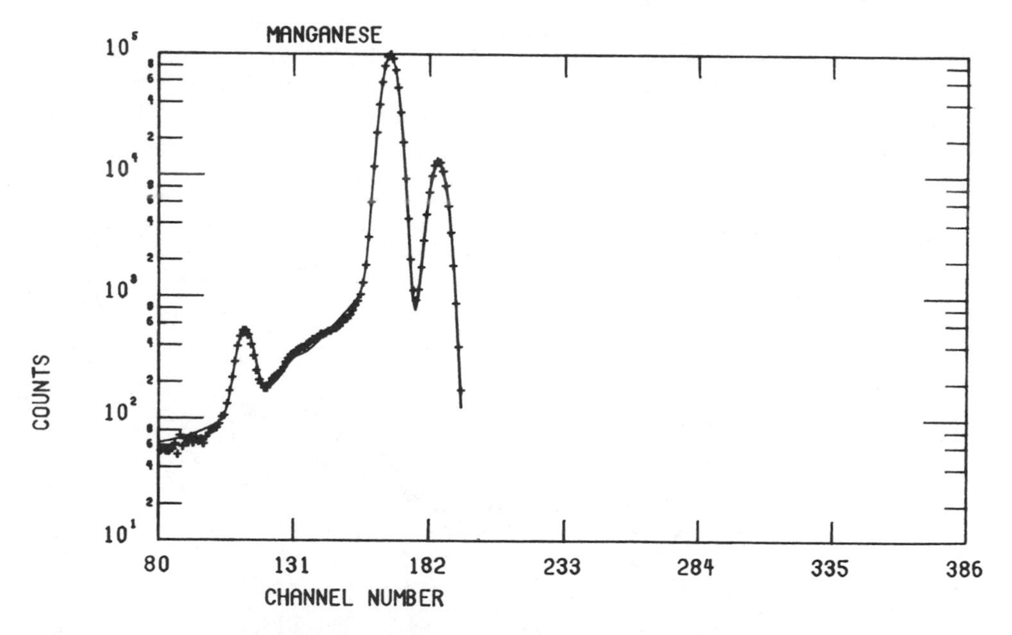

Figure 2. The response function model compared to the experimental pulse-height spectrum for manganese.

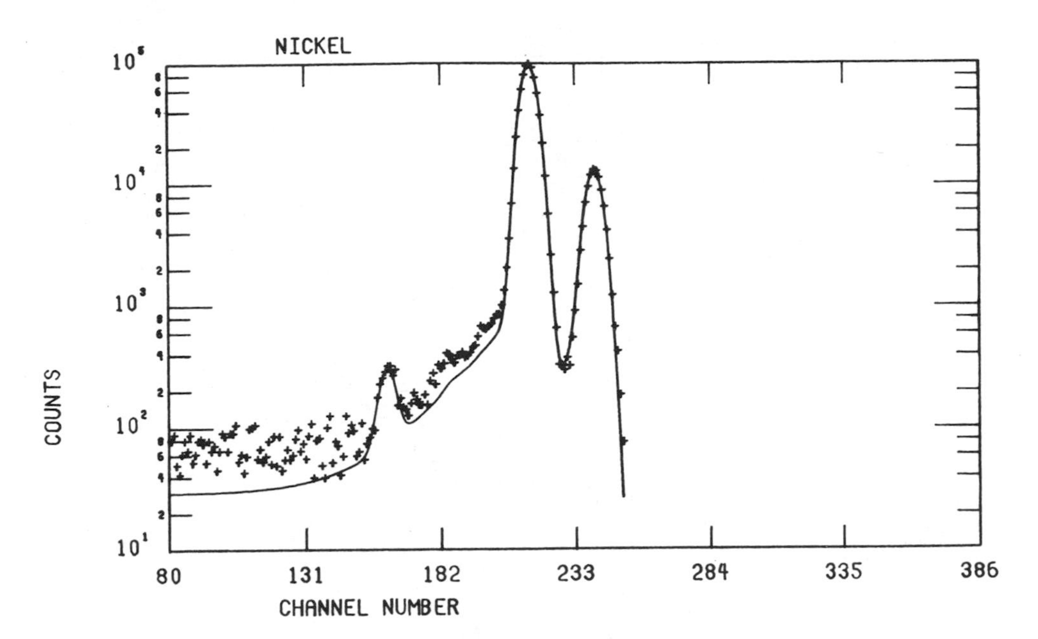

Figure 3. The response function model compared to the experimental pulse-height spectrum for nickel.

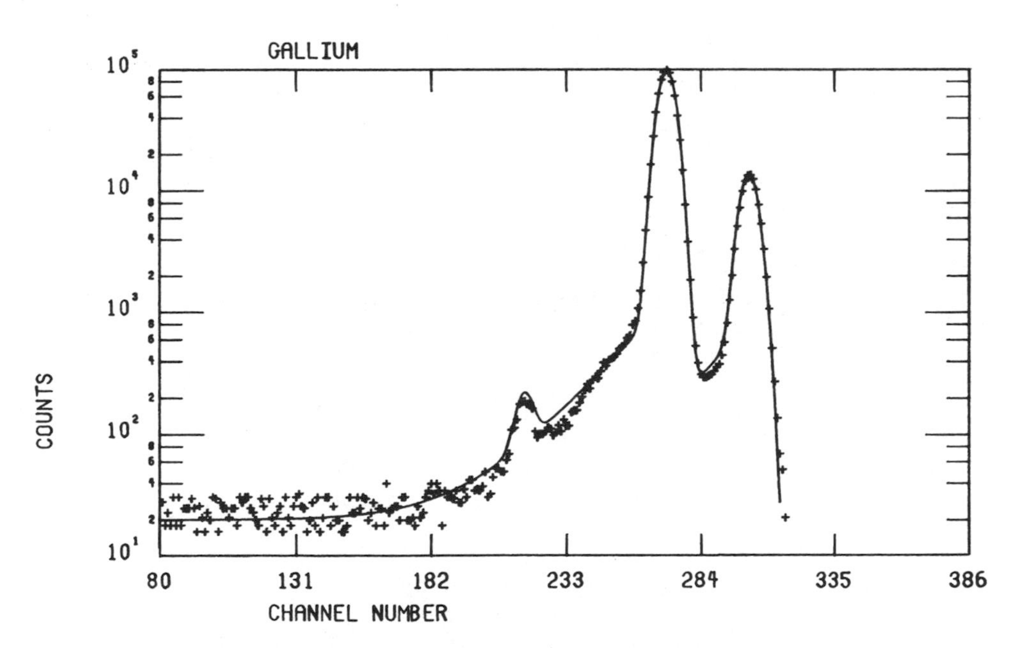

Figure 4. The response function model compared to the experimental pulse-height spectrum for gallium.

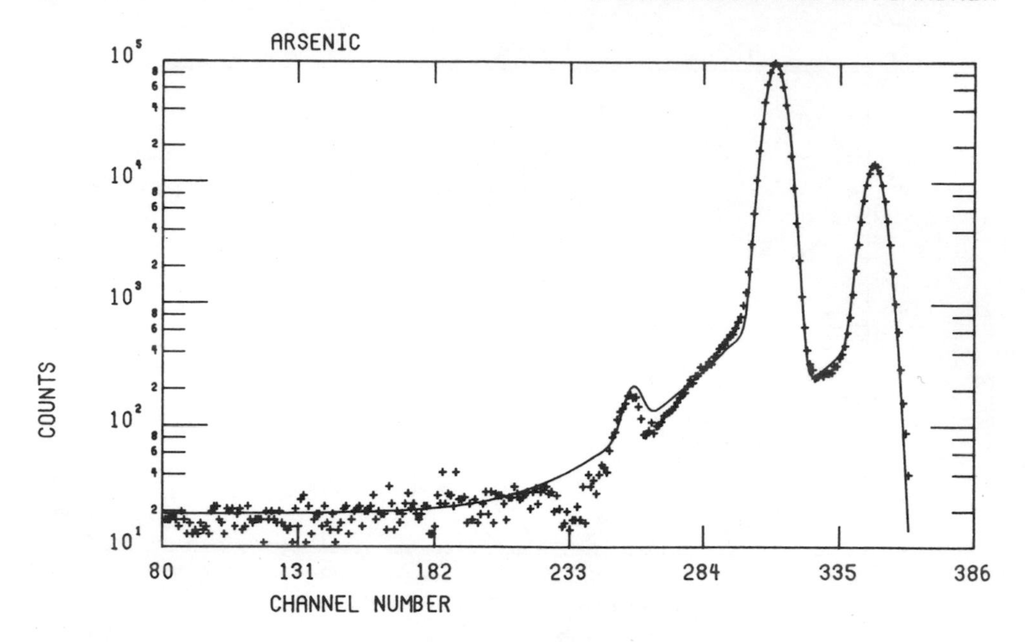

Figure 5. The response function model compared to the experimental pulse-height spectrum for arsenic.

The same spectral features have been used to describe a simple top hat detector system for a limited number of elemental spectra. No major changes in the features have been noted for this system. This indicates that the same spectral features should be useful in describing all Si(Li) detectors from the simplest to the most complex.

The elemental response function of Equation 7 has been incorporated directly into a least-squares computer program to calculate the X-ray intensities from the spectra of unknown samples. The results obtained agree with the results obtained using experimental elemental library spectra.

One important feature of the response function approach is that it could be used separately for each characteristic X-ray of each element in a least-squares analysis of an unknown spectrum. This would alleviate the problem of using different standards for thin and thick samples.

The response function approach as outlined here eliminates the need for the large amounts of computer storage that are required by the library least-squares method. This should allow the use of least-squares computer programs in the small minicomputers supplied with many X-ray analyzer systems. In addition to this the method should prove to be much easier to implement since pure element standards are not required, it should be more accurate, and simple mathematical methods can be used to adjust the models for changes in detector gain and zero shifts and even detector resolution changes.

ACKNOWLEDGEMENT

The authors gratefully acknowledge the partial financial support of the Environmental Protection Agency under Grant No. R-802759.

REFERENCES

1. P. Quittner, Gamma-Ray Spectroscopy, Adam Hilger Limited, London (1972).

2. P. Van Espen, H. Nullens, and F. Adams, "A Method for the Accurate Description of the Full Energy Peaks in Nonlinear Least-Squares Analysis of X-Ray Spectra", Nuclear Instruments and Methods, 145, 579-582 (1977).

3. F. Arinc, R. P. Gardner, L. Wielopolski, and A. R. Stiles, "Application of the Least-Squares Method to the Analysis of XRF Spectral Intensities from Atmospheric Particulates Collected on Filters", in R. W. Gould, C. S. Barrett, J. B. Newkirk, and C. O. Ruud, Editors, Advances in X-Ray Analysis, Vol. 19, p. 367-379, Kendall Hunt Publishing Company (1976).

4. F. Arinc, L. Wielopolski, and R. P. Gardner, "The Linear Least-Squares Analysis of X-Ray Fluorescence Spectra of Aerosol Samples Using Pure Elemental Library Standards and Photon Excitation", in T. G. Dzubay, Editor, X-Ray Fluorescence Analysis of Environmental Samples, p. 227-240, Ann Arbor Science Publishers, Ann Arbor (1977).

X-RAY FLUORESCENCE ANALYSIS OF STAINLESS STEELS AND LOW ALLOY STEELS USING SECONDARY TARGETS AND THE EXACT[a] PROGRAM

J. C. Harmon[b], G.E.A. Wyld, T. C. Yao[c] and J. W. Otvos[d]

Shell Development Company

Houston, Texas

ABSTRACT

EXACT is a mini-computer based fundamental parameters program which is utilized for matrix corrections in energy-dispersive X-ray analyses. We have previously shown this technique to work well with radioactive sources. However, due to the limited selection of isotopic sources available and their inherent low X-ray flux, we have investigated the use of Fe, Sn, and Dy secondary-targets as sources of monochromatic X-rays. Results to date indicate that the secondary-targets provide X-ray radiation which has sufficient monochromaticity for our technique to remain valid.

Calibration constants for over 50 elements have been determined and the data supports the theory that the calibration constant is independent of the matrix. In addition, calibration constants obtained from aqueous standards were applied to the analyses of NBS stainless steels and low alloy steels. The results further verify the matrix independency of the calibration constants. Also due to the many photon absorption and enhancement interactions in stainless steel, the results on these samples provide strong support for the validity of the EXACT fundamental parameters program.

---

a) Energy-Dispersive X-Ray Analysis Computation Technique.
b) Author to whom correspondence should be sent.
c) Present address: Kevex Corporation, Foster City, Ca.
d) Present address: University of California, Lawrence Berkeley Laboratory, Berkeley, California 94720

INTRODUCTION

In X-ray flourescence, several quantitative models are available[1-4] which provide means for converting raw X-ray intensities into quantitative results. Generally, these methods are empirically based and require a large number of standards. Further, the standards chosen should be similar in composition to the unknown sample. The fundamental parameter approach is based on physical constants, such as attenuation coefficients, jump ratios, and fluorescence yields. The equations as derived by Gillman and Heal[5], Sherman[6], and Shiraiwa and Fujino[7] for monochromatic excitation can be employed to yield exact matrix corrections for absorption and enhancement effects. Criss and Birks[8] have also reported on the use of the fundamental parameter approach with non-monoenergetic X-ray sources. Gardner, Wielopolski, and Doster have used Monte Carlo calculations and secondary targets in fundamental parameter programs[9]. We have previously reported on our use of radioactive isotopes[10] as monoenergetic excitation sources and an energy dispersive system for fundamental parameter matrix corrections. We have employed the attenuation coefficients from Vieigele[11], fluorescence yields from Fink and Ras[12] and jump ratios from Birks[13].

Recently we have extended our use of the fundamental parameters approach to secondary-target excitation methods and found that the excitation produced with these sources is of sufficient monochromaticity for our approach to remain valid. The wide application of the method is illustrated by the fact that aqueous standards may be used for the analysis of stainless steels. No additional calibrations are necessary, since the model is explicit and employs only fundamental constants in the enhancement term.

THE MODEL

We have employed the Shiraiwa[7] equation as a model for our fundamental parameter approach and derived a general equation for the use of several monoenergetic sources. Both absorption and enhancement effects are treated. It is further desirable to correct for changes in source intensity that have occurred since the calibration constants were obtained. The general equation for our approach[10] (using array notation) is:

$$W_i = \frac{I_i\ (\gamma_{sm}\ G + \gamma_{im})}{K_i\ \mu_{si}\ G(1 + Z)} \cdot \frac{M_s}{M'_s} \qquad (1)$$

where Z is the expression which corrects for enhancement and has the following relation:

$$Z = \sum_{k} Q_{ki} W_k \quad \frac{Ln(1+(\gamma_{sm} \csc E)/\gamma_{km})}{\gamma_{sm} \csc E} + \frac{Ln(1+(\gamma_{im} \csc F)/\gamma_{km})}{\gamma_{im} \quad \csc F} \tag{2}$$

(note that if Z=0, there is no enhancement)

$$Q_{ki} = \frac{\mu_{ki} \quad \mu_{sk} \quad \omega(1-1/J)}{2 \ \mu_{si}} \tag{3}$$

and

$$\gamma im = (\mu im \ W_m), \text{ etc.} \tag{4}$$

and

$$G = \frac{\text{Cosecant (angle of incident radiation)}}{\text{Cosecant (angle of fluorescent radiation)}} \tag{5}$$

The subscripts are array indices referring to the lists of elements as follows:

i = the elements being determined
m = all of the elements present in the sample
k = the elements causing enhancement
s = the elements used as sources

Thus, μ_{im} is the mass attenuation coefficient of element m at the fluorescent wavelength of element i, etc.

Other symbols used include:

m_s = monitor counts at time of calibration
m'_s= monitor counts at time of analysis
E = angle of incident radiation
F = angle of fluorescent radiation
W_i = weight fraction element determined
I_i = measured intensity of element determined
K_i = calibration constant for element determined
J = Jump ratio
ω = Fluorescent yield.

Equation (1) indicates that fluorescence intensities can be converted to weight percent with only a single calibration constant, K, necessary for each element. For the induced fluorescence effect no calibration is needed since this process takes place entirely within the sample and is dependent upon physical constants and composition only and not upon instrumental factors. Thus, in a system of n elements, only n constants must be determined, and these constants need not be determined in the same matrix as the unknown sample.

The model has limitations which are related to assumptions that are made in the derivation:

1. The exciting radiation must be monochromatic.
2. The sample should have "infinite" thickness. This condition is usually realized by samples a few millimeters thick. However, if K radiation is being measured from elements with atomic number greater than about 40 in a light matrix such as water, then "infinite thickness" may be as great as several centimeters. On the other hand, the sample thickness needs to be small relative to its distance from the source and detector. A separate calibration may be needed for "infinitely thick" samples which violate this rule since this changes the geometry of the analysis. Alternatively, a smaller sample (thin sample) may be employed as detailed in the computation section.
3. The sample must be homogeneous and its surface flat and smooth on a scale commensurate with the mean free path for the x-rays. Generally, we grind samples to pass through a 325 mesh screen. However, some samples may need to be pressed to provide an adequate surface.

We have tested the model on a multitude of samples in various matrices. When errors occurred we were able to trace the cause to a violation of one of the above requirements.

EXPERIMENTAL

Six secondary targets are routinely employed in our analyses. These include iron, dysprosium, tin, germanium, molybdenum and copper. Iron, dysprosium, and tin are employed for most analyses, and the remaining targets utilized for special applications were maximum sensitivity for a given element is needed. The sample changer and detector system is the Kevex Model 0810. The electronic accessories coupled with the detector are based on the Tracor-Northern Model 880 system. Spectra are stored on a dual floppy disk unit (DEC RX-11). The PDP 11/05 computer serves both as a multichannel analyzer and data reduction system. We have translated our fundamental parameters program (EXACT) into the Flextran language employed on the NS-880 system.

Equation (1) shows that only a single calibration constant is needed for each element. We usually employ aqueous standards (generally 1 to 2 percent by weight). These are prepared from high purity metals or standard compounds. The calibration constants can then be calculated from measured intensities, since the matrix and weight percent of the element of interest are known exactly. Changes in source intensity are compensated by measuring a platinum monitor before each set of analyses. The monitor intensity is then used to adjust the intensities of the elements determined. Background and peak overlap corrections are also utilized.

COMPUTATION METHOD

Our fundamental parameters computer program, EXACT, is an interactive program and requests the information necessary for the computation. The analyst specifies for each secondary target the count time, the monitor counts, the elements determined, and the measured intensities for each element. In addition, elements of known concentration and their weight percent are entered. One element or a compound which is considered as the remainder of the sample (normalizing element) may also be included in the determination.

The program obtains the appropriate attenuation coefficients, jump ratios, fluorescent yields, and calibration constants from the memory and the calculations are performed in an iterative manner. The final results are reported as weight percent.

Microsamples (i.e., where the sample size is smaller than the X-ray beam) can also be analyzed, however, with reduced accuracy. The fluorescence of all elements present must be measured or an element should be entered as a known weight percentage. During the iterations the intensities are gradually increased, while their relative amounts are held constant, until the sum of all weight fractions equals 1. In this microtechnique, no normalizing element can be used.

Thin samples can also be handled, however, the operator must specify the weight of the sample and its cross-sectional area. The magnitude of the correction is dependent on the composition of the sample and is modified with each iteration. The correction is of the form:

$$I_\infty = I / \left(1 - \exp\left[-g/cm^2 \cdot (\gamma_{sm} \csc E + \gamma_{im} \csc F)\right]\right) \qquad (6)$$

where I_∞ = the intensity that would have been observed for an infinitely thick sample

I = the observed intensity

g = grams of sample

cm^2 = cross sectional area.

RESULTS AND DISCUSSION

One of the major advantages of the fundamental parameters approach is that single calibration constants for each element of interest are sufficient. To verify this, we analyzed elements in different matrices and calculated the calibration constants. This data is shown in Figures 1 and 2. The excitation was with secondary targets and the elements and their associated matrices were chosen to vary widely in composition and concentration. The good agreement obtained indicates that the X-rays produced by secondary-target excitation are sufficiently monochromatic for our purposes. These curves serve as aids in identifying problem calibration constant which do not fit the curve and also provide approximate calibration constants for elements which have not previously been run. The data follow smooth curves and can be represented by the following equations:

Fe Target $K = 2.11 \times 10^6 + 3.48 \times 10^5 Z - 1.98 \times 10^3 Z^2 + 431 Z^3$

Correlation Coefficient = 0.997

Sn Target $K = 1.57 \times 10^6 - 1.94 \times 10^5 Z + 7.89 \times 10^3 Z^2 - 96 Z^3$

Correlation Coefficient = 0.995

where K = calibration constant

Z = atomic number.

Since K depends on the radiation source, the apparatus geometry, and the properties of each element, these equations will be different for different instruments. The above relations should hold as long as the same source, tube voltage, and geometry are employed. The tube current may be changed to keep the dead time below a certain level, however, all intensities should be normalized to the same reference tube current.

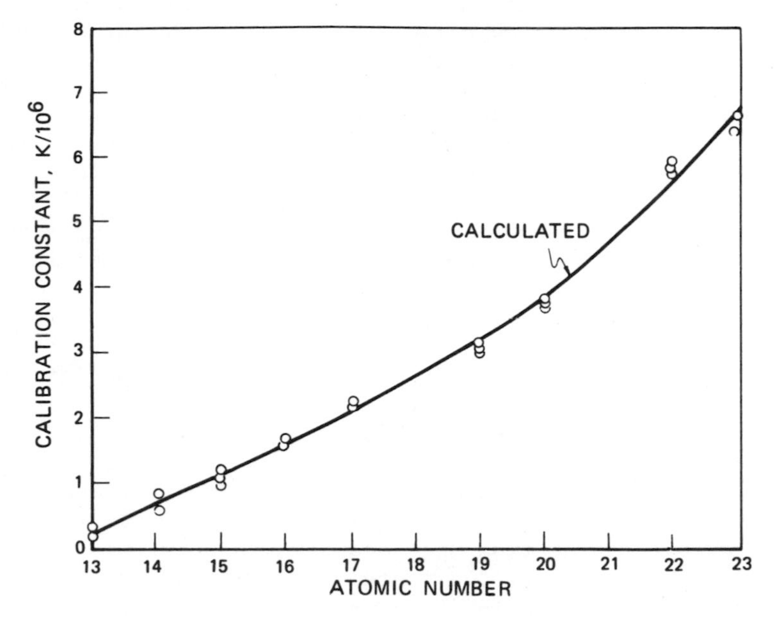

Figure 1 K-α series, Fe source, 20 kv, 45 Ma

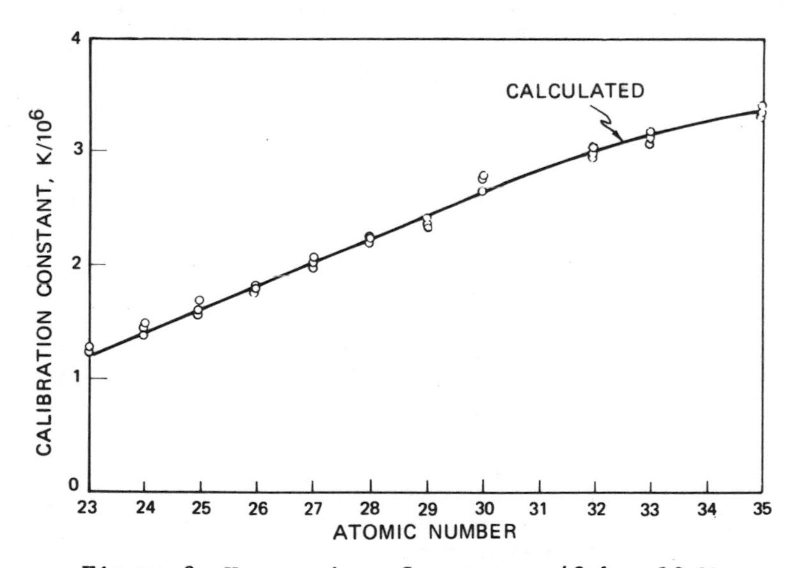

Figure 2 K-α series, Sn source, 40 kv, 10 Ma

In order to excite the K-lines of elements 44 through 60 we employ a dysprosium secondary-target. The calibration of these heavier elements has also provided a more rigorous test of the fundamental parameter computational model. The model assumes that the samples are infinitely thick, however, the K radiation from elements excited with the dysprosium source has sufficient energy to pass completely through the sample in many cases. This means, in effect, that these samples have to be regarded as thin samples. Thin samples are calculated with the incorporation of equation (6) into the generalized equation (1). Table 1 is an example of results obtained when the thickness correction is not applied. It is apparent that, in applying the thickness correction, the results further verify that the calibration constant is independent of the matrix. Our practice is to treat all samples run with the dysprosium secondary-target as "thin samples".

An important test of our method has involved the analysis of NBS alloys. The elements to be measured are caused to fluoresce primarily by excitation from the secondary-targets. In addition, they also may be caused to fluoresce by the X-rays emitted by other elements in the sample (enhancement). In order for enhancement to occur, these X-rays must be higher in energy than the excitation potential of the element being enhanced. In the case of alloy steels, nickel enhances iron, manganese, and chromium; iron enhances manganese and chromium; and manganese enhances chromium. In addition, fluorescent radiation from each of the above elements is attenuated by the other elements present in the bulk sample. Comparison of X-ray and chemical results on several National Bureau of Standards alloys are shown in Table 2. The results obtained using secondary-target excitation were based on calibration constants which were determined from aqueous standards of the element determined. This has provided a severe test of the model and provided strong evidence that the calibration constants are matrix independent and may be transferred from one matrix to another.

CONCLUSIONS

We have investigated the use of secondary-target X-ray sources as a means of providing monochromatic excitation in energy dispersive X-ray spectrometry. The results show that the X-rays produced are sufficiently monochromatic for fundamental parameter mathematical approaches, such as EXACT, to be employed for routine quantification of elements in many different matrices. This can be accomplished with a minimum number of standards, generally one standard for each element of interest. The secondary-targets can be selected to provide maximum sensitivity for a range of elements.

TABLE 1: THICKNESS CORRECTION FOR CESIUM

SAMPLE	CALIBRATION[a] CONSTANT/10^6	CORRECTED CALIBRATION[b] CONSTANT/10^6
CsCl	1.18	1.18
CsI	1.18	1.18
.96% Cs in H_2O	.74	1.17
1000 ppm Cs in H_2O	.64	1.20
Mean	.935	1.183
σ	.29	.013
Relative σ	30.6%	1.06%

a) Constants determined assuming that the sample were "infinitely thick", i.e., no correction.
b) Constants determined applying a "thin" sample correction wherein the sample weight and cross-sectional area were measured.

TABLE 2: ANALYSIS OF NBS METAL ALLOYS

ELEMENT	METHOD	NBS#160	NBS#121	NBS#1285	NBS#344
Ni	T-EDXS[a]	14.1	11.5	77.0	7.3
	Chemical Analysis	14.1	11.2	76.1	7.3
Fe	T-EDXS	63.1	69.5	6.9	74.8
	Chemical Analysis	Balance	Balance	6.8	Balance
Cr	T-EDXS	18.4	17.2	15.8	14.9
	Chemical Analysis	18.7	17.7	16.3	14.9
Mn	T-EDXS	1.8	1.7	.32	.74
	Chemical Analysis	1.6	1.5	.31	.57
Mo	T-EDXS	2.6	.06	-	2.3
	Chemical Analysis	2.8	.07	-	2.4

a) Tube-excited Energy Dispersive X-ray Spectroscopy (Secondary-Target)

REFERENCES

1. Lachance, G. R. and Traill, R. J., "A Practical Solution To The Matrix Problem in X-Ray Analysis", Can. Spectry., 11, 43(1966).

2. Claisse, F. and Quintin, M., "Generalization of The Lachance-Traill Method for The Correction of The Matrix Effect in X-Ray Fluorescence Analysis", Can. Spectry., 12, 129(1967).

3. Rasberry, S. D. and Heinrich, K. F. J., "Calibration For Interelement Effects in X-Ray Fluorescence Analysis", Anal. Chem., 46, 81(1974).

4. Locas Tooth, H. J. and Pyne, C., "The Accurate Determination of Major Constituents by X-Ray Fluorescence Analysis in The Presence of Large Interelement Effects", Advances in X-Ray Analysis, 7, 523(1964).

5. Gillman, E. and Heal, H. T., "Some Problems in The Analysis of Steels by X-Ray Fluorescence", Brit. J. Appl. Phys., 3, 353(1952).

6. Sherman, J., "The Theoretical Derivation of Fluorescent X-Ray Intensities from Mixtures", Spectrochimics Acta, 7, 283(1955).

7. Shirawiwa, T. and Fujino, N., "Theoretical Calculations of Fluorescent X-Ray Intensities in Fluorescent X-Ray Spectrometric Analysis",Jap. J. Appl. Phys., 5, 886(1966).

8. Criss, J. W. and Birks, L. S., "Calculation Methods for Fluorescent X-Ray Spectrometry", Anal. Chem., 40, 1080(1968).

9. Gardner, R. P., Wielopolski, L., and Doster, J. M., "Adaptation of The Fundamental Parameters Monte Carlo Simulation to EDXRF Analysis With Secondary Fluorescer X-Ray Machines", Advances in X-Ray Analysis, 21, 129(1977).

10. J. W. Otvos, G. Wyld, and T. C. Yao, "Fundamental Parameter Method for Quantitative Elemental Analysis with Monochromatic X-Ray Sources", 25th Annual Denver X-Ray Conference, 1976.

11. Bracewell, B., and Veigele, W. J., "Devel. In Appl. Spec.", 9, 357-400 (1971).

12. Fink, R. W. and Ras, P. V., "Handbook of Spec.", Vol. 1, pp. 219-223, CRC Press, Cleveland, Ohio (1974).

13. Birks, L. S., "Handbook of Spec.," Vol. 1, pp. 230, CRC Press, Cleveland, Ohio (1974).

THE RESOLUTION OF X-RAY FLUORESCENCE SPECTRA BY THE LEAST SQUARES METHOD

Colin G. Sanderson

Environmental Measurements Laboratory, U. S. Department of Energy, New York, New York 10014

INTRODUCTION

Least squares computer programs have been used for the resolution of complex gamma-ray spectra obtained from NaI(Tl) detectors for many years. With the addition of computer codes, which account for the differences between gamma-ray and x-ray fluorescence spectra, these same programs can be used to quantify energy dispersive x-ray fluorescence data.

WLSQXR is a modified version of our weighted least squares program (1,2) for the analysis of gamma-ray spectra. In order to analyze x-ray fluorescence spectra from large volume air samples, corrections for variations in primary beam flux, primary beam scatter and sample self-attenuation are required. Corrections for analyzer gain and zero energy channel (3) (ZEC) shifts which were part of the original gamma-ray program are also required.

The monthly air samples collected by the Environmental Measurements Laboratory (EML) consist of one to six 20 cm diameter Microsorban filters representing a total air volume of about 30,000 m^3. Discs 5 cm in diameter are cut from each filter, combined and compressed at 2500 kg/cm^3 into a single disc. The resulting discs, which vary from 25 to 75 mg/cm^2, are 0.25 to 0.75 mm thick. Because of this heavy sample loading, attenuation corrections are quite significant; corrections for Pb range from 10% to 40%; Fe, from 30% to 100%; V, from 50% to 300%.

The x-ray fluorescence spectrometer at EML is a Siemens SRS-1 with a 3 kilowatt x-ray generator. A Si(Li) x-ray detector has been mounted on the outside of the goniometer vacuum chamber at the 0° crystal axis so that energy dispersion data can be obtained. Data acquisition and reduction are performed with an ORTEC 6260 multichannel analyzer system thru a CAMAC-interfaced Digital Equipment Corporation PDP 11/04 computer. Analysis is performed at 45 kV and 40 mA with a tungsten tube.

SAMPLE ANALYSIS

Because of the long path length from the sample to the x-ray detector, the spectrometer is filled with helium. While helium equilibrium is being achieved and the x-ray generator is stabilizing, the $Fe_{K\alpha}$ and $Pb_{L\alpha}$ lines from a solid disc of iron and lead are monitored in a continuous series of one minute counts. When stability is reached, these count rates are recorded as a measure of the primary beam flux that will be used to excite the samples. The channel position of these two peaks are also used to compute the daily gain and ZEC of the system. The exact value of the system gain and ZEC is not required for each sample because one feature of the least squares program is an automatic iteritive energy calibration correction.

Samples and single element standards are initially counted while backed with the same Fe-Pb disc used to determine system stability in order to measure the sample self attenuation. After a one minute count, the data are transferred to the second half of the CAMAC memory. The Fe-Pb disc is removed and counting resumed in the cleared first half of the memory.

If the spectrum for a library standard is being obtained, the exact positions of the $Fe_{K\alpha}$ and $Pb_{L\alpha}$ peaks from the initial count are determined and new gain and ZEC are calculated.

When the analysis is completed, the spectrum is recorded on the system's floppy disc along with the system energy calibration, the integrated sum of the $Fe_{K\alpha}$ and $Pb_{L\alpha}$ peaks from the initial system stability determination and the sample self-attenuation coefficients.

Individual sample self-attenuation is calculated at 6.4 and 10.5 keV by equating the $Fe_{K\alpha}$ and $Pb_{L\alpha}$ count rates from the Fe-Pb disc with and without the sample in place. These measured attenuations are then used to develop a set of coefficients to describe sample attenuation over the energy range from 1.5 keV to 16 keV as follows. Equation 1 describes self attenuation for x-rays at a given energy based upon sample thickness (d), density (ρ), and the mass attenuation coefficient (μ).

$$I/I_o = \exp(-\mu\rho d) \tag{1}$$

The mass attenuation coefficient, which is proportional to energy raised to a power n, can be equated to energy with the addition of a proportionality constant,

$$\mu \approx (\text{Energy})^{-n} = k\ (\text{Energy})^{-n}. \tag{2}$$

The constants from equations 1 and 2 may be combined,

$$K = k\rho d \tag{3}$$

Thus yielding,

$$I/I_o = \exp(K \cdot \text{energy}^{-n}) \tag{4}$$

which expresses attenuation as a function of energy and two constants, K and n. Since we have made measurements at two energies, we have two equations which can be solved to yield the values of K and n.

The library standards and sample spectra are corrected channel by channel with this expression during execution of the least squares program so that the resulting spectra will appear to have been obtained from a weightless source.

A linear least squares analysis requires a set of standard library spectra that represent as closely as possible the spectra to be obtained from the samples. Because these standard spectra are generated experimentally - not computed from peak shape algorithms - their exact gain and base values must be known. They must also be consistent - all must have the same gain and base conditions. However, this may or may not be the case in actual practice. Therefore, the first step in the execution of the least-squares program is to shift each spectrum to the same nominal values using Schonfeld's shift routine from "alpha-m" (3).

A background spectrum taken with blank filter paper of the same type used to support the standard elements is also shifted, then subtracted from each standard. At this point, the channel by channel attenuation correction is applied.

Our experience has shown that gain shifts have never been greater than 0.5%. However, ZEC shifts have been as much as ± 0.5 channels. If the gain and ZEC are known, iteration is not necessary. If the gain is known to be stable but the ZEC is uncertain the program can be set to perform only ZEC iterations, or vice versa if the ZEC is stable but the gain shifts.

Since each sample will scatter the primary beam to varying degrees, a background spectrum from blank Microsorban is included with the standards so that the varying effect of scatter can be compensated for. However, the background spectrum must be corrected channel by channel with the same attenuation function used for the sample being computed.

RESULTS

Since the program is a weighted least squares procedure which uses the variance for weights, the inverse matrix can be used to estimate individual deviations. In Table I, the standard error obtained from six replicate analysis is shown with the standard deviation derived from a single least squares analysis. The sample analyzed was a Columbia Scientific Industries (CSI) multi-element filter paper standard 4a. The data were fitted for the five elements known to be present. The close agreement between the standard error and the standard deviation indicates realistic uncertainties are obtained by the least squares program.

The same CSI 4a standard was analyzed at different primary beam fluxes. Table II shows that the program can correct flux variations by as much as ± 20%.

The results of the analysis of CSI standard 3a are shown in Table III. The unattenuated values were obtained by analysis of the filter "as is". The correction factors listed indicate the amount of self attenuation the least squares program computed. The attenuated values were obtained when the 3a filter paper was analyzed sandwiched between blank Whatman 541 filter papers. The computed correction for Ti went from 1.50 to 2.50 yet yielded excellent agreements with the expected unattenuated value. The

Table I. Comparison of Standard Error and Standard Deviation for a Standard Filter

Element	Standard Error (%)	Standard Deviation (%)
V	2.2	2.6
Mn	0.8	1.6
Co	0.8	1.4
Cu	4.0	6.0
Pb	0.8	1.4

Table II. Flux Correction

Element	($\mu g/cm^2$) 40 mA	32 mA	48 mA	Standard Deviation
V	44.5	47.6	44.6	1.2
Mn	44.4	44.8	45.0	0.8
Co	44.5	44.3	44.7	0.6
Cu	44.5	44.0	46.2	1.6
Pb	44.5	45.0	44.1	0.7

Table III. Attenuation Correction

	Unattenuated		Attenuated	
Element	$\mu g/cm^2$	Correction	$\mu g/cm^2$	Correction
Ti	44.5	1.50	44.0	2.50
Cr	44.7	1.42	44.3	2.01
Fe	44.5	1.33	44.4	1.72
Ni	43.9	1.27	44.1	1.52
Zn	44.5	1.22	44.7	1.39
Background	1.01	-	1.34	-

background or blank filter included as a library standard was fitted at 34% over unity in the latter case because of the additional scatter produced by the three filter papers.

A further test of the program involved cutting 2 discs from the same 20 cm Microsorban air filter. Each disc was analyzed separately, then they were compressed into a single disc and re-analyzed.

For all the elements found the sum of the individual measurements statistically agree, at the 1 σ level, with the results obtained when the 2 discs were analyzed as one (Table IV). Attenuation correction factors for Ti went from 1.8 for the individual filters to 2.7 for the combined filters, Fe went from 1.4 to 1.8 and Zn went from 1.3 to 1.5.

Table IV. Analysis of Individual and Combined Filters ($\mu g/cm^2$)

Element	A	B	A + B	Combined A & B
Ti	2.3±4.6	2.4±4.5	4.7±6.0	9.8±9.5
V	9.2±4.0	4.9±3.6	14.1±5.7	14.7±6.0
Mn	2.3±1.7	3.6±1.9	5.9±2.3	6.1±3.4
Fe	74.4±4.4	75.8±4.3	150.2±6.2	148.6±8.6
Co	4.0±1.8	3.5±1.7	7.5±2.5	9.7±3.6
Zn	15.7±4.1	15.1±4.0	30.8±5.7	30.8±7.2
Pb	59.4±3.8	54.2±3.6	113.6±5.2	109.0±6.9

REFERENCES

1. "HASL Procedures Manual," J. H. Harley, editor, U. S. Department of Energy Report HASL-300, revised annually (1978).

2. I. Salmon, "Computer Analysis of Gamma-Ray Spectra from Mixtures of Known Nuclides by the Method of Least Squares," National Academy of Sciences Report NAS-NS-3107 (1962).

3. E. Schonfeld, "Alpha M - An Improved Computer Program for Determining Radioisotopes by Least-Squares Resolution of Gamma-Ray Spectra," Oak Ridge National Laboratory Report ORNL-3975 (1966).

THE REDUCTION OF MATRIX EFFECTS IN X-RAY FLUORESCENCE ANALYSIS BY THE MONTE CARLO, FUNDAMENTAL PARAMETERS METHOD

R. P. Gardner and J. M. Doster

Department of Nuclear Engineering
North Carolina State University
Raleigh, North Carolina 27650

ABSTRACT

A review of the application of the Monte Carlo, fundamental parameters method to XRF fluorescence analysis for the reduction of matrix effects is made. The analytical solutions arising from theoretical equations are given along with the restrictive assumptions that are necessary to this approach. The extensions of the fundamental parameters method by the Monte Carlo simulation to practical situations that require much less restrictive assumptions are outlined. The average angle approach to the use of the analytical solutions is investigated by comparison with the Monte Carlo method. Future extensions of the fundamental parameters method by the Monte Carlo approach are discussed.

INTRODUCTION

Until recently the fundamental parameters method for the elimination of matrix effects in XRF analysis has been applied solely by using the analytical solutions to the theoretical equations (1). These analytical solutions have been derived by a number of investigators (2-7) over the years. While these analytical solutions from theoretical treatments are aesthetically appealing as well as computationally efficient, they suffer from requiring that a number of restrictive assumptions be made in order to obtain them. Some of these restrictions (such as the assumption of narrow beam incidence and exit angles) have been overcome by using rather drastic simplifications (such as the use of one average angle of incidence and exit). Since no alternative solutions existed, the accuracy of these simplifications has not been determined.

Recently the authors and other co-workers began to apply and implement the Monte Carlo approach (8-16) for extending the fundamental parameters method to practical XRF analysis situations that are much less restrictive than the analytical solutions to the theoretical equations. This approach has been used to extend the fundamental parameters method to: (1) analyzers with a wide range of incidence and exit angles, (2) analyzers with a wide range of excitation photon energies, (3) samples that are less than infinitely thick, and (4) certain simple heterogeneous samples. Future extensions of the Monte Carlo approach that are presently being developed include: (1) extension to more complex heterogeneous samples, and (2) inclusion of the effect of scattering within the sample.

In this paper the Monte Carlo simulation is used to examine the validity and accuracy of the common practice of using a fixed average angle of incidence and exit when applying the analytical solutions of the theoretical equations to practical XRF analyzers that actually involve a wide range of these angles. Finally, a future extension of the Monte Carlo method to XRF analysis presently being developed is described which is based on obtaining the total spectral response to a sample.

THEORETICAL TREATMENTS WITH ANALYTIC SOLUTIONS

Analytic closed form solutions for the fluorescent X-ray intensities from homogeneous mixtures were obtained by Sherman (4,5) for primary, secondary and tertiary interactions. To obtain these solutions, the necessary simplifying assumptions are: (1) fixed angles of excitation incidence and characteristic radiation exit, (2) homogeneous samples, (3) thick samples, (4) excitation by monoenergetic photons, (5) scattering within the sample can be neglected, and (6) the solutions are normalized to the pure element responses. Sherman reports the solutions for the fluorescence intensities as Primary Contributions

$$I_{f3} = \tau'_3/(\mu'+\mu''_{f3}) \tag{1}$$

Primary and Secondary Contributions

$$I_{f2} = \tau'_2/(\mu'+\mu''_{f2})+[\tau'_3\tau_{2f3}/(\mu'+\mu''_{f2})][(1/\mu')\log(1+\mu'/\mu_{f3})+ (1/\mu''_{f2})\log(1+\mu''_{f2}/\mu_{f3})] \tag{2}$$

Primary, Secondary, and Tertiary Contributions

$$I_{f1} = \tau'_1/(\mu'+\mu''_{f1})+[\tau'_2\tau_{1f2}/(\mu'+\mu''_{f1})][(1/\mu')\log(1+\mu'/\mu_{f2})+$$

$$(1/\mu''_{f1})\log(1+\mu''_{f1}/\mu_{f2})\,|+|\tau'_3\tau_{1f3}/(\mu'+\mu''_{f1})\,||(1/\mu')\log(1+\mu'/\mu_{f3})+$$

$$(1/\mu''_{f1})\log(1+\mu''_{f1}/\mu_{f3})\,|+|\tau'_3\tau_{2f3}\tau_{1f2}/(\mu'\mu''_{f1})\,||(1/\mu_{f2})\log(C_1/\mu_3)$$
$$+\ (1/\mu_{f3})\log(C_1/\mu_{f2})+(1/A_1)\,|(\tfrac{1}{2})\log^2(A_3/A_2)+(\tfrac{1}{2})\log^2(B_3/B_2)-$$
$$(\tfrac{1}{2})\log^2(A_3/B_2)+\psi(1)+\psi(-D_3/A_2)+\psi(-D_2/B_3)\,|+(1/\mu''_{f1})\,|\underline{(\tfrac{1}{2})\log^2(\mu_{f3}/A_2)}-$$
$$(\tfrac{1}{2})\log^2(B_3/B_2)-(\tfrac{1}{2})\log^2(C_1/\mu_{f3})-\psi(-D_2/B_3)-\psi(\mu_{f2}/C_1)\,|+(1/\mu')\,|$$
$$(\tfrac{1}{2})\log^2(\mu_{f2}/A_3)-(\tfrac{1}{2})\log^2(A_3/A_2)-(\tfrac{1}{2})\log^2(C_1/\mu_{f2})-\underline{\psi(D_3/A_2)}-\psi(\mu_{f3}/C_1)\,| \quad (3)$$

The notation for these equations and the definition of the ψ function are given by Sherman (4,5).

It should be noted that the underlined terms in the tertiary contribution are in error. The term given as $\frac{1}{2}\log^2(\mu_{f3}/A_2)$ should be corrected to $\frac{1}{2}\log^2(\mu_{f3}/B_2)$. The term given as $-\psi(D_3/A_2)$ should be corrected to $-\psi(-D_3/A_2)$. Note also that all secondary contributions should be multiplied by ½ and the tertiary contribution should be multiplied by ¼ as previously corrected by Shiraiwa and Fujino (6). The fluorescence equations incorporating the above corrections have been programmed as an IBM standard FORTRAN subroutine named SHERMA and is available from the authors.

The treatment of primary and secondary interactions has been extended by Sherman to samples of finite thickness and analytic solutions are available. Shiraiwa and Fujino (6,7) for the same geometrical considerations have derived analytical integral relationships for primary, secondary and tertiary fluorescence for the discrete and continuous spectra of photons emitted by a conventional X-ray tube. These integral equations require evaluation by numerical solution.

MONTE CARLO SIMULATION

A Brief Description of the Probabilistic Monte Carlo Method

The probabilistic Monte Carlo method is a technique for simulating the random processes involved in physical systems by sampling from probability distribution functions (pdf's) that describe the frequencies of the random events. In practice, a large number of particles (or photons) are followed from birth to death by a series of random walks. The fate of the particle along each path is determined by randomly choosing from the appropriate probability distribution function (pdf). The procedure is repeated until a sufficient number of particle histories have been observed to allow conclusions to be drawn about the phenomenon of interest.

Variance reduction techniques may be employed to increase the efficiency of the Monte Carlo simulation by giving better precision for the same number of particle histories. These variance reduction techniques usually consist of forcing the random walks within boundaries favorable to the successful completion of the history. Each variance reduction must be accompanied by an appropriate weight factor between zero and one to prevent the introduction of bias.

Random numbers distributed according to any probability distribution may be obtained in terms of the rectangular distribution from the expression

$$\int_{-\infty}^{y} g(y')dy' = \int_{0}^{x} f(x')dx' \tag{4}$$

where g(y) is the distribution of interest and f(x) is the rectangular distribution. This expression may be reduced to

$$\int_{-\infty}^{y} g(y')dy' = x \tag{5}$$

where x is a rectangularly distributed random number.

To illustrate this technique and also a possible variance reduction method, consider the exponential pdf used in determining path lengths between interactions

$$g(y) = \mu \exp(-\mu y) \qquad y \geq 0 \tag{6}$$

$$g(y) = 0 \qquad y < 0 \tag{7}$$

Note that the integral from minus infinity to infinity is one and this must be true for all pdf's. To obtain exponentially distributed random numbers substitute Equation 6 into Equation 5 to obtain

$$\mu \int_{0}^{y} \exp(-\mu y')dy' = x \tag{8}$$

The solution of this equation is

$$y = (-1/\mu)\ln(1-x) \tag{9}$$

To force the interaction between the points y_1 and y_2 one has

$$x = \mu \int_{y_1}^{y} \exp(-\mu y')dy' / \mu \int_{y_1}^{y_2} \exp(-\mu y')dy' \tag{10}$$

Solving this equation for y one has

$$y = -(1/\mu)\ln\{x[\exp(-\mu y_2)-\exp(-\mu y_1)]+\exp(-\mu y_1)\} \tag{11}$$

The corresponding weight factor is

$$w = \mu\int_{y_1}^{y_2} \exp(-\mu y)\,dy / \mu\int_0^{\infty} \exp(-\mu y)\,dy$$

$$w = \exp(-\mu y_1) - \exp(-\mu y_2) \qquad (12)$$

For more detailed information on the probabilistic Monte Carlo method the reader is referred to References 17, 18 and 19.

Application of the Monte Carlo Method to XRF Spectrometry

The Monte Carlo technique is subject to few of the assumptions that limit analytical solutions. System geometries, sample thicknesses, source distributions, matrix and scattering effects may be modeled precisely without adding significantly to the computational time. The nature of the XRF problem is such that complete variance reduction is possible. This means that every exciting photon from the source may be forced to result in the successful detection of a characteristic X-ray. Very efficient Monte Carlo programs may thus be written for the detailed simulation of XRF systems.

Anticipated and demonstrated uses of the Monte Carlo method to XRF analysis include: (1) extension of the fundamental parameters method to less restrictive assumptions, (2) penetration depth calculations for the electron excitation of samples, (3) the calculation of the response to photon excitation source backscattering from samples, (4) detector characteristics, (5) penetration depth calculations for the heavy charged particle excitation of samples, (6) the calculation of the response to heavy charged particle excitation source backscattering from samples, (7) the generation of total spectral responses, and (8) XRF spectrometer design. Only the first of these is discussed in detail in the present paper.

Review of Existing XRF Monte Carlo Applications

The Monte Carlo method has been applied to two typical XRF systems. The first of these (8) consists of an annular radioisotope monoenergetic source; a combination shield - detector-collimator, and sample holder; a sample of finite thickness; and a circular detector as shown in Fugure 1. The second system modeled (14) is the secondary fluorescer analyzer as shown in Figure 2.

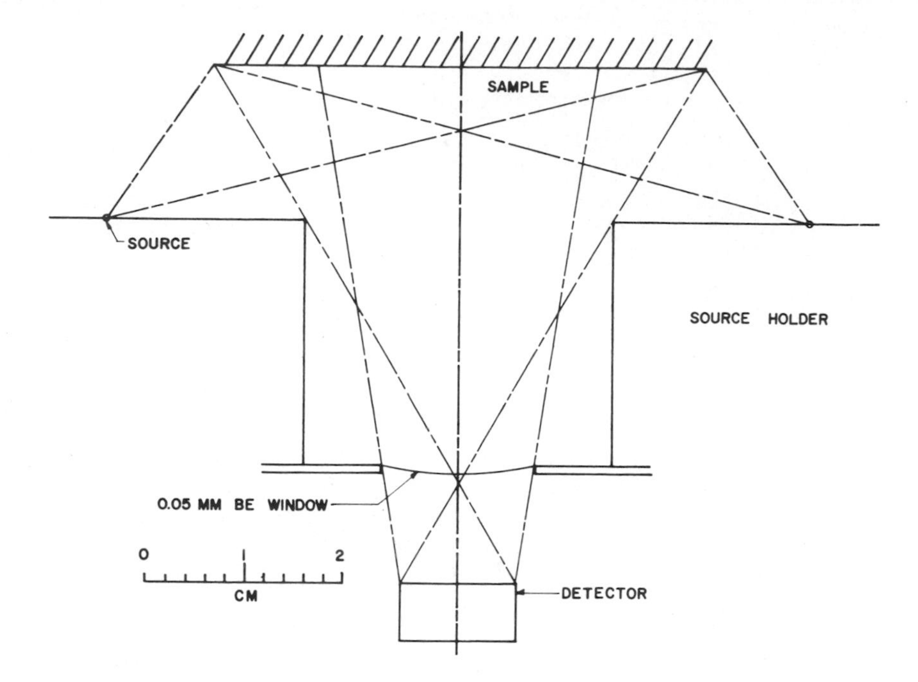

Figure 1. Drawing of radioisotope annular source X-ray analyzer system.

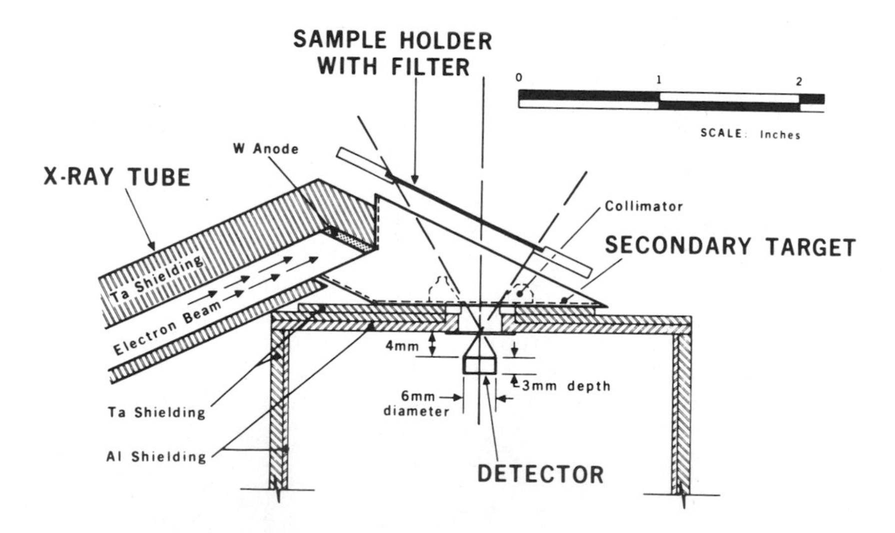

Figure 2. Drawing of a typical secondary fluorescer X-ray machine excitation analyzer system.

The Monte Carlo analysis of this system must include capabilities for a distributed source, various sample tilt angles and the source intensity distribution of the fluorescer target. Both models were developed for homogeneous samples of up to three elements. The pertinent computer programs are available from the authors as IBM standard FORTRAN subroutines named X RAY and SXRAY, respectively.

The Monte Carlo method has been further extended to the case of excitation from the continuous and discrete spectra (9) of photons obtained from X-ray tubes. The results have been shown to compare favorably with the analytic integral solutions proposed by Shiraiwa and Fujino (6,7) when restricted to the thick sample, fixed entrance and exit angle case.

EVALUATION OF THE FIXED AVERAGE ANGLE ASSUMPTION

Results obtained with the analytic closed form solutions of Sherman (4,5) for fixed average angles are compared to those obtained by Monte Carlo simulation for two typical X-ray analysis systems. The two systems of interest are: (1) the EDXRF analyzer that uses an annular ring radioisotope exciting source with circular coaxially mounted sample as shown in Figure 1 and (2) the secondary fluorescer X-ray machine analyzer as shown in Figure 2. Many of these two types of systems have geometrical designs that incorporate wide ranges of incidence and exit angles. These systems have been previously modeled by a fundamental parameters Monte Carlo simulation method (8,14). The Monte Carlo technique has been shown to compare accurately with the analytical formula results when restricted to the narrow beam geometry assumed in the analytical formula derivations. Due to the flexibility of the Monte Carlo method, it is used here as the basis of the comparison.

Multiple Discrete Exciting Energies

Both X-ray systems are capable of excitation with polychromatic beams of multiple discrete energy photons. The resulting response may be obtained by the expression

$$R = \sum_{i=1}^{n} R_i W_i \qquad (13)$$

where n is the number of discrete energy photons, W_i is the fractional abundance of the ith photon, R_i is the response due to the ith photon, and R is the expected total response.

This averaging technique was applied to a Ni-Fe-Cr ternary fixed entrance and exit angle system with two discrete energy photons. Figure 3 shows the excellent agreement between the Monte Carlo and Sherman results.

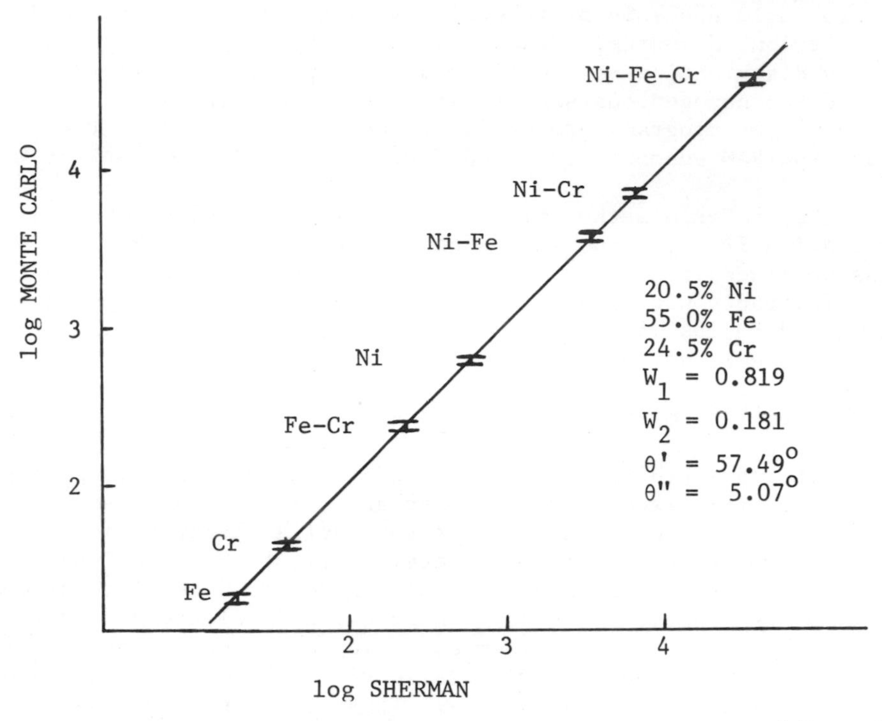

Figure 3. Comparison of the Monte Carlo and Sherman model predictions for X-ray intensities.

Wide Angle Geometry

Incident and exit angles necessary for the evaluation of Sherman's solutions were obtained by two methods. The first of these is by utilizing the Monte Carlo programs to generate response weighted average angles according to the relations

$$\text{Incident:}\quad \bar{\theta}' = \sum_{i=1}^{n} I_i \theta'_i \Big/ \sum_{i=1}^{n} I_i \tag{14}$$

$$\text{Exit:}\quad \bar{\theta}'' = \sum_{i=1}^{n} I_i \theta''_i \Big/ \sum_{i=1}^{n} I_i \tag{15}$$

where n is the total number of histories, I_i is the product of all sequential probabilities necessary for the successful detection of the ith history, θ'_i is the incident angle with respect to the normal for the ith history, and θ''_i is the exit angle with respect to the normal for the ith history. These parameters are convenient by products of the intensity calculations and result in no significant increase in computational time. The second method for obtaining the incident and exit angles is by taking the arithmetic mean of the minimum and maximum possible angles specified by the system geometry. A point that should be noted is that the response weighted average angle depends on the sample composition. This dependence is related directly to the absorption characteristics of the matrix and is therefore completely neglected in the geometry determined average angles.

Application of the Theoretical Formulas to Secondary Fluorescer Machines

The secondary fluorescer machine is characterized by: (1) photon emission from any point on the fluorescer target, (2) multiple discrete exciting photons, and (3) variable sample tilt angles. The Monte Carlo model is shown in Figure 4 with both average angle representations. In addition, the minimum and maximum incidence and exit angles are outlined. The predicted intensities from the Monte Carlo simulation as compared to those predicted by the Sherman equations for a Ni-Fe-Cr ternary were obtained for three sample tilt angles and both average angle techniques. A table of these values is available from the authors.

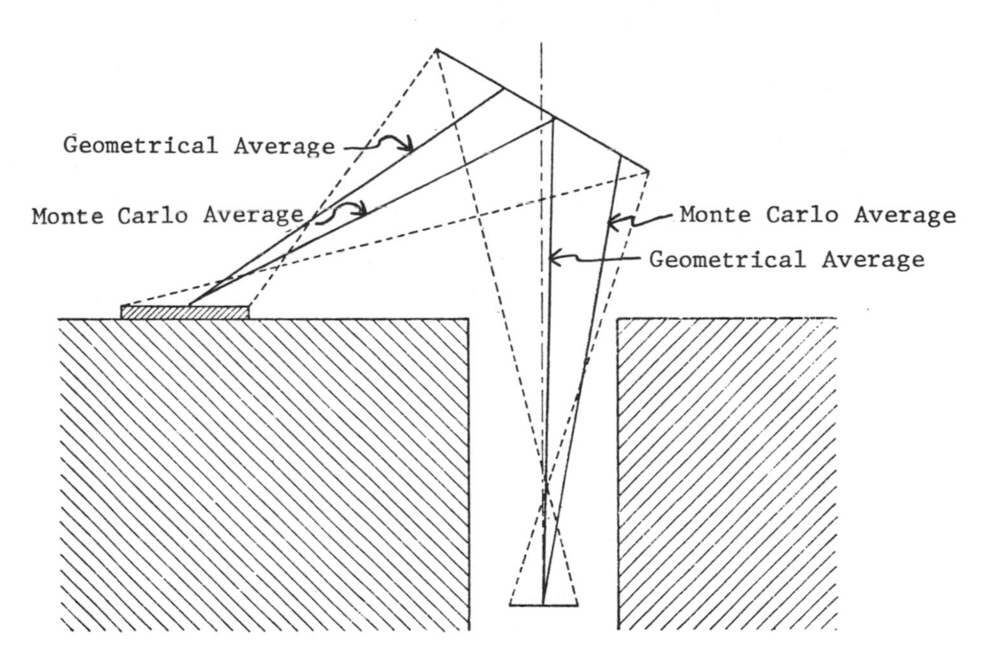

Figure 4. Illustration of the two methods of choosing the average incidence and exit angles for a secondary fluorescer XRF analyzer.

Application of the Theoretical Formulas to Radioisotope Ring Source Systems

The radioisotope ring source system is characterized by: (1) multiple discrete energy exciting photons and (2) source-sample-detector geometry such that excitation may be represented as from a point isotropic source. The equivalent Monte Carlo model is shown in Figure 5 again illustrating the incident and exit angle representations. The sample has the same Ni-Fe-Cr composition as the previous one. The Monte Carlo and Sherman results were compared for various sample to detector distances to determine the effect of increased broadening of the incident and exit beams. A table of these values is also available from the authors.

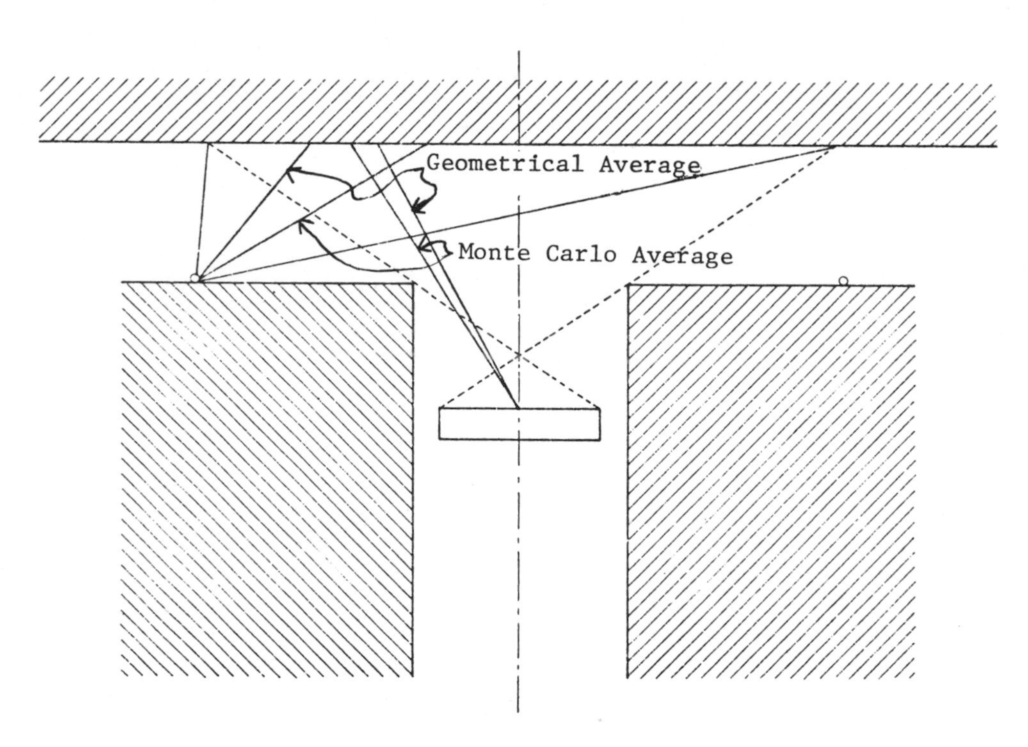

Figure 5. Illustration of the two methods of choosing the average incidence and exit angles for a radioisotope annular source XRF analyzer.

Results

A measure of the applicability of the theoretical formulas to each of the situations simulated may be obtained by a reduced chi-squared value. This value is evaluated from the expression

$$\chi^2_\nu = (1/n) \sum_{i=1}^{n} (R_{Mi}-R_{Si})^2/\sigma^2_{Mi} \tag{16}$$

where n is 7, the number of calculated intensity contributions which includes Ni, Fe, Cr, Ni-Fe, Ni-Cr, Fe-Cr, and Ni-Fe-Cr; R_{Mi} is the Monte Carlo calculated contribution, R_{Si} is the contribution calculated from the Sherman equations, and σ_{Mi} is the standard deviation of the Monte Carlo contribution. Reduced chi-squared values between zero and one indicate agreement within the uncertainty of the Monte Carlo calculations. Table I lists the total x^2_ν value for each case and the partial x^2 value for each calculated intensity. The results obtainable with the Monte Carlo predicted average angles show excellent agreement in all but one case (z=7cm,radioisotope ring source). Use of geometry predicted average angles yield results within the Monte Carlo uncertainty in only two of six cases. It is of interest to note for both average angle techniques the relative amounts that the calculated intensities contribute to the x^2_ν values. As a result of the response weighting, the Monte Carlo calculations predict average angles that will tend to minimize errors in the primary intensities. The contrast between the two average angle techniques becomes even more apparent when this effect is considered.

Conclusions

Responses due to multiple discrete exciting source energies may be obtained with the Sherman (4,5) analytic solution by relative abundance averaging. The Sherman equations for first, second and third order fluorescence effects are applicable to XRF systems with wide beam geometrics when used with Monte Carlo predicted average incident and exit angles. Average incident and exit angles predicted purely from system geometry may be substituted for Monte Carlo predicted angles where the maximum possible deviation from the mean angle is small. It should be kept in mind however that these average angles fail to reflect changes in sample composition and a large fraction of the observed error stems from the primary intensities.

FUTURE WORK

The authors are presently working on the development of a Monte Carlo method to simulate the entire spectral response obtained from a sample. This response could then be used with a suitable non-linear least-squares method with experimental spectra to obtain ele-

TABLE I. Chi-square Values for Monte Carlo Predicted and Geometry Predicted Average Angles of Incidence and Exit

	Case 1		Case 2		Case 3	
	MC	GH	MC	GH	MC	GH
Ni	0.004	4.998	0.096	0.016	0.006	0.331
Fe	0.061	2.448	0.012	0.000	0.001	0.151
Cr	0.060	0.096	0.001	0.002	0.004	0.006
Ni-Fe	0.003	1.146	0.012	0.381	0.832	0.112
Ni-Cr	0.006	0.002	0.010	0.200	0.498	0.018
Fe-Cr	0.357	0.042	0.519	0.063	0.575	0.093
Ni-Fe-Cr	0.357	0.750	0.016	0.344	1.255	0.162
χ^2_ν	0.087	1.355	0.095	0.144	0.453	0.125

	Case 4		Case 5		Case 6	
	MC	GH	MC	GH	MC	GH
Ni	0.817	3.509	0.024	3.247	0.035	1.833
Fe	0.864	2.342	0.240	2.392	0.118	1.131
Cr	0.105	0.151	0.711	0.548	0.112	0.071
Ni-Fe	2.530	3.620	2.493	4.972	3.874	6.069
Ni-Cr	1.016	0.605	0.093	0.005	1.220	0.923
Fe-Cr	0.367	0.024	0.869	0.157	4.343	3.134
Ni-Fe-Cr	1.037	0.137	2.864	0.645	9.946	7.071
χ^2_ν	0.962	1.484	1.042	1.709	2.807	2.890

MC - Monte Carlo predicted average angles
GH - Geometry predicted average angles

Case 1 - Secondary Fluorescer, tilt angle = 0.0 degrees
Case 2 - Secondary Fluorescer, tilt angle = 30.0 degrees
Case 3 - Secondary Fluorescer, tilt angle = 60.0 degrees
Case 4 - Ring source, sample to detector = 2.75cm
Case 5 - Ring source, sample to detector = 5.0cm
Case 6 - Ring source, sample to detector = 7.0cm

mental amounts. The primary advantage of such a method would be that all of the spectral information obtained could be used in the analysis.

ACKNOWLEDGMENT

The authors gratefully acknowledge the partial financial support of the Environmental Protection Agency under Grant No. R-802759.

REFERENCES

1. J. W. Criss and L. S. Birks, "Calculation Methods for Fluorescent X-Ray Spectrometry", Analytical Chemistry, 40, 1080-1086 (1968).

2. K. Glocker and H. Schreiber, "Quantitative rontgen-spekralanalyze mit Kalterregung des spectrums", Annals of Physics, 85, 1087(1928).

3. E. Gillam and H. T. Heal, "Some Problems in the Analysis of Steels by X-Ray Fluorescence", British Journal of Applied Physics, 3, 353 (1952).

4. J. Sherman, "The Theoretical Derivation of Fluorescent X-Ray Intensities from Mixtures", Spectrochimica Acta, 7, 283-306(1955).

5. J. Sherman, "Simplification of a Formula in the Correlation of Fluorescent X-Ray Intensities from Mixtures", Spectrochimica Acta, 15, 466-470(1959).

6. T. Shiraiwa and N. Fujino, "Theoretical Calculation of Fluorescent X-Ray Intensities in Fluorescent X-Ray Spectrochemical Analysis", Japanese Journal of Applied Physics, 5,886-899(1966).

7. T. Shiraiwa and N. Fujino, "Theoretical Calculation of Fluorescent X-Ray Intensities of Nickel-Iron-Chromium Ternary Alloys", Bulletin of the Chemical Society of Japan, 40, 2289-2296(1967).

8. R. P. Gardner and A. R. Hawthorne, "Monte Carlo Simulation of the X-Ray Fluorescence Excited by Discrete Energy Photons in Homogeneous Samples", X-Ray Spectrometry, 4,138-148(1975).

9. A. R. Hawthorne and R. P. Gardner, "Monte Carlo Simulation of X-Ray Fluorescence from Homogeneous Multielement Samples Excited by Continuous and Discrete Energy Photons from X-Ray Tubes", Analytical Chemistry, 47, 2220-2225(1975).

10. A R. Hawthorne and R. P. Gardner, "Monte Carlo Models for the Inverse Calculation of Multielement Amounts in XRF Analysis," Transactions of the American Nuclear Society, Supplement No. 3, 21, 38-39(1975).

11. A. R. Hawthrone, R. P. Gardner, and T. G. Dzubay, "Monte Carlo Simulation of Self-Absorption Effects in Elemental XRF Analysis of Atmospheric Particulates Collected on Filters", in R. W. Gould, C. S. Barrett, J. B. Newkirk, and C. O. Ruud, Editors, Advances in X-Ray Analysis, 19, p.323-337, Kendall/Hunt Publishing Company (1976).

12. A. R. Hawthorne and R. P. Gardner, "Fundamental Parameters Solution to the X-Ray Fluorescence Analysis of Nickel-Iron-Chromium Alloys Including Tertiary Corrections", Analytical Chemistry, 48, 2130-2135(1976).

13. A. R. Hawthorne and R. P. Gardner, "Monte Carlo A-plications to the X-Ray Fluorescence Analysis of Aerosol Samples" in T. G. Dzubay, Editor, X-Ray Fluorescence Analysis of Environmental Samples, p. 209-220, Ann Aobor Science Publishers, Inc., Ann Arbor (1977).

14. R. P. Gardner and L. Wielopolski, "Adaptation of the Fundamental Parameters Monte Carlo Simulation Method to EDXRF Analysis with Secondary Fluorescer X-Ray Machines", in C. S. Barrett, D. E. Leyden, J. B. Newkirk, and C. O. Ruud, Editors, Advances in X-Ray Analysis, 21, p.129-142, Plenum Press (1978).

15. R. P. Gardner, L. Wielopolski, and K. Verghese,"Application of Selected Mathematical Techniques to Energy-Dispersive X-Ray Fluorescence Analysis", Atomic Energy Review, 15,(4),701-754(1977) (1977).

16. R. P. Gardner, L. Wielopolski, and K. Verghese, "Mathematical Techniques for Quantitative Elemental Analysis by Energy Dispersive X-Ray Fluorescence", Journal of Radioanalytical Chemistry, 43,611-643 (1978).

17. N. M. Shaeffer, Editor, Reactor Shielding for Nuclear Engineers, U.S. Atomic Energy Commission, Office of Information Services (1973).

18. J. M. Hammersly and D. C. Handscomb, Monte Carlo Methods, John Wiley & Sons, Inc. (1964).

19. L. L. Carter and E. D. Cashwell, "Particle-Transport Simulation with the Monte Carlo Method,TlD-26607 Technical Information Center, ter, Office of Public Affairs, US Energy Research and Development Administration (1975).

'PAREDS' - AN INTERACTIVE ON-LINE SYSTEM FOR THE INTERPRETATION OF EDXRF DATA

Wolfhard Wegscheider[1] and Donald E. Leyden

Department of Chemistry, University of Denver

Denver, Colorado 80208

ABSTRACT

Multielement data as they are obtained from energy-dispersive x-ray spectrometry can only be interpreted efficiently if multivariate statistical techniques are employed. It is shown that these and pattern recognition techniques can easily be implemented on the dedicated mini-computer system that is usually supplied with the spectrometer. These algorithms are helpful in interpreting multielement data and in gaining insight in and understanding of those aspects of the data structure that are inherent to the data. Such a program was tested using ground water data. Further extensions of the concept of integrated data interpretation are discussed.

INTRODUCTION

One of the strongest advantages of energy dispersive x-ray fluorescence analysis (EDXRF) is the rapid production of a large amount of information. This information, first being acquired as intensity data, is then projected onto the space of concentration (or mass). On practically all commercial EDXRF-systems, software of varying complexity is supported to arrive at the analytical results in almost real time either from standards or from fundamental parameters. Taking into account the multielement capability of EDXRF, huge masses of data are produced and need to be interpreted. No significant software is offered on commercial systems

[1]On leave from: Institute for General Chemistry, Micro- and Radiochemistry, Technical University, Graz, Austria.

to assist the spectroscopist with the reduction of these multidimensional data and with its adequate representation in terms of the desired information. Indeed, it is a very rare situation that concentration data provide the necessary information directly, while it appears to be more common that some other property or quality is of primary interest and multielement data only relate to it in some more or less complex manner. As a result of this unfortunate situation, one tries to resort frequently to one or a few "key" elements and tends to neglect a large amount of additional information without even considering its usefulness.

This study describes a package of programs that is based on classical multivariate statistics as well as on pattern recognition methods and can be used to interpret EDXRF-data. The major objective of this work was to investigate which routines can be efficiently implemented on mini-computers of the type provided with standard x-ray systems. A long range goal of this investigation is to sort out those programs that are less useful for interpreting multielement data and to explore the feasibility to incorporate new algorithms.

Even though most of the work in pattern recognition for chemical data analysis has been done on large computers,(12) it is thought that the experience gained in this field can be extrapolated to minicomputers without serious deficiencies. In the field of multielement data interpretation programs of this type have proven useful in the classification of archaeological items,(3) in the field of forensic analysis,(4,5) in predicting the performance of material(6) as well as in source identification of airborne particulate matter(7,8) and of oil spills.(9) The current investigation was carried out for trace element (and supplementary) groundwater data from Colorado.(10) This data set, although stemming mainly from atomic absorption results, could have easily been obtained with energy-dispersive x-ray fluorescence analysis utilizing a preconcentration technique.(11,12)

EXPERIMENTAL

The work was performed on an EDXRF system consisting of a Spectrace 440 spectrometer from Nuclear Semiconductor equipped with a NS-880 analyzer from Tracor-Northern interfaced to a 24K PDP 11/05 computer system. The peripherals consisted of a Texas Instruments Silent 700 terminal for input and output, a CRT-screen for output, a tape unit and a dual floppy disk system for storage of intermediate and permanent files. All the programming was done in FLEXTRAN, a compact and powerful higher programming language, supplied by the manufacturer. All the instructions are entered in mnemonic English using a small main program that allocates the variables, creates the overlay locations (COMMON blocks) and calls the subroutines. The package is designed to leave enough memory core for the assembler routines used for data acquisition plus 1024 channel storage for spectral data. The subroutines are grouped in three sections: one for preprocessing, one to solve the eigenvalue problem using Jacobi's procedure and one for the classification of the data. The display routines are contained in the preprocessing package to give the operator the possibility for visual inspection throughout the entire preprocessing phase.

RESULTS AND DISCUSSION

Pattern recognition systems for data interpretation can be used on a wide variety of different data and consist invariably of the same subunits; these are depicted schematically in Figure 1.(13) In the case of energy-dispersive x-ray fluorescence analysis (EDXRF) the input comes either from the pulse height analyzer in form of raw data, or after applying suitable corrections from the computer in form of mass or concentration data. The transducer stores these data now in matrix form so that the elements of each row represent the spectroscopic information for one sample (object). The columns of this matrix hold one and the same measurement (e.g. the concentration of Cu) for all the objects. Several objectives have to be met in the data preprocessing:(13)

- remove that part of the information which is irrelevant for the problem and which may even be confusing

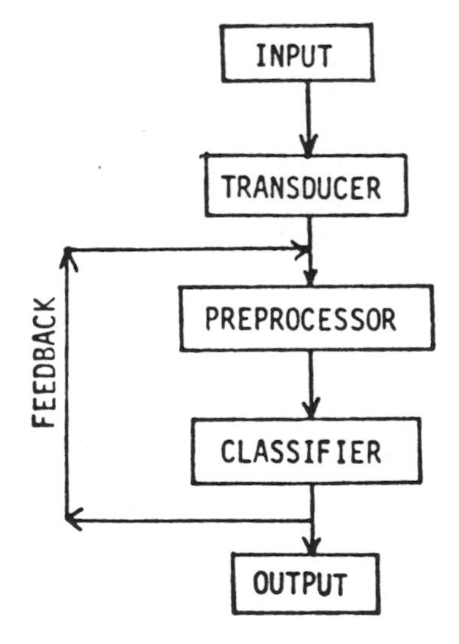

Figure 1. Basic components of a pattern recognition system

-retain enough information to emphasize the invariant properties that are inherent to a certain class of objects, and

-transform that information so that it can easily be interpreted in the next step.

Especially the preprocessing step leaves much room for imaginative experimentation as it is very much dependent on the data set.

The classification of the data can best be exemplified using a two-class problem. A decision function is sought that is applied to the transformed data set and has the following properties

$$f(\vec{x}) \leq 0 \text{ for } \vec{x} \text{ being an element of class 1} \quad (1)$$

$$> 0 \text{ for } \vec{x} \text{ being an element of class 2}$$

The process that leads to $f(\vec{x})$ is generally called training or learning. In certain cases where there is no a priori knowledge of class membership, another type of procedure, called unsupervised learning or cluster analysis, can be applied to learn something about the intrinsic properties of the data.

In the current system the data matrix is filled either by linking it to an integration routine, to a program giving concentration (or mass) data or by entering the data by hand. The latter is frequently necessary for supplementary (non-elemental) information. The first row of each vector is reserved for property data that can either be continuous or discrete. As many as 120 objects with up to 10 different elements can be analyzed at the same time. Two situations need to be discussed separately. First, in the case of missing data, these have been coded especially to recognize them later in the data analysis process and treat them separately. Secondly, data below the detection limit have been substituted by a value equivalent to one half the detection limit.

The groundwater data set that was used to show the operation of the program package consists of 149 samples (objects) with up to ten measurements made on each sample. The four different origins of these samples served as properties (Table I); from this set, 100 samples were chosen randomly, but it was seen that the 4 categories were represented about equally. This gave a sample to measurement ratio of 10 and mean ratio per class of 2.5. In the first step basic statistical information is obtained on the whole data set as summarized in Table II. As can be seen from the skewness data, the distribution of most elements are left skewed

Table I

Categories for Groundwater Data

Class #	Origin	
1	outside the Mineral Belt	without plumbing contamination
2	from the Mineral Belt	without plumbing contamination
3	outside the Mineral Belt	with plumbing contamination
4	from the Mineral Belt	with plumbing contamination

Table II

Basic Statistics of Groundwater data

Measurement	# of Missing Data	Mean	Standard Deviation	Skewness	Kurtosis
T	5	9.52[1]	3.6	-0.22	2.9
SC[5]	46	292[2]	572	5.20	32.7
pH	5	6.47	9.8	-3.38	21.3
Eh	5	419[3]	99.1	0.81	7.1
Cd	0	6.74[4]	22.7	4.84	30.1
Cu	0	55.4	122	3.32	14.0
Fe	5	605	2160	4.26	20.6
Mn	5	554	2300	5.71	39.0
Pb	0	2.04	1.64	1.55	5.8
Zn	0	1660	4500	4.62	2.9

[1]in °C [2]in µmhos/cm [3]in mV [4]all elemental data in ppb
[5]specific conductivity

suggesting a log-normal distribution. The only right skewed dis-distribution is found for the pH. To help identify these statistical properties histogram routines are supplied for the CRT-display as well as for a hard copy; these histograms can also be plotted for each class separately in case any bimodality is suggested from the initial histograms.

To assist the spectroscopist in data interpretation, several display options for the properties and measurements have been created. First, it is possible to examine pairs of measurements, as well as one measurement vs. property for continuous properties by plotting the two variables versus each other. In a number of cases, it has been found valuable to plot elemental ratios versus each other.(14) To accomplish this the original data matrix can easily be transformed to give the desired ratios since the floppy disk system permits rapid restoration of the original data if needed.

The high dimensionality of the data space, however, will give in most cases severe overlap of data points. To retain a maximum amount of information in the display process, a diagonalization of the dispersion matrix can be performed so that:

$$D = X^T X \qquad (2)$$

D...dispersion matrix
X...data matrix

Solving the eigenvalue problem

$$(D - \lambda I) = 0 \qquad (3)$$

will retain the maximal variance in the first λ_i's. If the corresponding eigenvectors V are obtained, it is possible to create a linearly transformed data matrix

$$X' = XV^T \qquad (4)$$

This has been done for the groundwater data and the result is shown in Figure 2; the two dimensions shown are the first two abstract eigenvectors and about 88% of the total information can be preserved that way. Another, but more expensive way of transformation is the "non-linear mapping"(14) that will be implemented in future editions of this program package.

Even though it is possible to map a maximum of information in just two dimensions using this routine, that information need not be pertinent to the class identification problem. It was pointed out before that the representation of the data is of great impor-

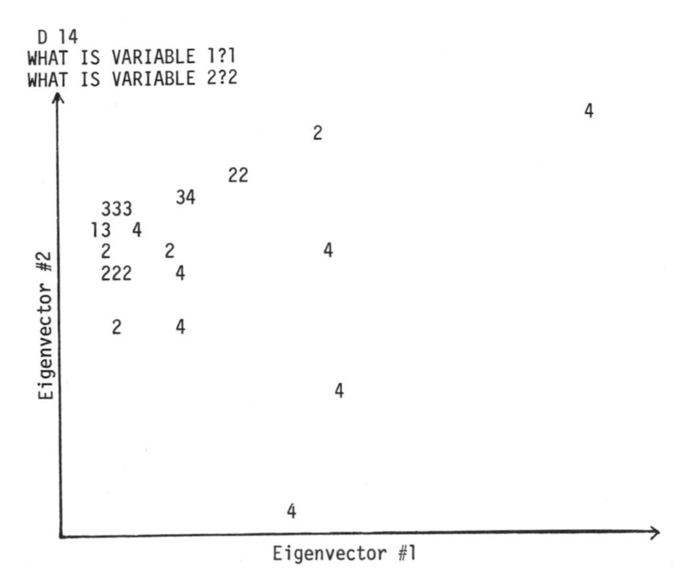

Figure 2. Two dimensional linear mapping of groundwater data retaining the maximum of variance.

tance in extracting the necessary information. A major step towards that goal usually is the scaling of data. Several options are available and they are given in Table III; the same program contains an additional option for log-transformation. In the case of the current data set, only autoscaling was employed giving a mean of 0 and a standard deviation of 1 for all data. The extraction of information in the course of preprocessing (Figure 1) is initiated using weighting options, the between to within variance ratios and the Fisher weights.(3) A partial reproduction of the printout is given in Table IV. This program part can also easily be applied to the data ratios as these are stored intermediately in the matrix of the original data. A feature vs. feature plot of the two variables with the largest overall Fisher weights is shown in Figure 3. Even though the separation of classes is far from optimal, some trends can easily be read from these plots. One of the major problems in taking the "raw" weights lies in the high correlation of the measurements; correlation and covariance matrix can easily be created in another part of the routine. These data can be visualized using a method of cluster analysis to identify related groups (Figure 4). In this case two distinct clusters can be identified and it is suggested that these contain the measurement vectors with qualitatively differing information content. Feature selection that is also essential to maintain a ratio of the number of dimensions to the number of objects of about 3 can be accomplished using a simple procedure devised by B.R. Kowalski and co-workers.(13) If feature

Table III

Scaling Options

Autoscaling	$X'_{i,j} = \dfrac{X_{i,j} - \overline{X}_i}{S_i}$
Range Scaling	$X'_{i,j} = \dfrac{X_{i,j} - X_{min_i}}{X_{max_i} - X_{min_i}}$
Mean Subtraction	$X'_{i,j} = X_{i,j} - \overline{X}_i$
Variance Normalization	$X'_{i,j} = \dfrac{X_{i,j}}{\sum_{j=1}^{m} X^2_{i,j}}$
Mean Normalization	$X'_{i,j} = \dfrac{X_{i,j}}{\overline{X}_i}$

$X'_{i,j}$ = transformed datum

$X_{i,j}$ = original datum

S_i = the standard deviation of measurement i

m = number of objects

decorrelation is desired a more complex procedure is required that will be created in later versions of the package.(1) Since the number of orthogonal dimensions for the current set is small compared to the number of objects it is possible to work even without any feature selection routine. The last step in any pattern recognition problem consists of the classification. As indicated in Figure 1 this is frequently accomplished by a recursive procedure creating a weight vector to divide either two classes, or one class from all the others, or the classes pairwise whatever seems desirable to the operator. This is accomplished using the Linear Learning Machine that is shown to converge if a solution exists.(15) The weight vector can then be used to make predictions on objects with unknown class membership. If that can be done a true intrinsic property of the objects has been identified using multielement data.(1)

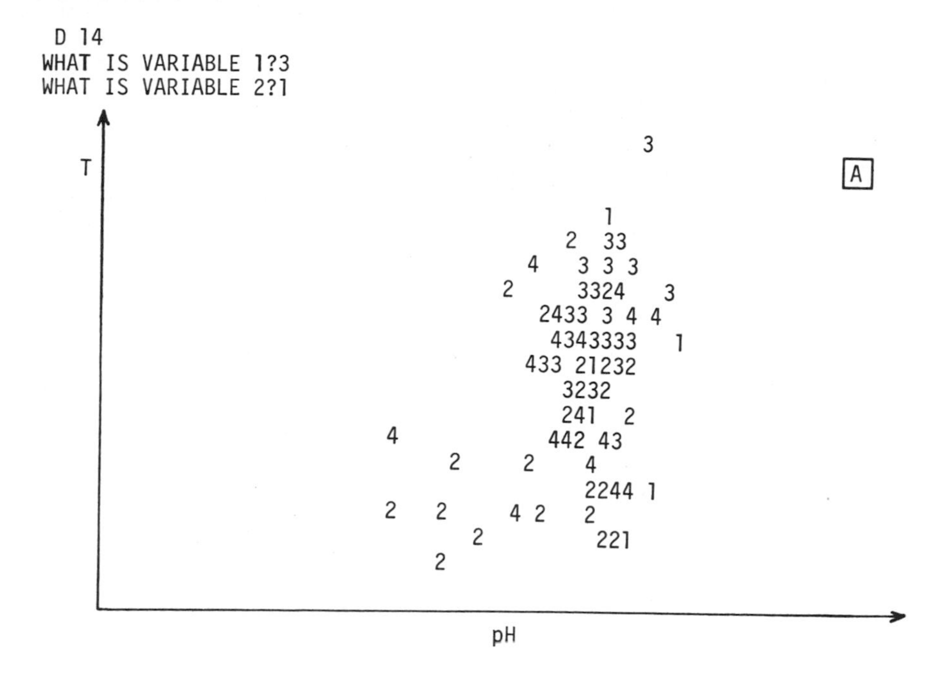

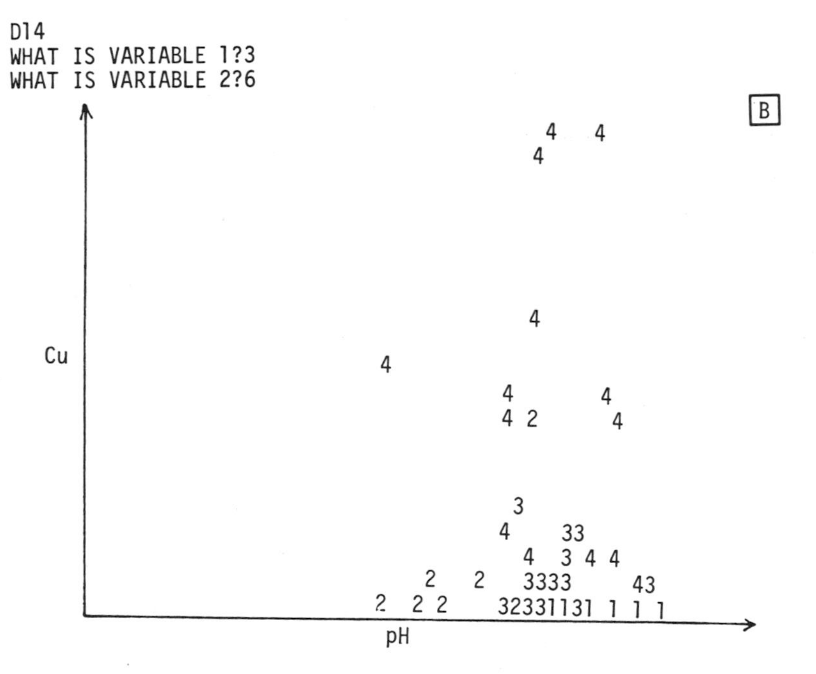

Figure 3. Display of the best variables for discriminating between groundwater from different origin; A) variance weights. B) Fisher wieghts

Table IV

Partial Reproduction of Output of the Weighting Routine

Feature	Categories	# of Items	Variance Weight	Fisher Weight
1	1/2	6/31	0.241	2.023
1	1/3	6/29	5.076	0.366
1	1/4	6/28	0.158	0.012
1	2/3	31/29	20.597	0.720
1	2/4	31/28	2.650	0.094
1	3/4	29/28	12.166	0.441
1			2.249	0.279

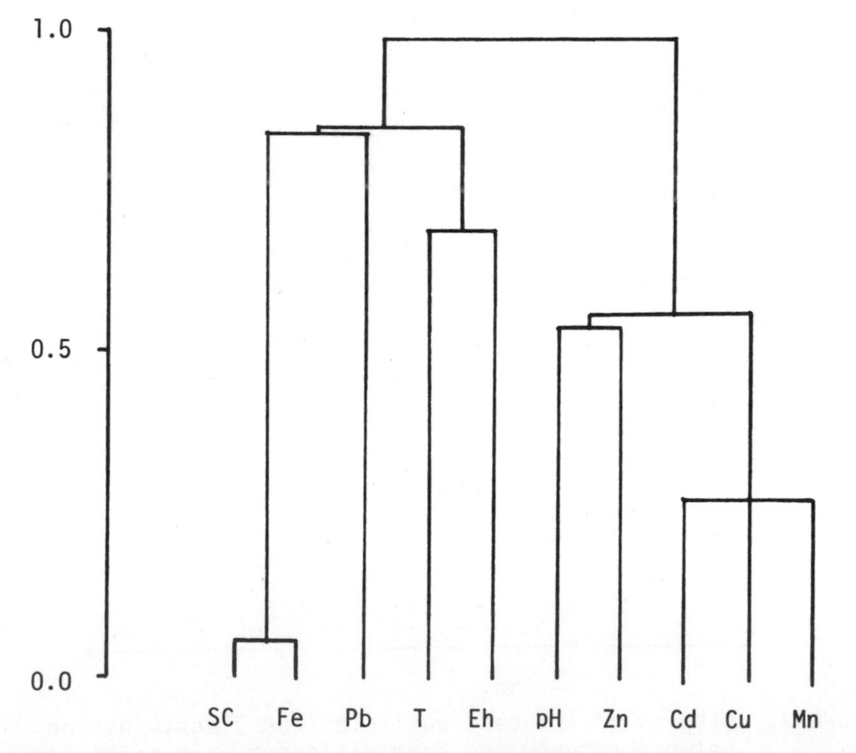

Figure 4. Hierarchical cluster analysis on the measurement correlations.

Another procedure that is widely used in literature for the classification of data is the K-Nearest-Neighbor technique (KNN); this method does not require the linear separability of classes, but can be rather time consuming if large masses of data have to be compared. "Near", is frequently defined by an inverse of the Euclidean distance or the more general Mahalanobis distance.(1,13) The class membership of an unknown is then assigned according to the majority membership of the K-Nearest-Neighbors. It is thought that this routine will provide a worthwhile addition to the current package.

In conclusion, it can be stated that the standard mini-computer systems are capable of handling a pattern recognition polyalgorithm and that this polyalgorithm facilitates on-line data interpretation with heavy operator interaction. Further work will have to sort out those algorithms that are more valuable for multi-element data than others before general recommendations can be made.

ACKNOWLEDGMENTS

This research was supported in part by the grant CHE-7618385 from the National Science Foundation. One of the authors (W.W.) thanks the Austrian-American Educational Commission for support under the Fulbright-Hays Act.

REFERENCES

1. B.R. Kowalski, "Measurement Analysis by Pattern Recognition", Anal. Chem., 47, 1152A (1975).

2. H.J. Stuper and P.C. Jurs, "ADAPT: A Computer System for Automated Data Analysis Using Pattern Recognition Techniques", J. Chem. Inf. Comp. Sci., 16, 99 (1976).

3. B.R. Kowalski, T.F. Schatzki and F.H. Stross, "Classification of Archaeological Artifacts by Applying Pattern Recognition to Trace Element Data", Anal. Chem., 44, 2176 (1972).

4. D.L. Duewer and B.R. Kowalski, "Forensic Data Analysis by Pattern Recognition Categorization of White Bond Papers by Elemental Composition", Anal. Chem., 47, 526 (1975).

5. P.J. Simon, B.C. Giessen and T.R. Copeland, "Categorization of Papers by Trace Metal Content Using Atomic Absorption Spectrometric and Pattern Recognition Techniques", Anal. Chem., 49, 2285 (1977).

6. B.R. Kowalski, "Pattern Recognition Techniques for Predicting Performance", Chemtech., May, 1974, p. 300.

7. H.E. Neustadter, J.S. Fordyce and R.B. King, "Elemental Composition of Airborne Particulates and Source Identification: Data Analysis Techniques", J. Air Pollut. Contr. Assoc., 26, 1079 (1976).

8. P.D. Gaarenstroom, S.P. Perone and J.L. Moyers, "Application of Pattern Recognition and Factor Analysis for Characterization of Atmospheric Particulate Composition in Southwest Desert Atmosphere", Environ. Sci. Technol., 11, 795 (1977).

9. D.L. Duewer, B.R. Kowalski and T.F. Schatzki, "Source Identification of Oil Spills by Pattern Recognition Analysis of Natural Elemental Composition", Anal. Chem., 47, 1573 (1975).

10. R.W. Klusman and K.W. Edwards, "Toxic Heavy Metals in Groundwater of a Portion of the Front Range Mineral Belt", Final Completion Report, No. 72, OWRT Project No. A-023-COLO, Environmental Resources Center, Colorado State University (1976).

11. D.E. Leyden, "Advances in the Preconcentration of Dissolved Ions in Water Samples", in Advances in X-ray Analysis, Vol. 20, H.F. McMurdie, C.S. Barrett, J.B. Newkirk and C.O. Ruud, Editors, p. 437 (Plenum Press, New York-London: 1977).

12. D.E. Leyden, W. Wegscheider, W.B. Bodnar, E.D. Sexton, and W.K. Nonidez, "Critical Comparison of Preconcentration Methods for Trace Ion Determination by Energy and Wavelength Dispersive X-ray Spectrometry", in preparation.

13. P.C. Jurs and T.L. Isenhour, "Chemical Applications of Pattern Recognition" (Wiley New York: 1975).

14. A.M. Harper, D.L. Duewer, B.R. Kowalski and J.L. Fashing, "ARTHUR and Experimental Data Analysis: The Heuristic Use of a Polyalgorithm", in Chemometrics: Theory and Application, B.R. Kowalski, Editor, Vol. 52, ACS Symposium Series, p. 14 (American Chemical Society, Washington, D.C.: 1977).

15. N.J. Nielsson, "Learning Machines", (McGraw-Hill, New York: 1965).

VARIATION IN INTENSITY RATIOS

USED TO IDENTIFY ASBESTOS FIBERS

John C. Russ

EDAX Laboratories

P.O. Box 135, Prairie View, IL 60069

Identification of asbestos fibers and other small particulate matter observed in TEM and STEM is often based in whole or part on the ratios of elemental intensities[1,2,3]. The underlying principle in this method is the linear relationship between concentration ratio and intensity ratio, which has been proposed by many authors, some of whom measure the proportionality factors[4] and others calculate them from quasi-theoretical relationships[5].

To be able to routinely apply this method to confidently identify particles, it is necessary to evaluate the magnitude of variation that may be encountered, and the extent to which it may allow for confusion. We have analyzed a series of asbestos minerals and others which potentially could be mistaken for them. Table 1 lists the concentrations of the major elements which would normally be chosen for analysis, and Figure 1 shows the concentration ratios Mg/Si and Fe/Si, indicating that the materials are indeed all distinct.

It is important to note that the samples available did not encompass the complete range of elemental composition for a given asbestos mineral. (eg., crocidolite from Bolivia contains much more magnesium than amosite from South Africa, although Table 1 shows higher magnesium levels in amosite than in crocidolite). The additional variation possible should be borne in mind, as it will increase the uncertainty to be described.

Figure 2 shows representative electron-excited spectra of particles at 80 kV in the STEM. From such spectra, measured for 400 seconds to obtain good counting statistics, we determined the factors k_{xSi} defined by the relationship:

$$\frac{Conc_x}{Conc_{Si}} = k_{xSi} \frac{Inten_x}{Inten_{Si}}$$

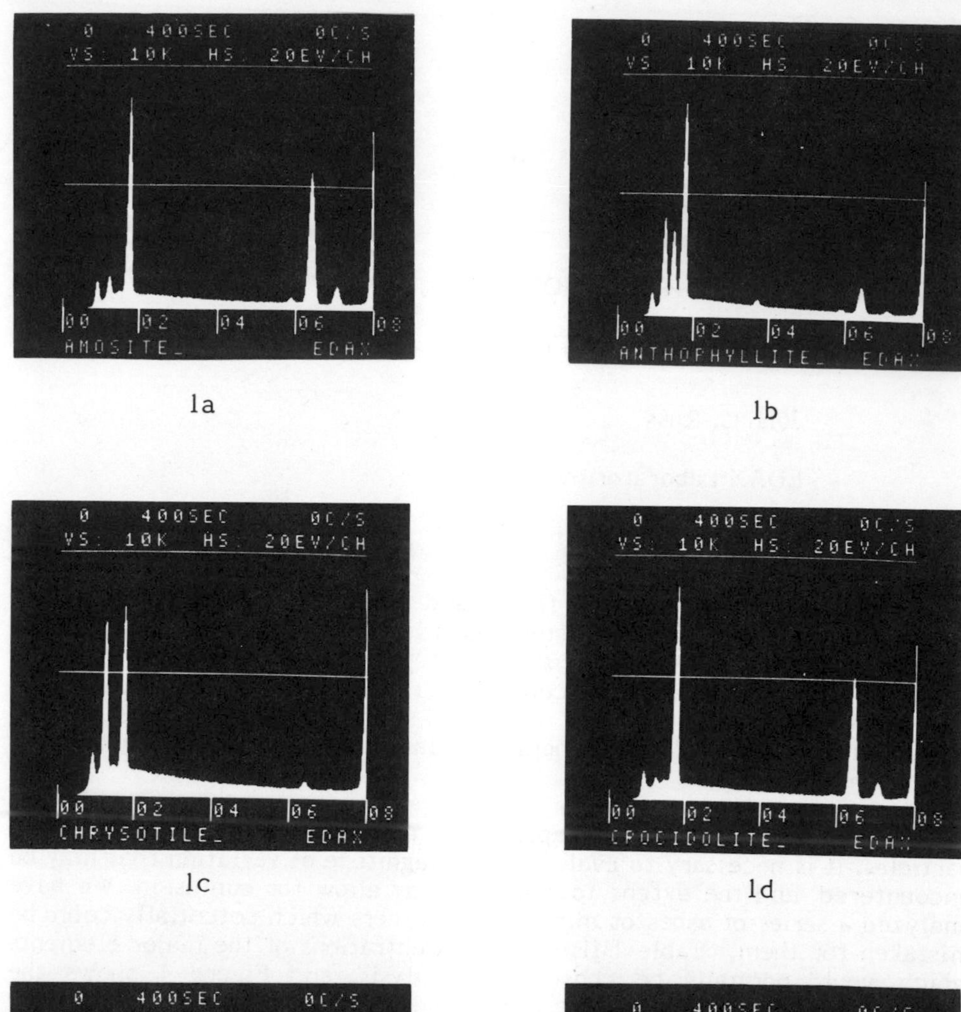

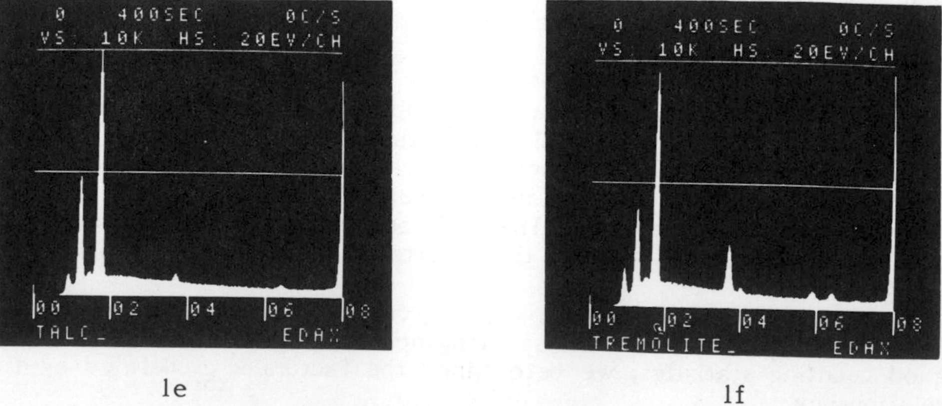

Figure 1 (a-f) Representative spectra from various mineral particles.

Table 1
Bulk concentration (w/o) of major elements (by XRF)

mineral	% Mg	% Si	% Fe
amosite	2.4 - 4.0	22.4 - 23.1	28.4 - 34.2
anthophyllite	12.7 - 17.2	21.5 - 27.1	4.3 - 8.4
chrysotile	24.0 - 25.9	18.3 - 19.6	1.4 - 2.4
crocidolite	0.6 - 1.60	22.7 - 23.8	27.9 - 29.5
talc	18.3 - 19.2	28.1 - 29.6	0.3 - 1.0
tremolite	14.4 - 14.8	26.5 - 29.6	1.0 - 3.3

Since the k_{xSi} factor is not truly a constant, but depends to some extent on particle size, orientation and matrix composition, we measured a series of particles ranging from approximately 0.1 to 5µm. and for the larger particles measured spectra both with the beam centered on the fiber or particle and with the beam scanning to cover the entire area (or for some of the longer fibers, a representative section). The total peak area above a fitted background curve was integrated and the mean bulk composition used to calculate k_{xSi} factors, which ranged as shown in Table 2.

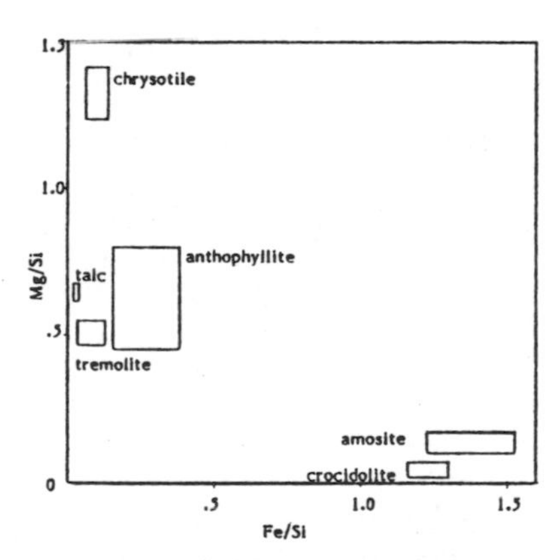

Figure 2 Plot of concentration ratios for minerals (see Table 1).

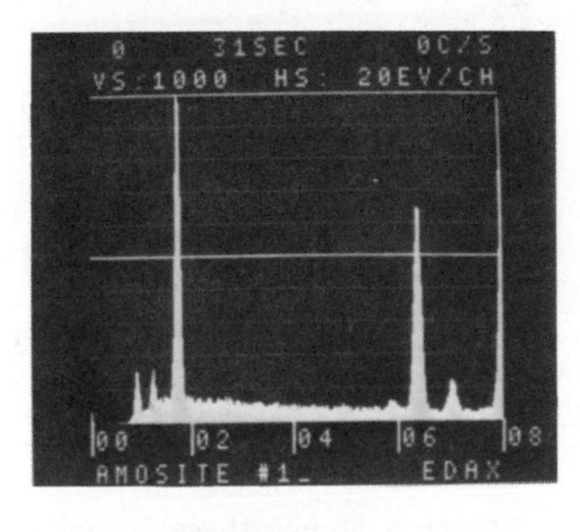

Figure 3 Repetitive short time measurement on fiber of amosite, showing poor peak statistics.

Table 2
Calculated k_{xSi} factors for different minerals and particle sizes

mineral	k_{MgSi}	k_{FeSi}
amosite	not determined	1.12 - 1.57
anthophyllite	1.30 - 1.58	1.09 - 1.61
chrysotile	1.29 - 1.42	not determined
crocidolite	not determined	1.22 - 1.59
talc	1.31 - 1.71	not determined
tremolite	1.31 - 1.65	not determined

In general, the higher values of k_{MgSi} and lower values of k_{FeSi} represent the larger particles. The overall mean values for the factors were k_{MgSi} = 1.41 and k_{FeSi} = 1.48. These values include fundamental constants such as fluorescent yield, the influence of microscope parameters such as the accelerating voltage influence on ionization cross-section, and X-ray detector parameters such as beryllium window thickness, and so will vary from one experimenter and apparatus to another.

Since practical identification of mineral particles in the microscope normally involves many particles, it is necessary to restrict the analyzing time per particle. As an apparently practical compromise, we chose to count until the silicon peak reached 1000 full scale (at which point its integrated area was 5500-6000 counts). This required times varying from less than 20 seconds for the larger particles to more than 100 seconds for the smaller ones.

For a series of ten such measurements, all on a single particle of each type (each approximately 0.5 μm in particle or fiber diameter), we then integrated the Mg, Si and Fe peaks and divided to obtain ratios. Table 3 shows the range of variation due to the statistics of counting, and Figure 3 shows a representative spectrum.

Table 3
Range of intensity ratios for ten measurements on same particle

mineral	Mg/Si	Fe/Si
amosite	0.023 - 0.181	0.917 - 1.008
anthophyllite	0.410 - 0.489	0.132 - 0.202
chrysotile	0.946 - 1.036	0.021 - 0.104
crocidolite	0.010 - 0.129	0.822 - 0.916
talc	0.402 - 0.511	0.006 - 0.051
tremolite	0.307 - 0.413	0.018 -0.111

Figure 4 shows these intensity ratio ranges graphically, and Figure 5 shows the same ranges enlarged by the error due to assuming the k_{xSi} factor is constant. It is clear that while many of the minerals can be readily distinguished, not all can. For amosite and crocidolite, some additional means of identification such as the presence of other elements, or electron diffraction must be added to the simple comparison of intensity ratios of the major elements. Identification of more (preferably all) of the elements present in a mineral can in principle end in positive identification of the mineral. For example, the sodium in crocidolite and manganese in amosite are distinctive. However, in this case the L peak of copper (from the sample grid) obscures the presence of any sodium peak, and in any case both peaks are smaller than the principal elements already used, so that longer analysis times would be needed to obtain statistically useful numbers. It is not practical to use a longer time to reduce the statistical spread, as the overlap would persist.

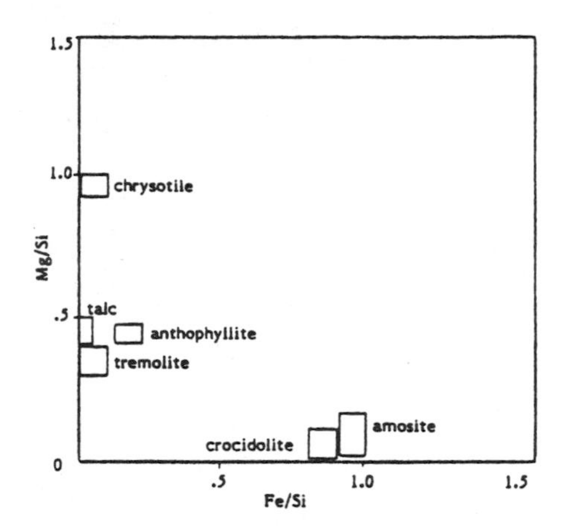

Figure 4 Plot of intensity ratios for minerals (see Table 3).

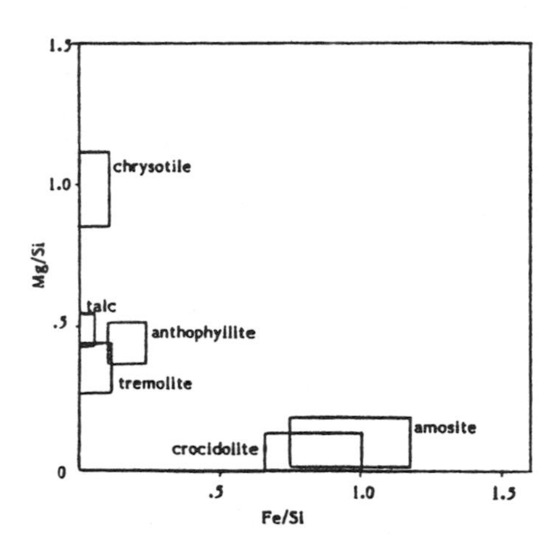

Figure 5 Plot of total range of intensity ratios, including variation in k_{xSi} due to particle size, geometry and matrix.

Individual correction of the k_{xSi} factors for particle size, and conversion to concentration, is only useful for spectra having good counting statistics, and so is too time consuming for routine application to large numbers of analyses. Nevertheless, the use of fully corrected results (i.e. the conversion of spectral data to oxide chemistry with total propagated errors due to statistics) is even more true if different investigators' results are to be compared, since the k_{xSi} factors are dependent on the accelerating voltage used. The calculation can also include atomic ratios of the cations present, which may be useful for phase identification as well.

An additional caution is that since analysis of samples of this type requires collecting data from large numbers of particles or fibers, the problem of total counting calculation time becoming large to obtain more accurate results can be prohibitive. On the other hand, the temptation to assume that the errors described in the simple use of intensity ratio matching will tend to average out when large numbers of fibers are analyzed must be resisted. This does not happen, as the errors are generally likely to be strongly biased, and not random. Confused identification is most likely to result when there are mixtures of mineral types present, a frequent occurrence in samples of real interest.

In general, the experimenter should be very careful in applying this fast and simple method. It is essential to first consider all probable overlaps of identification with other minerals which might be present.

ACKNOWLEDGEMENT

The author is particularly grateful to Dr. F.D. Pooley for advice and comments.

REFERENCES

1. D.R. Beaman and D.M. File, "Quantitative Determination of Asbestos Fiber Concentrations", Analytical Chemistry, Vol. 48, No. 1, January 1976, p. 101-110.

2. F.D. Pooley, "The Use of an Analytical Electron Microscope in the Analysis of Mineral Dusts", Phil. Trans. R. Soc. Lond. A. 286, p. 625-638, 1977.

3. F.D. Pooley, "The Identification of Asbestos Dust with an Electron Microscope Microprobe Analyzer", Norelco Reporter, Special Issue 1976, p. 5-9.

4. G. Cliff and G.W. Lorimer, "Quantitative Analysis of Thin Foils Using EMMA 4 - The Ratio Technique", Proc. Fifth Europ, Congr. on Electron Microsc. Manchester, p. 140-141, London: Inst. of Physics.

5. J.C. Russ, "The Direct Element Ratio Model for Quantitative Analysis of Thin Sections", Microprobe Analysis as Applied to Cells and Tissues, T. Hall, P. Echlin and R. Kaufmann, 1974, p. 269-276.

6. F.D. Pooley, private communication.

USE OF A NEW VERSATILE INTERACTIVE REGRESSION ANALYSIS PROGRAM FOR X-RAY FLUORESCENCE ANALYSIS

W. N. Schreiner	R. Jenkins
Philips Laboratories	Philips Elect. Instruments
Briarcliff Manor, NY	Mahwah, NJ

INTRODUCTION: XRF can be a powerful tool for quantitative elemental analysis - it can also be a big headache. The problem is, of course, that to perform a proper quantitative analysis one needs to first develop a good matrix correction model to convert the measured line intensities to concentrations. For empirical models, aside from the inherent difficulties in optimizing a set of interelement correction coefficients, the commercially available computer controlled spectrometers have in the past either provided rather poor regression analysis programs with their software or none at all. The reason for this has been the rather limited core size of the on-line computer and the lack of fast efficient mass storage facilities such as floppy disks. Thus, model selection and evaluation were difficult for the spectroscopist. Either he had to use large scale off-line computers which were inconvenient and time consuming or he blindly believed the on-line computer results because proper evaluation of regression results was difficult or even impossible.

The advent of sophisticated disc operating systems for mini-computers now provides the requisite capability to design high-speed interactive fitting routines which provide adequate error analysis to properly evaluate matrix correction models. Today I want to describe such a program to you and show how its error analysis output allows the spectroscopist to rapidly select the optimum correction terms and to determine their values based on a non-linear least squares fit to a set of standards. This program is called RUNFIT (1).

MODEL FITTING WITH RUNFIT

RUNFIT is part of the software package of the AXS (2) (Automated X-Ray Spectrometer) sold by Philips Electronics Instruments. The principle software program modules are shown in Fig. 1. SETHDWR defines which accessories and peripherals are configured in a given installation. SETELMT defines the conditions of measurement for elements. SETJOB selects the basic correction model, various measurement parameters and output requirements for on-line analysis. SETUP allows the user to run multiple samples under individually selected jobs. RUN governs the automatic operation of the instrument while MEASURE takes the actual measurements and CALCON computes the concentrations from the selected model. The data is stored on disk files and may be accessed for editing by DFEDIT, for plotting by PLOT and/or for determination of the interelement correction coefficient by RUNFIT.

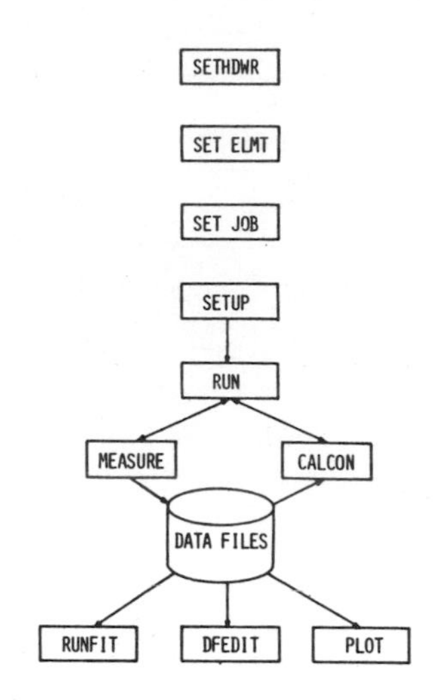

Fig. 1. AXS Software.

The system RUNFIT/CALCON have the following four models built into them:

a) Straight Line
$$C_i = A\ I_i + B$$
b) Lucas Tooth/Pyne (3)
$$C_i = (A\ I_i + B) \quad (1 + \alpha\ I_j + \ldots)$$
c) Lachance/Traill (4)
$$C_i = (A\ I_i + B) \quad (1 + \alpha\ C_j + \ldots)$$
d) Rasberry/Heinrich (5)
$$C_i = (A\ I_i + B) \quad (1 + \beta \frac{C_j}{1+C_i} + \ldots)$$

The basic straight line model can be used when there is no interference from other elements. The other three allow interelement corrections to be made by intensity or concentration or for secondary flourescence. Second degree terms in the analyte can also be included but these generally make the resulting equations unstable. The equations are arranged such that (with the exception of (b) the left hand bracket contains only instrument dependent factors and the right hand bracket contains only matrix dependent factors. By dividing through by I_i (where $B \approx 0$) any significant matrix coefficients will appear as non-zero slopes in a plot of C_i/I_i for the analyte vs. I_j, C_j or $C_j/(1+C_i)$ of the interfering element.

For this talk I chose to present fits for Stephenson's oxide mixtures (6) to demonstrate how parameter selection can be carried out using RUNFIT and PLOT.

Table I shows the concentration ranges for the Stephenson oxide mixtures. We will look at the elements Zr and Si. Fig. 2 shows a plot of C vs. I for Zr. The line through the points is the straight line least squares fit to the data points (triangles) and the dots near the line are the one standard deviation error limits for the location of the true regression line. Most of the data points fall outside these limits indicating strong matrix effects. The next question is which elements are responsible for these effects.

TABLE I. STEPHENSON'S THIRTY-SIX OXIDE MIXTURES

ELEMENT	% COMPOSITION RANGE
Zr	35 - 84
Al	5 - 54
Si	1 - 11
Ca	5 - 15
Ce	5 - 15

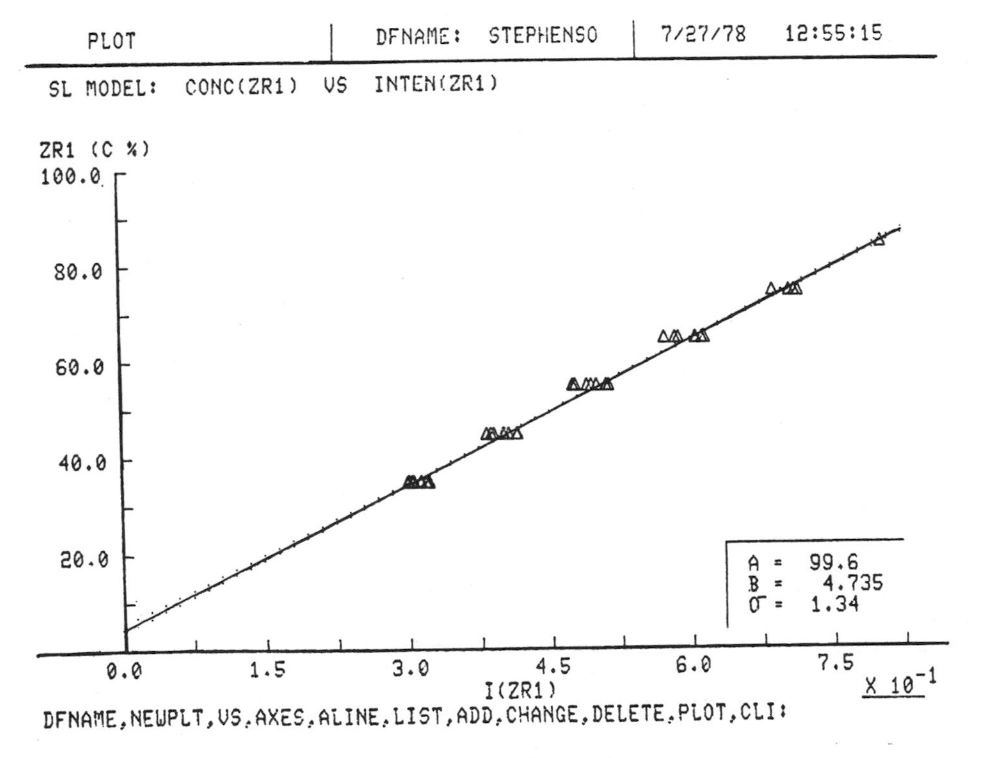

Fig. 2. Zr Data of Stephenson

With PLOT we can systematically look at each element in turn to get a feel for possible candidates before going to RUNFIT and actually performing a least square analysis. Fig. 3a-d show C/I for Zr vs. I for the elements Al, Si, Ca, Ce respectively. Due to the large scatter of points only Ca appears to have a statistically significant slope. Note, particularly in Fig. 3b, how a poor choice of standards could appear to be an interelement effect. If only the groupings of standards at the lower left and upper right were used, a very sloping line would be obtained.

RUNFIT confirms these observations based on statistical tests rather than the "eyeball method". In Fig. 4 the operator calls RUNFIT into execution and asks for program #3, an intensity correction model. He then asks for the analyte Zr and the data file STEPHENSO. The current parameter values are listed in response and the program then waits at the command line. Next he asks for intensity corrections to be IN(cluded) for Al, Si, Ca, Ce and subsequently issues the command to SOLVE for the least squares solution. The iterations proceed with the current weighted sum of squares being printed at each step, until a solution is reached, whereupon a summary of the fit is printed. A, B (the slope and intercept) and σ are listed for two fits, one without corrections and one with them. The correction coefficients and their errors are given along with a statistical t-test on the coefficients. The only coefficient which appears significant is Ca which is more than 4 standard deviations different from zero. The next step is to eliminate the fitting parameters one at a time, beginning with the least significant one first. The operator also asks to see the equations evaluated for the standards (not shown) to see if there are any outliers and to note the K value from the equation $\sigma=K\sqrt{\langle c\rangle + 0.1}$. K is a measure of the fit which effectively removes $\langle c\rangle$ from the σ. Thus one can compare fits made over different concentration ranges.

Table II shows the subsequent fitting process with the least important correcting element being omitted at each step until the final correction model is reached which only includes Ca as originally anticipated. Note that the fit is not significantly worse by excluding Si, Al and Ce but the model is much better determined since t for Ca increases from 4 to 7. Also RUNFIT converges in 2-4 iterations for this data. This is typical of a well behaved case.

We now look at a more difficult case: Si. Actually Si is no worse than Zr except that the interelement effects are weak and there are many pitfalls in trying to fit poorly determined parameters. The Si data is shown in Fig. 5 (ignoring for the moment the dashed parabola). Notice that it comes in 2 groups at 1 and 11% and that there are matrix effects. None of the plots of C/I for Si vs. Zr, Al, Ca, Ce (Fig. 6a-d) exhibit significant interelement effects with the possible exception of Zr. However, in fitting all

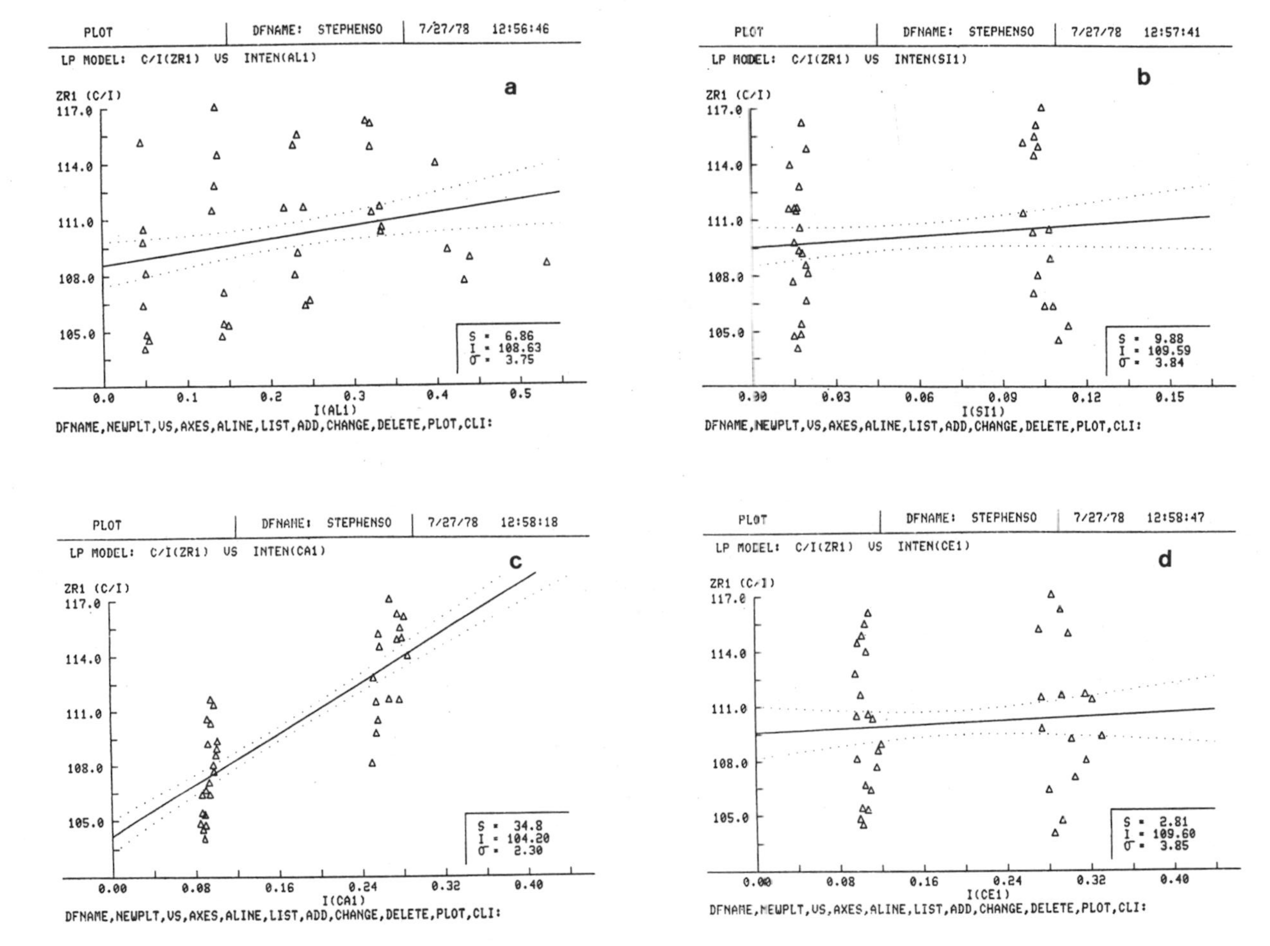

Fig. 3a-d. Possible Corrections to Zr

possible correction elements as before (Table III) we notice that all the t's are large compared to what we might expect from the plots.

```
RUNFIT                                          CALCULATE CONCENTRATIONS? Y
RUNFIT - REV 0.10
ENTER PROGRAM # 3
PROGRAM  3      LP MODEL - STEPHENSON OXIDE MIXTURES
VALID ANALYTES ARE:  ZR1 AL1 SI1 CA1 CE1
ENTER ANALYTE: ZR1
CURRENT DATA FILE IS:  STEPHENSO
USE DATA FROM FILENAME:
ELEMENTS MEASURED:   ZR1 AL1 SI1 CA1 CE1
ANALYTE - ZR1    A =    0.00000    B =    0.00000
CORR EL     ZR1        AL1        SI1        CA1        CE1
COEF     0.00000  0.00000  0.00000  0.00000  0.00000
ERROR    0.00000  0.00000  0.00000  0.00000  0.00000
PROG#,MODEL,ANALYTE,DFNAM,PARM,ZERO,FIX,INCL,EXCLD,DELETE,SOLVE,RESULTS,
 LSTDEV,CLI: IN AL1 SI1 CA1 CE1
VALID CORR ELEMENTS ARE: ZR1 AL1 SI1 CA1 CE1
ENTER CORR ELEMENT: AL1 SI1 CA1 CE1
ENTER CORR ELEMENT: SOLVE
PROG#,MODEL,ANALYTE,DFNAM,PARM,ZERO,FIX,INCL,EXCLD,DELETE,SOLVE,RESULTS,
 LSTDEV,CLI: SO
FITTING: AL1 SI1 CA1 CE1
WGT'D SSQ =  0.10802E  6  61.440       20.445       18.005
  18.005
CONVERGED.  SIG =    0.72771    INCR = 10    NITR,NSSC =    4  15
PROGRAM  3      LP MODEL - STEPHENSON OXIDE MIXTURES
PARAMETERS FITTED ON      7/27/78      AT     13: 5
ANALYTE = ZR1              STRAIGHT LINE FIT              CORRECTIONS FIT
  SIG/DF                      1.34660/ 34                    0.72929/ 34
  ACONST                 99.61731 +-    1.62256        98.62999 +-    0.82265
  BCONST                  4.73538 +-    0.81754         1.66954 +-    0.41101
CORR EL     TYPE            CORR COEFFICIENT           T FOR H0: COEF=0
  1  ZR1    ALPH          0.00000
  2  AL1    ALPH          0.05997 +-    0.08052           0.745
  3  SI1    ALPH          0.08566 +-    0.09737           0.880
  4  CA1    ALPH          0.31474 +-    0.07359           4.277
  5  CE1    ALPH          0.02348 +-    0.04831           0.486
PARMS FIT TO DATAFILE: STEPHENSO          5 STANDARDS DELETED:
 37  38  39  40  41
```

Fig. 4. RUNFIT Output

TABLE II. FITTING Zr TO LP MODEL (STEPHENSON OXIDE MIXTURES).

Corr Element	Value	t	SIGMA	K	Slope	NITR
Ca	0.315	4.3	0.731	0.100	98.6	4
Si	0.086	0.9				
Al	0.060	0.8				
Ce	0.023	0.5				
Ca	0.288	5.7	0.732	0.101	98.6	4
Si	0.056	0.7				
Al	0.029	0.6				
Ca	0.269	7.3	0.737	0.101	98.6	2
Si	0.033	0.5				
Ca	0.266	7.3	0.740	0.102	98.4	3

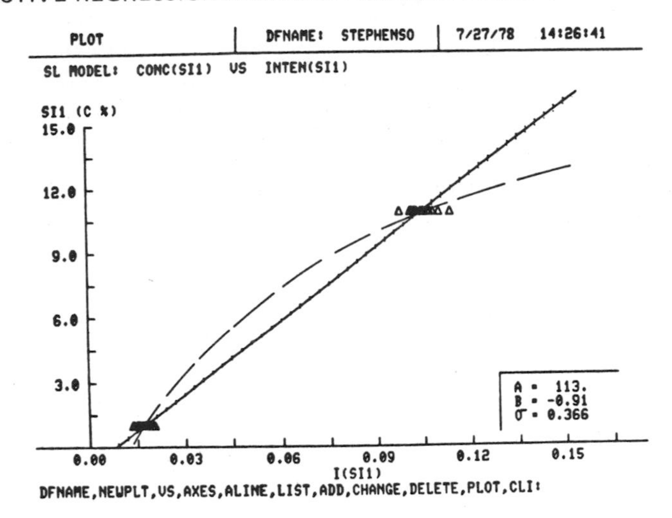

Fig. 5. Si Data of Stephenson.

TABLE III. FITTING Si TO LP MODEL (STEPHENSON OXIDE MIXTURES).

Corr Element	Value	t	SIGMA	K	Slope	NITR
Zr	-0.80	7.9	0.214	0.140	326.7	38
Al	-0.72	6.6				
Ca	-0.54	5.1				
Ce	-0.34	3.8				
Zr	-0.01	1798	0.007	0.003	- 5.1	97
Al	-0.01	1258				
Ca	-0.01	392	(BKGND=100.5)			
Ce	-0.01	392				
Zr	-0.19	2.2	0.311	0.133	123.6	3
Ce	0.25	1.7	0.323	0.138	107.8	3
Ca	0.18	1.1	0.347	0.148	109.5	5
Al	0.06	0.5	0.362	0.155	111.6	3
Zr	-0.16	1.8	0.286	0.122	117.9	4
Ce	0.17	1.3				
Si	3.85	4.0	0.295	0.126	195.7	12

Also notice that the slope is 327 (not near 100) and that RUNFIT took a long time to converge - 38 iterations. What is happening here is that since there are no strong interelement effects, I = C to within a proportionality factor and therefore in effect the answer, C(Si), is being given away in the form of 100%

minus the concentrations of all the other elements. To show that this is the case, I attempted to make a concentration correction for all the elements and the results are given in the 2nd fit of Table III. Note the enormous t values and the "excellent fit". However, all correction coefficients = -0.01 (the conversion factor from concentration to weight faction) and the slope is essentially 0 with a background of 100. In effect

$$C_{Si} = (-5I_{Si} + 100)\ [1 - 0.01\ (C_{Zr} + C_{Al} + C_{Ca} + C_{Ce})]$$
$$\simeq 100 - \sum_{j \neq Si} C_j$$

Since this constraint equation was not built into the model, it appears as an exact fit. Many people do not realize this when they attempt to correct for every element in sight.

A better way is to begin by fitting each correction element one at a time and keep only the most significant one. The remaining elements are again tried one at a time and the most significant one kept. This is repeated until little or no further improvement is achieved by adding additional correction elements. The next 5 fits of Table III show the results of this method. The final model includes only correction terms for Zr and Ce.

One final example is an attempt at a self absorption correction for Si on Si (last fit in Table III). Here t looks alright but RUNFIT had more difficulty in converging than usual and the coefficients look somewhat abnormal. This fit is shown as the dashed line in Fig. 5. It is a parabola so oriented that it tends to go horizontally through the grouping of points at 11%. Clearly any sample at ≈ 6% will be in gross error if estimated with this fit. Second order corrections are notoriously unstable and should be used with extreme caution. In general, when RUNFIT has difficulty converging, the user should beware that his correction model and/or selection of standards is probably unsatisfactory.

CONCLUSIONS

It is important to note that RUNFIT (or any other program) cannot guarantee against misapplication of matrix correction models by the spectroscopist, which unfortunately is a rather common occurence. He is still responsible for choosing an appropriate mathematical model for the physical system at hand. This implies understanding the assumptions and approximations made by alternative models and if these are acceptable in any given circumstance. Where RUNFIT is useful with its error analysis output, is in allowing easy and systematic determination of which corrections are significant and which may be eliminated to result in an optimal model for the description of the system under study.

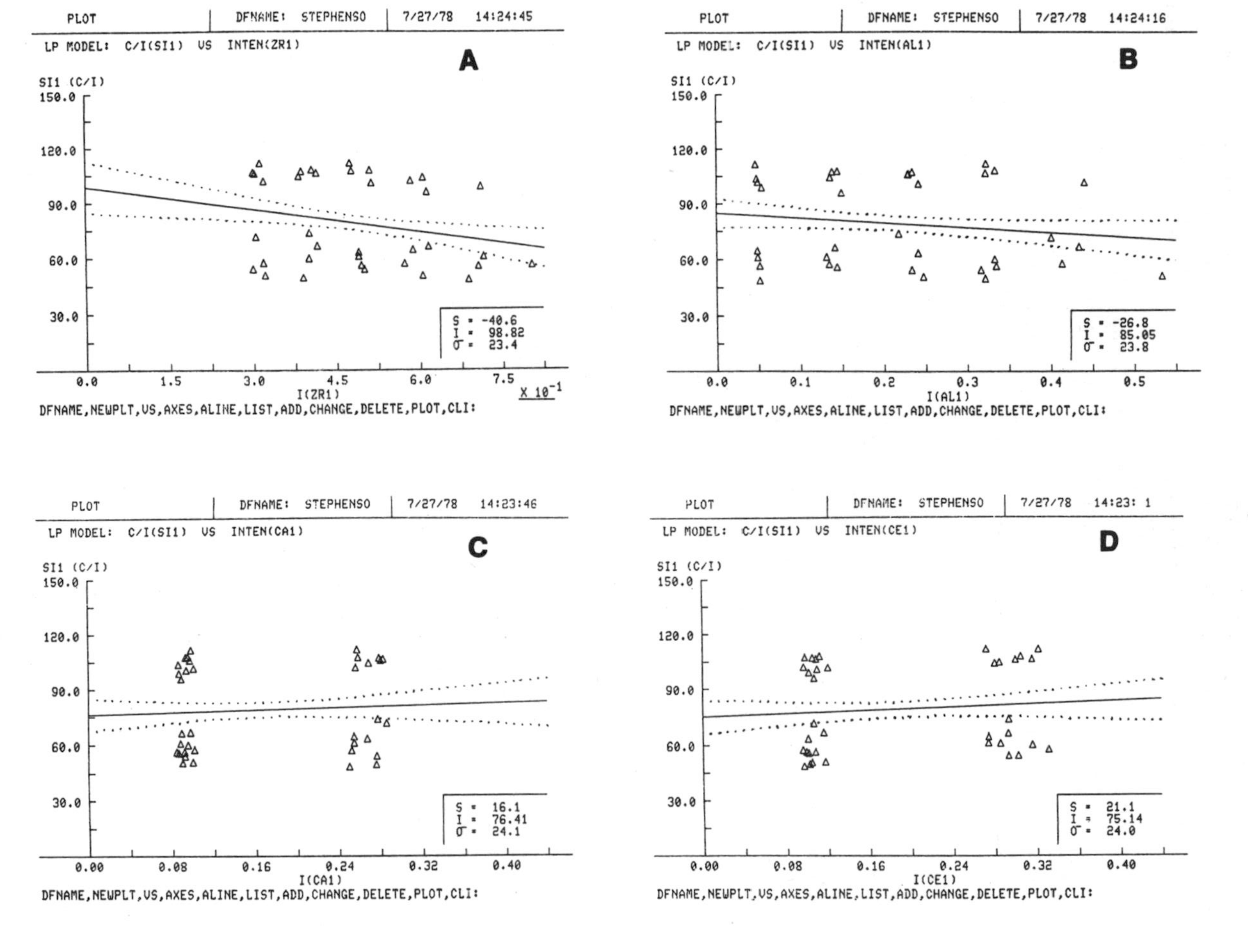

Fig. 6a-d. Possible Correction to Si.

REFERENCES

1. W.N. Schreiner and R. Jenkins, "RUNFIT - A Non-linear Least Squares Fitting Routine for Optimizing Empirical XRF Matrix Correction Models", X-Ray Spectrometry (1978) to be published.
2. R. Jenkins, Y. Hahm, D. Myers, F.R. Paolini and W.N. Schreiner, Norelco Reporter, 24, 30 (1977).
3. H.J. Lucas-Tooth and C. Pyne, Advan. X-Ray Anal., 7, 523 (1964).
4. G.R. Lachance and R.J. Traill, Can. Spectros., 11, 43 (1966) and 11, 63 (1966).
5. S.D. Rasberry and K.F.J. Heinrich, Anal. Chem, 46, 81 (1974).
6. D.A. Stephenson, Anal. Chem., 43, 310 (1971).

MODELLING INTENSITY AND CONCENTRATION IN ENERGY DISPERSIVE X-RAY FLUORESCENCE

R.B. Shen, J.C. Russ, and W. Stroeve

EDAX International, Inc.

P.O. Box 135, Prairie View, IL 60090

It might seem at first glance that quantitative models relating intensity to concentration should be identical for energy - or wavelength - dispersive fluorescence analysis. In both cases the interelement effects that complicate the use of simple linear calibration curves occur in the specimen, at which time the X-rays are not yet aware of which kind of detector will be used to count them. This can be true in some cases, but not in general.

The first, rather simple difference is that energy dispersive counts do not normally require any calculated dead time correction, since the counting electronics do that automatically. This is accomplished by stopping the system clock during the time interval required to process each pulse and store the count, time during which the electronics are "dead" and cannot accept another X-ray pulse. The clock is also stopped after the rejection of piled-up pulses. These are detected in a separate amplifier with a short time constant, and if two events have occurred (two X-rays entered the detector) within a time interval less than the width of the amplified pulse, so that the pulses are overlapped and cannot be measured accurately to determine the X-ray energy, they are rejected. Then while the sytem clock is stopped, two additional X-ray pulses are processed and stored to take their place. Based on the assumption that the X-rays are truly randomly generated and statistically represent the composition of the sample, this process gives representative stored counts in the spectrum. The measured elemental intensity, if expressed in counts per (live) second, is linearly proportional to tube current even up to count rates at which the system is "dead" most of the time, and a preset analysis time of (for example) 100 seconds may require ten minutes of elapsed or clock time.

Since in some energy dispersive X-ray systems the generator current is under computer control, it may also be practical to increase the current to the point of maximum stored count rate, for most efficient analysis including the effects of dead time. In this case, the elemental intensities are expressed as counts per (live) second per microamp of tube current, which is generally different for different samples and compositions. If long term stability is achieved by measuring one or more reference standards from time to time, to compensate for aging of the tube or other components, the intensity would then be the ratio of counts (per second-microamp) on the unknown or analytical standard to the counts (per second-microamp) on the reference. In any case, the word intensity will be used henceforth to denote any of these quantities or ratios.

However, the ED spectrum, with its rather broad peaks, presents the additional problems of substantial scattered background under the peaks and varying degrees of peak overlap. It is possible to process the spectrum to obtain "pure" intensity values before relating intensities to concentration, but this is often neither easy nor, in fact, necessary. Sometimes intensities can include more than just the X-rays from a single characteristic peak.

Simple integration of an intensity "window" or "region of interest" centered on a major peak of each element (and including the underlying background and any overlapping peaks from other analyzed elements) will yield "intensity" values I_i which can be related to the concentration C_i by a family of equations of the form [1]:

$$C_i = \underbrace{I_i (K_i}_{a} + \underbrace{\Sigma\alpha_{ij} C_j)}_{b} + \underbrace{B_i}_{c} + \underbrace{B_{ij} I_j}_{d} \qquad (1)$$

Term a) represents the general sensitivity, or slope of the calibration curve. Term b) describes the effects of other elements present on the slope of the curve, and can readily be replaced by the Lucas-Tooth $\Sigma K_{ij} I_j$ or Rasberry-Heinrich $\Sigma\alpha_{ij} C_j + \Sigma\beta_{ik} C_k/(1 + C_i)$ instead of the Traill and Lachance form shown.

The user can build a proper intensity concentration model using any combination of these terms on an element-by-element basis. The Lucas-Tooth or Delta-I is particularly useful when some elements present in the sample have an important interelement effect (either because they vary significantly or are major elements) but are not themselves of interest and may not be well analyzed in the standards. The Traill-Lachance or Delta-C is most commonly used when large concentration ranges are encountered, and the Rasberry-Heinrich adds an additional term useful when strong fluorescence is expected. All of these models, of course, degenerate to a straight calibration line if no interelement terms b) are added.

The c) term is the intercept of the calibration curve, which corresponds approximately to the scattered intensity underlying the peak. Term d) can adjust the intercept somewhat for the influence of elemental concentrations on scattering, but its main purpose is to remove the overlaps due to other elements. The overlap factors can relate either the portion of a neighboring peak lying in the integration region of the element of interest to the integral of that peak itself, or the ratio of an interfering minor line to the principal line.

To illustrate the use of these models, we analyzed a set of nine Brammer alloy steel standards for iron, chromium and nickel. The chromium strongly absorbs the iron X-rays, while the nickel fluoresces iron atoms in the sample. The intensity and concentration values were used with the various models to determine the "goodness of fit". Table 1 shows the measured intensities and concentrations for the elements. Note that the iron Kβ was measured as well as the Kα; it will be used for subsequent calculations - only the Kα intensity is used in the examples immediately following. The intensity values listed below are given as ratios to the reference standard. The iron intensities of two million total count (collected in 40 live seconds) should give statistical precision better than 0.1% relative, so that the "goodness of fit" of the various models can be meaningfully compared.

Table 1

Sample	Fe			Cr		Ni	
	%	int.Kα	Kβ	%	int.	%	int.
BS-51D	96.672	2.0964	0.40371	.20	0.017520	1.75	0.019871
BS-52C	97.612	2.0946	0.39638	.08	0.019559	.125	0.003038
BS-55D	96.364	2.0191	0.38937	1.50	0.081260	.095	0.002739
BS-56C	98.357	2.1117	0.40287	.020	0.007812	.027	0.001997
BS-57D	99.091	2.1389	0.40246	.020	0.008214	.016	0.001653
BS-58B	94.319	2.0095	0.38110	1.25	0.065656	3.25	0.034230
BS-59A	96.911	2.0561	0.38418	.98	0.052931	.215	0.003678
BS-64A	98.438	2.1189	0.40070	.098	0.011874	.039	0.002169
BS-66A	97.5554	2.0919	0.38358	.050	0.009088	.019	0.001737

A linear fit of iron intensity to concentration gives a mean absolute deviation between actual and calculated concentrations of 0.443% (percent iron), and an "rms" deviation of 0.188% (relative error). The results of the other models are listed below. The use of a Delta-I correction for Ni and Cr actually degrades the results (because introducing two additional parameters reduces the number of degrees of freedom). The Delta-C results are not better (indeed, slightly worse) than the simple linear fit. The Rasberry-Heinrich model shows a substantial and significant improvement. To verify that this apparent improvement is real, and not an accidental mathematical coincidence, the R-H model was also applied with the nickel introduced as an absorption term and chromium as a fluorescence term. This of course does not correspond to the actual physical processes that occur, and this model, as shown, gives the poorest results.

In addition to the calculations using the Fe Kα intensity, the linear and R-H models were also applied to the Fe Kβ peak which is smaller by a factor

Model		Average absolute deviation	rms deviation
Linear		0.443	0.188
LTP	Cr, Ni slope correction	0.680	0.230
LCT	Cr, Ni slope correction	0.582	0.198
RH	Cr absorption correction Ni fluorescence correction	0.110	0.044
RH	Cr fluorescence correction Ni absorption correction	0.820	0.276
Kβ-linear		0.829	0.330
Kβ-RH-	Cr absorption correction Ni fluorescence correction	0.266	0.107

of about 6. The same relative improvement is noted when the proper R-H model, with chromium absorption and nickel fluorescence terms, is used. The difference in "goodness of fit" is poorer than the Kα by a ratio of very nearly $\sqrt{6}$, which we would expect simply on the basis of the effect of the smaller intensity values. This further confirms our results.

The proper use of the peak overlap term (d in the equations shown above) can be shown for the same steels by the analysis of sulfur, which is interfered with by the L line of molybdenum. The measured intensities (cps) for sulfur K (including overlap) and molydbenum K are listed below, with the concentration data.

Table 2 - Mo and S in steels

Spec. #	Mo %	Mo int.(cps)	S %	S int.(cps)	S net int.
1	.013	139.08	.005	41.93	3.24
2	.070	172.00	.092	84.25	31.10
3	.205	226.63	.062	106.73	29.57
4	.400	308.13	.020	121.58	8.61
5	.008	136.53	.028	49.88	12.31
6	.005	134.10	.018	46.33	9.82
7	.120	189.68	.011	64.03	3.10
8	.175	210.73	.021	79.13	8.96
9	.029	146.98	.028	62.10	9.94
10	.011	137.80	.009	42.03	3.90
11	.007	137.30	.099	80.35	42.44
12	.215	233.48	.014	84.68	4.51

If the data are plotted, the molybdenum points fit a straight line but the sulfur points do not. After adding the overlap term to the equation, the corrected net sulfur intensities give a much better calibration curve, as shown in Figure 1.

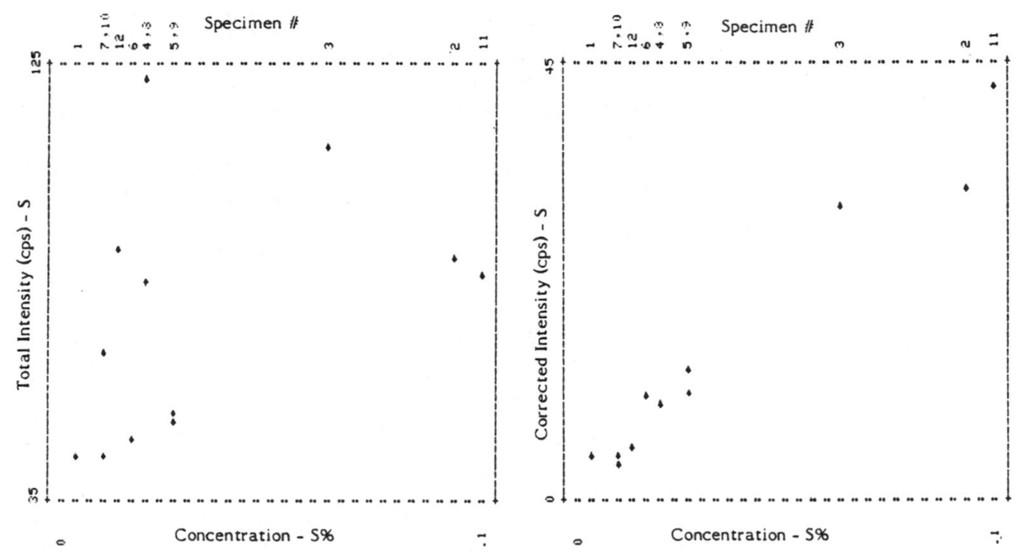

Plot of raw S intensity vs. concentration.

Plot of S intensity corrected for Mo overlap vs. S concentration.

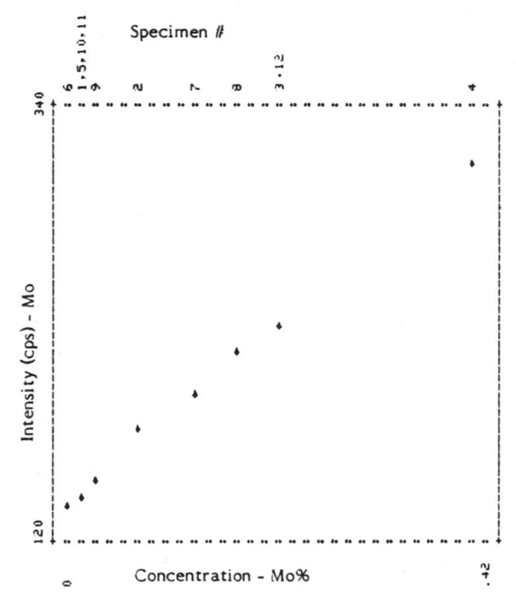

Plot of Mo intensity (cps) vs. concentration.

Figure 1: Intensity - concentration plots for Mo and S, showing effect of overlap correction.

This approach has the advantage of being easy to calculate by simple linear methods, requiring little computer time or size. Indeed, it is often used and gives excellent results. One difficulty is that terms are not "clean" but attempt to lump together several effects. For example, if we consider the case of correcting for the overlap between the Kβ line of chromium with the Kα line of manganese, the overlap factor in term d) presumes that the β/α ratio for chromium (since the Cr Kα line is used for the chromium intensity) is constant. It may not be, if the variation of matrix composition is enough to change the matrix absorption coefficient for the two different energies. This effect is much more important if different shells are involved, such as using an overlap factor to adjust the overlap of the molybdenum L lines on sulfur, based on the molybdenum Kα intensity. In this case the energies are very different, and the excitation of the different shells may be affected by the concentration of other elements. Nevertheless, it may give an adequate correction as was shown in the example above.

Another objection in some cases may be the implicit assumption that the I_i value is itself free from background and overlaps, so that in effect the element j must have a fairly major, isolated peak. This is also true, of course, if I_j values are used in term b.

A more serious objection is that coefficients have no firm theoretical basis to allow checking for "reasonableness" or supplementation by independent calculation from fundamental parameters. Also they cannot be transferred from one instrument to another nor be made completely independent of long term changes in tube output, etc.

The "intensity" values, even though they include counts from several different sources (characteristic lines of one or more overlapping elements, scattered continuum radiation, plus the peak of the element of interest), cannot be improved on a statistical precision basis by separation of the different components of the spectra. Any spectrum fitting process is limited in ultimate statistical precision to give a propagated error σ_N in the net number of counts N which is $\sqrt{N + 2B + 2S}$ where B is the subtracted underlying background and S the stripped away overlapping peak(s). To this error, which approaches the simpler value of $\sqrt{N}$ only for very large, isolated peaks, must be added any additional error in the quality of the deconvolution method. Hence, for minor peaks it may be desirable to use the simple total integral value for intensity.

To obtain net peak intensities and so eliminate terms c) and d) requires first the removal of the background. In some cases (usually involving the analysis of trace elements in a reasonably constant matrix) this can be done by subtracting a blank spectrum, which may need to be normalized to fit the unknown at some selected point(s). More often it requires fitting a curve to the spectrum at selected points. This curve may be either empirical (eg. a polynomial) or quasi-theoretical. Since the latter method requires the matrix composition, the resulting iterative calculation becomes prohibitively cumbersome and is rarely used.

We generally use a fast and simple curve fitting method, which is significantly better than fitting straight line segments and able to handle the gradually changing scattered continuum away from tube lines (either coherently - or Compton-scattered) or absorption edges due to filters interposed in front of the X-ray tube. A series of background points is selected either by the user or by automatic testing to find points (60-100 eV wide) in the spectrum which are neither significantly higher than either neighbor (and thus could be peaks) nor lower than both (and thus could be the vally between two peaks). These points are then connected by a second degree polynomial which is fit in "sliding" fashion: the first three points determine a curve fit to points one and two; then points 2, 3, and 4 are used to get the curve from points two to three, and so forth.

The next step is to fit the peak(s), preferably using a non-Gaussian shape that accurately describes the response of the detector. However, the peak position may also vary by a significant amount with time. If the peak heights and positions are all independently adjustable, a solution can only be reached by iterative methods such as a Simplex approach[2]. This is not only slow, but for real peaks with statistical uncertainties does not always converge to the "true" answer, but rather the mathematically "best" one.[3]

We prefer to use two well separated major isolated peaks to determine (by linear least squares parabolic fits to ascertain their centroids)[4] the spectrum calibration (eV per channel) and do all further fits using assumed peak positions, with non-Guassian weighted least squares fits[5]. The post-analysis recalibration of the spectrum determines the actual eV per channel, which is usually close to the nominal value (eg. 19.972 eV/ch instead of 20). The centroid energies of the elemental peaks are then used to fit the modified Gaussian shapes, which are used to obtain the individual net intensities as well as to display the individual component peaks as a confirmation of the results. The use of all the major and minor peaks for each element serves to show the overlaps and warn of the presence of unexpected peaks (diffraction or other elements) that may interfere.

The same "recalibration" method is equally useful when total intensities are used. In that case the integration window must be adjusted by interpolation between channels to compensate for slight peak shifts[6]. This is necessary to get precise, reproducible intensity values in real systems where shifts of several eV may occur due to thermal or electrical effects over a time period of hours, and where without such peak tracking the intensity values would vary by several percent.

In any case, once a file of measured intensities and known concentrations on the standards has been compiled, it is necessary for the user to build the appropriate analytical model using the appropriate terms of equation (1) with summations for interelement slope corrections and/or background effects carried out for the important elements. Displayed graphs of intensity verses concentration points and lines, and ratios of C_j/I_j help the user to select the elements of importance. As an example, the

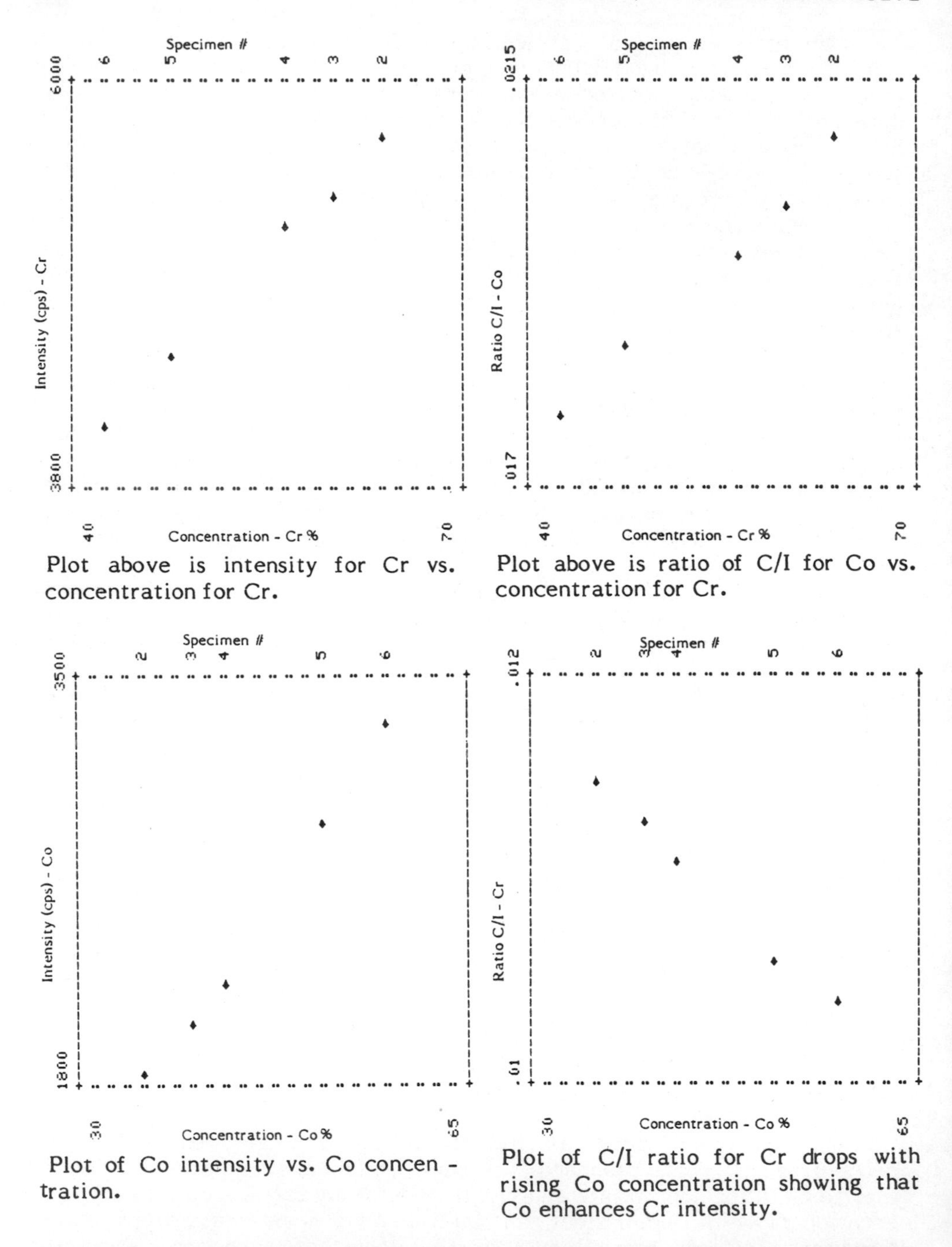

Plot above is intensity for Cr vs. concentration for Cr.

Plot above is ratio of C/I for Co vs. concentration for Cr.

Plot of Co intensity vs. Co concen - tration.

Plot of C/I ratio for Cr drops with rising Co concentration showing that Co enhances Cr intensity.

Figure 2 Variation in C/I ratio S for elements with change in concentration of other element shows that Cr has absorption effect on Co and that Co enhances Cr.

table below lists intensities from a series of binary Cr-Co alloys. Plots of the ratio C/I for each element against concentration (or intensity) of the other show linear trends, with slope equal to the coefficient in term b). The graphs shown in Figure 2 illustrate these relationships.

Table 3 - Intensity (cps) and concentration of Cr-Co alloys

Spec.	Cr		Co	
#	int.	%	int.	%
1	7388.55	100		
2	5673.23	65	1680.40	35
3	5322.83	60	1994.95	40
4	5157.23	57	2203.13	43
5	4452.90	47	2875.63	53
6	4061.83	42	3279.05	58
7			7478.28	100

The user can also choose to eliminate suspect stray points from his calculations. As this is an interactive process, the ultimate responsibility to select the right model and terms lies with the user, but the programs can assist by showing the results of different fits graphically and by calculated "goodness of fit" parameters. Once the best model relating concentrations and intensities has been found, it can be stored so that repetitive analysis of unknowns using this calibration model can be carried out by relatively unskilled operators.

REFERENCES

1. J.C. Russ, "Processing of Energy-Dispersive X-Ray Spectra", X-Ray Spectrometry, January 1977, p. 37-55.

2. C. Fiori, National Bureau of Standards, private communication.

3. P. Statham, "Reliability in Data Analysis Procedures for X-Ray Spectra", 8th Int. Cong. on X-Ray Optics & Microanalysis, Boston, Mass., August, 1977, p. 95A.

4. C. Nockolds, "Computer Assisted Analyzer Calibration", EDAX EDITor Vol. 6, No. 3, July 1976, p. 57.

5. J.C. Russ, "Resolving Spectrum Interferences Using Non-Gaussian Peaks", Canadian Journal of Spectroscopy, January-February, 1978.

6. J.C. Russ, "Getting Accurate Intensity Values From Energy-Dispersive X-Ray Spectra Using Fixed Energy Windows", Adv. in X-Ray Anal., Vol. 21, p. 221-237, Plenum Publ. corp., N.Y. (1978).

QUANTITATIVE ANALYSIS OF 300 AND 400 SERIES STAINLESS STEEL BY ENERGY DISPERSIVE X-RAY FLUORESCENCE

Bradner D. Wheeler and Nancy Jacobus

EG&G ORTEC

Oak Ridge, Tennessee 37830

ABSTRACT

Recent developments in analytical techniques and software have allowed the accurate quantitative determinations of both the major and minor elements in stainless steels by energy dispersive x-ray fluorescence. The successful analysis of 300 and 400 series stainless steel is reported utilizing this technique. The analysis of this type of material represents one of the most severe tests of the method due to numerous peak overlaps and interelement effects such as absorption and enhancement.

Sixteen standards of ASTM 300 series and ten 400 series were prepared by polishing on a 220 grit aluminum oxide belt and subsequently washing the surface in absolute methanol. Analyses were performed with an EG&G ORTEC 6110 Tube Excited Fluorescence Analyzer utilizing a dual anode (Rh/W) x-ray tube. Peak deconvolutions and interelement corrections were made with a 16K PDP-11/05 computer utilizing the program FLINT (1). Utilization of spectral deconvolutions and interelement corrections yields a relative accuracy of approximately 1% of the concentrations of the major elements.

INTRODUCTION

Quantitative analysis of 300 and 400 series stainless steels has been traditionally performed by wet chemical techniques, optical emission, atomic absorption, and wavelength x-ray fluorescence. Recent developments in solid state detectors and interelement correction programs (1), (2), (3), (4), (5), and (6) has allowed energy dispersive x-ray fluorescence to be applied in the rapid and accurate analysis of these alloys. The major and minor constituents can be determined with an accuracy equal to conventional chemical or optional instrumental techniques. The x-ray technique also has the significant additional advantage of being non-destructive and generally less time consuming.

INSTRUMENTATION

An EG&G ORETC 6110 Tube Excited Fluorescence Analyzer was utilized for this study and operated under the following instrumental parameters:

Element:	Al	Si,P,S,Ti,V,Cr, Mn,Fe,Co,Ni,Cu	Nb,Mo	Sn
Anode:	W	Rh	W	W
Anode Voltage:	5kV	15kV	45kV	45kV
Anode Current:	400 A	20 A	200 A	200 A
Filter:	None	None	Cd	Mo
Energy Scale:	0-10keV	0-20keV	0-40keV	0-40keV
Atmosphere:	Vacuum	Vacuum	Air	Air
Counting Time:	200 sec	100 sec	40 sec	40 sec

SAMPLE PREPARATION

The standards utilized in this study were obtained from the British Bureau of Standards, National Bureau of Standards, and Carpenter Steel Corporation. Each sample was prepared by first grinding on a 60 mesh belt, then final grinding on an aluminum oxide belt of 220 mesh size. Following the final grinding, the sample was washed in anhydrous ethanol and ether.

EXPERIMENTAL PROCEDURE AND RESULTS

Accurate quantitative chemical analysis by x-ray fluorescence is primarily a comparative technique in which it is desirable that the standards and samples be of a similar composition. Minor variations require that mathematical corrections be applied to the intensities of specific elements in order to correct compositional differences. Systems such as stainless steel may or may not display intensities which are directly proportional to the concentration due to the effect of another element within the sample. This is particularly true in the materials analyzed due to the presence of periodic neighbors in the energy range of 4.5 to 9 keV consisting of titanium through copper. Titanium, for example, creates a severe overlap due to the TiKβ occurring at the same energy level as VKα. A further complication is present as a result of enhancement and absorption effects. An example of this effect would be the absorption of nickel by iron since the NiKα occurs just on the high energy side of the FeK edge. A similar problem exists at the low Z end of the spectra but is also aggravated by the presence of L lines of higher energy elements such as niobium and molybdenum.

Since severe spectra overlaps occur in the region of titanium through copper, the individual peaks involved must be spectrally stripped or deconvoluted prior to any interelement correction procedure for absorption and enhancement. Zemany (6) determined vanadium, chromium, and manganese in low alloy steels by a correction method utilizing the spectrally free peak of VKα and stripping the VKβ interference from the CrKα and then subsequently stripping the corrected CrKβ from the MnKα. The spectral deconvolution routine utilized in this analysis is as follows:

$$C_{Ti} = a + bI_{TiK\alpha} \qquad (1)$$

$$C_{V} = a + b\left[I_{VK\alpha+TiK\alpha\beta}(TiK\beta/TiK\alpha)I_{Ti(1)}\right] \qquad (2)$$

where:

C_V = concentration of vanadium

a = x intercept

b = slope of calibration curve

$I_{VK\alpha+TiK\beta}$ = measured intensity of VKα and the interferring TiKβ

$(TiK_{\beta}/TiK_{\alpha})$ = ratio of TiKβ:TiKα as measured on pure titanium

Ti(1) = interference free TiKα intensity from equation (1)

$$C_{Cr} = a + b\left[I_{CrK\alpha+VK\beta}-(VK\beta/VK\alpha)I_{V(2)}\right] \qquad (3)$$

$$C_{Mn} = a + b\left[I_{MnK\alpha+CrK\beta}-(CrK\beta/CrK\alpha)I_{Cr(3)}\right] \qquad (4)$$

$$C_{Fe} = a + b\left[I_{FeK\alpha+MnK\beta}-(MnK\beta/MnK\alpha)I_{Mn(4)}\right] \qquad (5)$$

$$C_{Co} = a + b\left[I_{CoK\alpha+FeK\beta}-(FeK\beta/FeK\alpha)I_{Fe(5)}\right] \qquad (6)$$

$$C_{Ni} = a + b\left[I_{NiK\alpha+CoK\beta}-(CoK\beta/CoK\alpha)I_{Co(6)}\right] \qquad (7)$$

$$C_{Cu} = a + b\ I_{CuK\beta} \qquad (8)$$

$$C_{Si} = a + b\left[I_{SiK\alpha+WM\alpha}-(IWM\alpha/IWL\alpha)IWL\alpha\right] \qquad (9)$$

$$C_{S} = a + b\left[I_{SK\alpha+NbL\alpha+MoL\alpha}-(NbL\alpha/NbK\alpha)I_{NbK\alpha} -(MoL\alpha/MoK\alpha)IMoK\alpha\right] \qquad (10)$$

One of the most severe effects with spectral overlaps occurs in the analysis of sulfur where the NbLα and the MoLα are at the same energy level as the SKα. Figure 1 illustrates the improvement in the calibration following the spectral deconvolution routine as described in equation (10).

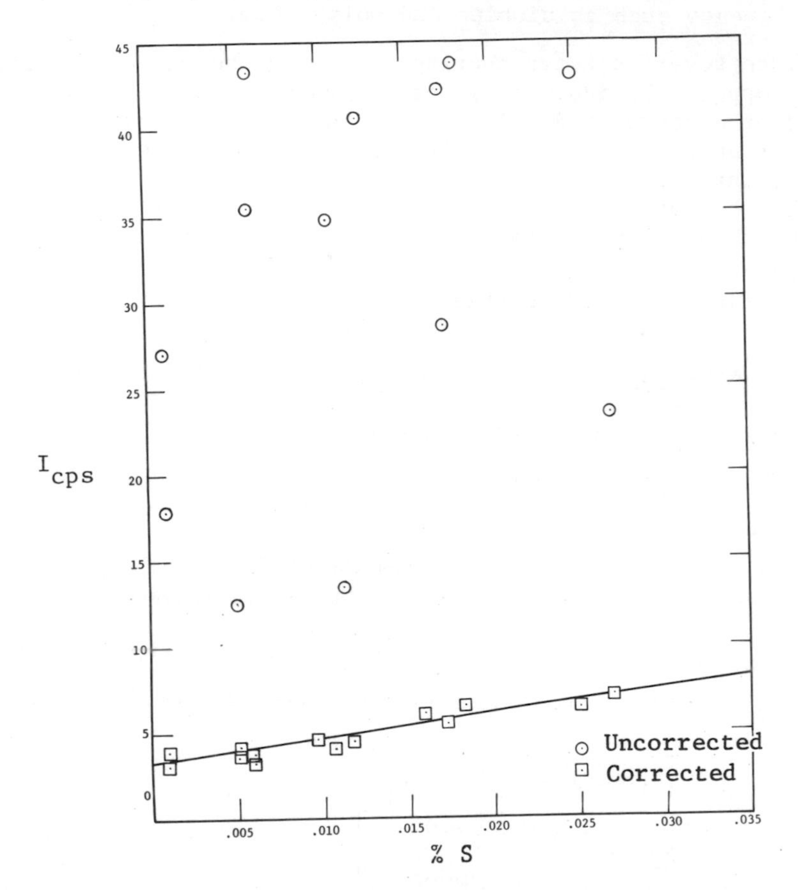

Figure 1. Sulfur Calibration Curve.

The previously stated equations only relate to spectral deconvolution and do not correct for absorption and enhancement effects which are also prevalent in these types of alloys. Referring to Figure 1, it is apparent that several elements are seriously affected -- an example being the enhancement of the PKα by the MoLα and NbLα due to their proximity to the PK edge. If the analysis is confined to samples of similar composition, the observed intensity will be approximately linear. A generalized linear calibration would be as follows:

$$C_i \approx a + bI_i \tag{11}$$

where: C_i = the concentration of the ith element

a = the x intercept of the calibration curve

b = the slope of the calibration curve

Intensities will often be reduced by absorption or increased by enhancement due to the presence of other elements in the matrix. The concentration of the ith element in this model is given by the following equation:

$$C_i = a + b'I_i \exp^{m_{ij}(I_j - N_j)} \tag{12}$$

where: m_{ij} = the interaction coefficient for element j on element i

I_j = the intensity of the jth element

N_j = the average jth element intensity of the standard

The interaction coefficients are determined by a non linear multiple least squares fit of the standard concentration -- intensity data which requires a minimum of n + 3 standards where n is the number of interfering elements. Utilization of this technique provides results summarized on Table 1.

Table 1. Accuracies of 300 and 400 Series Stainless Steel

Element	300 SERIES			Element	400 SERIES		
	Concentration Range	Average Conc.	Average Error		Concentration Range	Average Conc.	Average Error
Al	0.004- 0.210	0.065	0.006	Al	0.004 - 0.015	0.013	0.003
Si	0.35 - 0.76	0.49	0.02	Si	0.13 - 0.80	0.42	0.01
P	0.005- 0.035	0.023	0.004	P	0.010 - 0.026	0.017	0.001
S	0.001- 0.329	0.050	0.005	S	0.004 - 0.36	0.077	0.006
V	0.034- 0.130	0.060	0.011	V	0.025 - 0.09	0.047	0.008
Cr	17.11 -27.02	19.56	0.15	Cr	11.70 -25.19	15.58	0.08
Mn	0.41 - 1.78	1.47	0.04	Mn	0.38 - 1.22	0.60	0.02
o	0.06 - 0.50	0.20	0.03	Co	0.019 - 0.16	0.051	0.008
Ni	4.20 -30.65	12.41	0.15	Ni	0.24 - 8.22	1.93	0.02
Cu	0.067- 0.49	0.23	0.01	Cu	0.052 - 2.32	0.442	0.014
Nb	0.005- 0.66	0.146	0.001	Nb	0.001 - 0.67	0.238	0.001
Mo	0.022- 2.34	0.83	0.015	Mo	0.027 - 0.80	0.228	0.003
Sn	0.002- 0.014	0.008	0.001	Sn	0.004 - 0.013	0.007	0.002
				Pb	0.0001- 0.001	0.0007	0.0003
				Ag	0.0002- 0.0013	0.0005	0.0004

SUMMARY

Energy dispersive x-ray fluorescence has been shown to be a rapid and accurate method of analysis for stainless steels. Obvious advantages are that the method is non-destructive and that little sample preparation is required. Numerous peak overlaps are present but can easily be deconvoluted and resolved. Absorption and/or enhancement effects can then be solved utilizing the deconvoluted peaks.

REFERENCES

1. J. A. Cooper, B. D. Wheeler, D. M. Bartell, and D. A. Gedcke, Advances in X-Ray Analysis, Vol. 19, p. 213, (1976).

2. H. J. Lucas-Tooth and B. J. Price, "A Mathematical Method for the Investigation of Inter-Element Effects in X-Ray Fluorescent Analysis", Metallurgia, Vol. LXIV, No. 383, Sept. 1961.

3. G. R. LaChance, "A Simple Method for Converting Measured X-Ray Intensities into Mass Compositions", Geological Survey of Canada, Paper 64-50, 1964.

4. M. F. Hasler and J. W. Kemp, "Suggested Practices for Spectrochemical Computations, ASTM E-2 SM2-3.

5. S. D. Rasberry and K. F. J. Heinrich, "Calibration for Interelement Effects in X-Ray Fluorescence Analysis", Analytical Chemistry, Vol. 46, No. 1, Jan. 1974.

6. P. D. Zemany, "Line Interference Corrections for X-Ray Spectrographic Determination of Vanadium, Chromium, and Manganese in Low-Alloy Steels" Spectrochimica Acta, Vol. 16, pp. 736-741, 1960.

ENERGY-DISPERSIVE X-RAY ANALYSIS FOR CARBON ON AND IN STEELS

R. G. Musket

Kevex Corporation

Foster City, California 94404

ABSTRACT

Alpha-particle induced X-ray emission studies of steel samples have provided quantitative, non-destructive determinations of the carbon content in surface layers and in the bulk. The analyses were performed using the ALPHA-X TM technique of alpha-induced X-ray emission. Absolute carbon surface densities were determined directly from the ratio of the carbon X-ray count-rate for the unknown layers to that for standard free-standing carbon films. Assessment of the bulk detection limit was made using a steel standard with 0.5 weight-percent carbon. The three-sigma minimum detection limits for twenty minute analyses were 0.1 μg carbon/cm^2 (equivalent to $\simeq$5 Å) in the layers and 1000 ppm carbon in the bulk.

INTRODUCTION

Analysis for carbon on and in steels is important for a variety of technological reasons. Contamination of steel surfaces by very thin hydrocarbon and/or other carbonaceous layers (equivalent to as little as 20 Å of carbon) is generally recognized to seriously degrade the adhesion of primers and paints. Thus, quantitative, non-destructive determination of the surface carbon on steels is required to develop and evaluate steel cleaning, handling, and storage procedures. Since the bulk carbon concentration can have a marked influence on a number of the steel properties (e.g., brittleness, hardness, strength), quantitative assessment of the carbon content is quite important for the processing and use of steels.

In general, X-ray analysis of carbon on and in steels is limited by the low carbon-K fluorescence yield ($\omega_k \approx 0.003$), and strong absorption by the higher atomic number elements in the steel (e.g., for iron μ = 13,900 cm^2/g (1)). Due to the low fluorescence yield ω_k, the excitation mode (i.e., electrons, X-rays, or ions) must have a large ionization cross-section σ_I for the carbon K-shell to yield a reasonable X-ray production cross-section ($\sigma_X = \sigma_I \omega_k$). For thin surface layers, the optimum electron energy is about 1 keV, the X-ray energy should be just above the carbon K-edge of 284 eV (e.g., rhodium Mγ X-rays at 496 eV), and the best alpha-particle energy is about 5 MeV. The corresponding ionization cross-sections are approximately 8×10^{-19}, 3×10^{-19}, 4×10^{-18} cm^2 for 1 keV electrons (2), Rh(Mγ) X-rays (1), and 5.8 MeV alpha particles (3). The high absorption cross section for C(K) X-rays means that, even in the case of uniform ionization of the carbon atoms, ninety percent of the detected C(K) X-rays originated from within about 2000 Å of the surface of the steel. Of the above, only the alpha-particle excitation will yield essentially uniform ionization of such bulk carbon. The incident electron energy would have to be raised to achieve somewhat uniform excitation with the average effective cross section being reduced; the incident Rh(Mγ) X-rays would be strongly absorbed by the steel (μ=7,200 cm^2/g in Fe (1)) with the actual depth sampled (i.e., 90% depth) being only about 1400 Å. From these cross section and sampling-depth considerations, alpha particles should be preferred over electrons or X-rays for the excitation of carbon on or in steel.

To date, X-ray analysis of carbon in steel has been confined to wavelength dispersive X-ray spectrometry using either electron (4) or X-ray (5) excitation. However, low overall detection efficiencies for C(K) X-rays and interferences from multiple order diffractions of Cr(Lα) and Ni(Lα) X-rays and, in the case of X-ray excitation, scattered primary X-rays severely restrict the minimum detection limit (MDL). Since the bremsstrahlung radiation concommitant with alpha-particle excitation is greatly reduced compared to that for electron excitation and there is no background from scattered primary X-rays, alpha-particle excitation and energy-dispersive X-ray spectrometry with a windowless, high-efficiency (e.g., $\simeq$25% for C(K) X-rays (6))Si(Li) detector should be useful for the analysis of carbon on and in steels. The present study was designed to define the degree of usefulness of the ALPHA-X version of the alpha-induced X-ray emission technique (7).

Although detailed descriptions of the physical concepts relevant to ALPHA-X are available (7), a brief discussion of these concepts is included here to provide a basis for interpretation and evaluation of the results of the present study. For the alpha-particle energies of interest, alpha-particles interact almost exclusively with the inner and outer shell electrons of atoms in the bombarded sample. The alphas are, in effect, continuously slowed by the relatively small energy losses suffered in such Coulombic collisions. For example, 5.8 MeV alpha-particles (from the curium-244 source used in ALPHA-X) incident on an iron sample lose energy with $dE/dx \approx 340$ keV/μm (8). Thus, in traversing the first 2000 Å of iron or low alloy steel, these alpha particles lose only about 1.2 percent of their initial energy. Coupled with the relatively weak energy dependence of the ionization cross section σ_I for the carbon K-shell by 5.8 MeV alpha particles (3), this small energy loss over the 90% depth for C(K) X-rays assures essentially uniform ionization of the sampled carbon. Thus, for carbon in steel the 5.8 MeV alpha-particles probe depths limited almost entirely by the absorption of the carbon X-rays.

EXPERIMENTAL CONSIDERATIONS

In the study of carbonaceous films on low alloy steel (500 ppm carbon in the bulk), two samples were analyzed: one (Steel "L") had a lower level of carbon contamination than the other (Steel "H"). Commercially available free-standing carbon foils with known thicknesses of 5 to 100 $\mu g/cm^2$ were used as thin film standards (9). Evaluation of the MDL for carbon in the bulk of steel was made using British Chemical Standard Steel 407 (a low alloy steel with 94 w/o iron and 0.5 w/o carbon).

For the analyses, the as-received Steel "L" and Steel "H", the carbon foil standards, and the as-polished BCS Steel 407 were mounted in a twelve-sample carrousel and the carrousel was installed in an ion-pumped vacuum chamber with an ALPHA-X instrument (10), an argon gas inlet manifold, and an ion-sputtering gun (11) attached. ALPHA-X is a compact unit consisting of (a) an annular curium-244 radioactive source mounted in (b) a collimator assembly closely coupled to (c) a high-efficiency, windowless Si(Li) detector/cryostat (7). Figure 1 is a representation of the ALPHA-X probe arrangement. The collimator assembly limits the area analyzed to a 6 mm diameter spot centered on the probe axis and minimizes the background from extraneous radiation. The only adjustable parameter in an ALPHA-X analysis is the distance from the end cap of the probe to the sample; typically, the distance used was about 1 mm with the count rate decreasing about 20% per mm for additional distance. Of course, precise quantitative results of the type discussed in the present work require measurements with identical geometries for standards and unknowns. ALPHA-X mounts on the chamber via a 2-3/4" O.D. flange on the 1-1/2" I.D. gate valve which permits isolation of the retractable probe unit during specimen changes. The left side of Fig. 2 shows ALPHA-X attached to the ion-pumped specimen vacuum system. Typically, the pressure during analysis was $<10^{-7}$ Torr. The sputtering

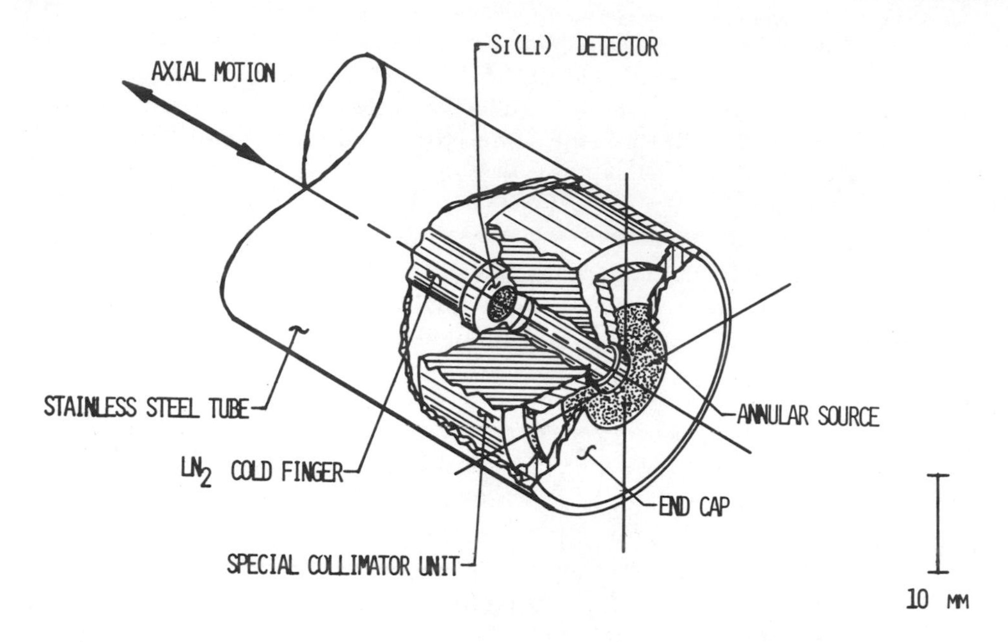

Figure 1. ALPHA-X probe arrangement.

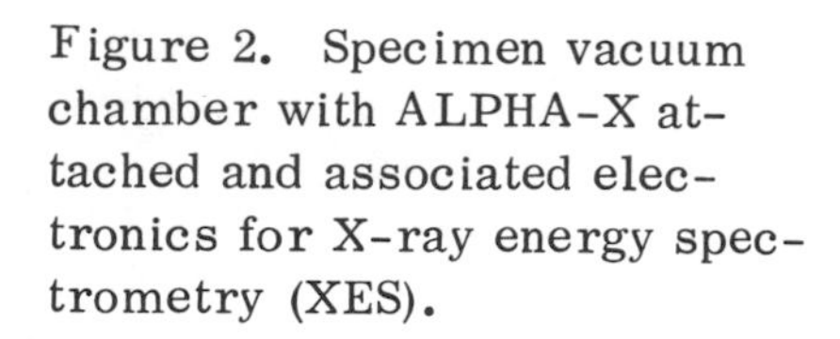

Figure 2. Specimen vacuum chamber with ALPHA-X attached and associated electronics for X-ray energy spectrometry (XES).

capability provided the means to sputter-clean specimens in situ and thereby determine the relative bulk and surface contributions to the C(K) X-ray count rate. For sputtering, the ion energy was 0.5 or 2 keV with average ion current densities of $\simeq$5 or 10 $\mu A/cm^2$ at a static argon pressure of $\simeq 6 \times 10^{-5}$ Torr. Since the sputtering yields for iron bombarded by 0.5 and 2 keV argon ions are 1.3 and 2 (12), iron would be eroded at rates of about one and four monolayers/min, respectively. However, uncertainties in estimating the sputtering yields for carbonaceous surface layers dictated a procedure of alternately sputtering and analyzing until additional sputtering produced no further reduction in the C(K) count rate. Final "clean" surface measurements on Steel "L" and BCS Steel 407 were made after sputtering for a total of 135 and 96 min, respectively.

Electronics for X-ray energy spectrometry (XES) consisted of a pulsed-optical feedback preamp, a pulse processor, an energy-to-digital converter, and an analytical spectrometer system, which provided on-line data acquisition, storage, and processing (10) (right side of Fig. 2). The data were acquired with a 10 eV/channel energy scale with the raw spectral data being stored on mini-disks. Analysis times were either 300 or 1200 seconds. A sloping-background subtraction program was used to obtain the net C(K) X-ray yields.

RESULTS AND DISCUSSION

Carbon Surface Layers on Steel

Figure 3 shows the calibration curve determined for the commercially available, free-standing carbon films (9). The horizontal error bars represent the ± 10% uncertainty in the supplier's film density measurements. Comparison of this calibration curve with the net carbon X-ray yield (i.e., net X-rays/sec) of a sample yielded the carbon surface density on the sample, provided the contribution from carbon in the bulk was negligible. No carbon X-ray absorption corrections were required for the thin layers encountered in this study of carbon surface layers. Measurement of at least one standard was made on each day that measurements were made on the samples to correct for any change in the detector efficiency for C(K) X-rays. However, over the one month duration of the experiments the efficiency was constant within one percent.

First analyses of Steel "H" and Steel "L" were made approximately 18 hours after the samples had been pumped down to $<10^{-6}$ Torr. Fig. 4A-B shows the X-ray spectrum for Steel "H". Note the high peak-to-background ratio that permits detection and quantification of very small characteristic

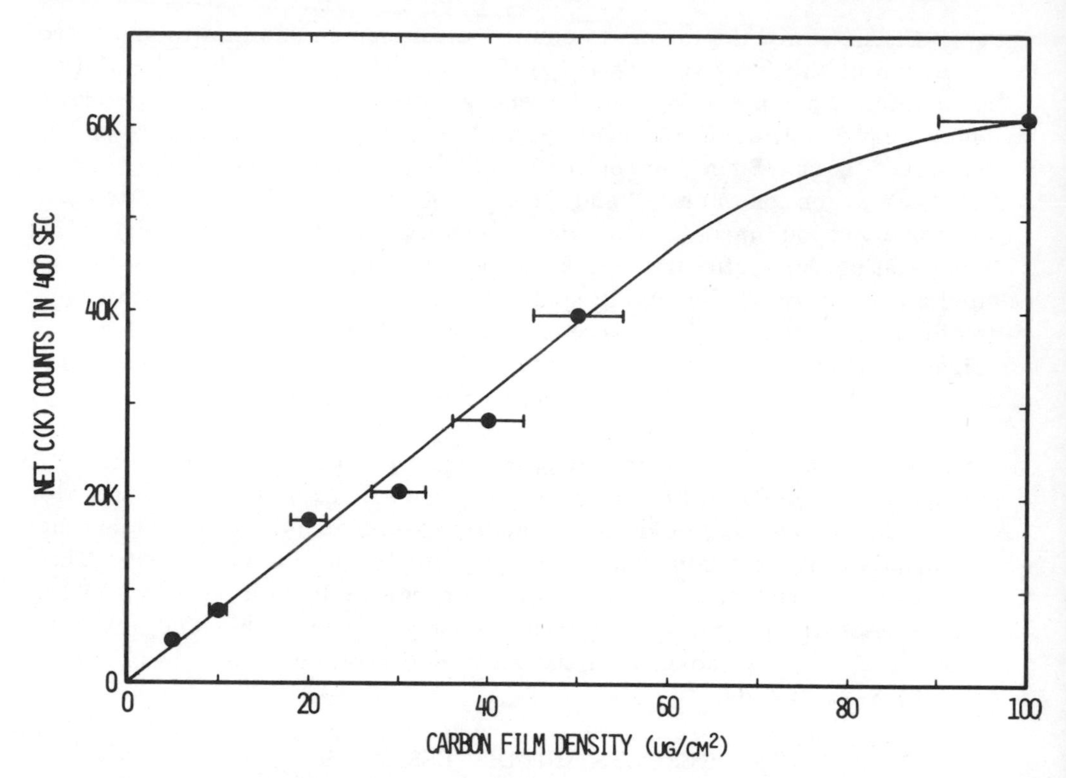

Figure 3. Calibration curve for the free-standing carbon films.

peaks corresponding to very low surface densities. The C(K) and Fe(L) X-rays resulted from ionizations by the 5.8 MeV alpha-particles, but the Fe(K) peaks were produced mainly through ionizations by the 12 to 20 keV X-rays emitted during about 11 percent of the curium-244 disintegrations. Fig. 5A-B compares the low-energy spectra for Steel "H" and Steel "L". Note how well the spectra agree everywhere except at the C(K) peak. The 1200 second analyses showed that Steel "H" and Steel "L" had 0.88 ± 0.09 and 0.25 ± 0.05 μg carbon/cm^2, respectively. The uncertainties reflect the specified accuracy of the standards (± 10%) and the counting statistics. Equivalent carbon thicknesses were 39 and 11 Å, assuming a carbon density of 2.26 g/cm^3.

Except for brief air exposure the samples were maintained at $<10^{-7}$ Torr for 17 days. After that time Steel "L" was sputter-cleaned and analyzed and Steel "H" was re-analyzed. Fig. 6A-B is a comparison of the results from the two samples. In this case, the spectra agree everywhere

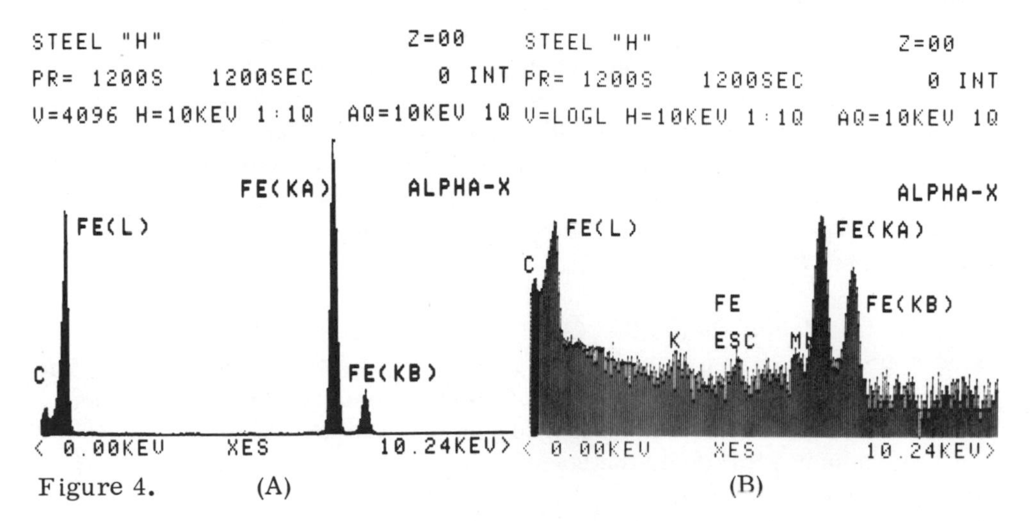

Figure 4. (A) (B)

X-ray spectrum for Steel "H" with count per channel scale (A) Linear and (B) Logarithmic.

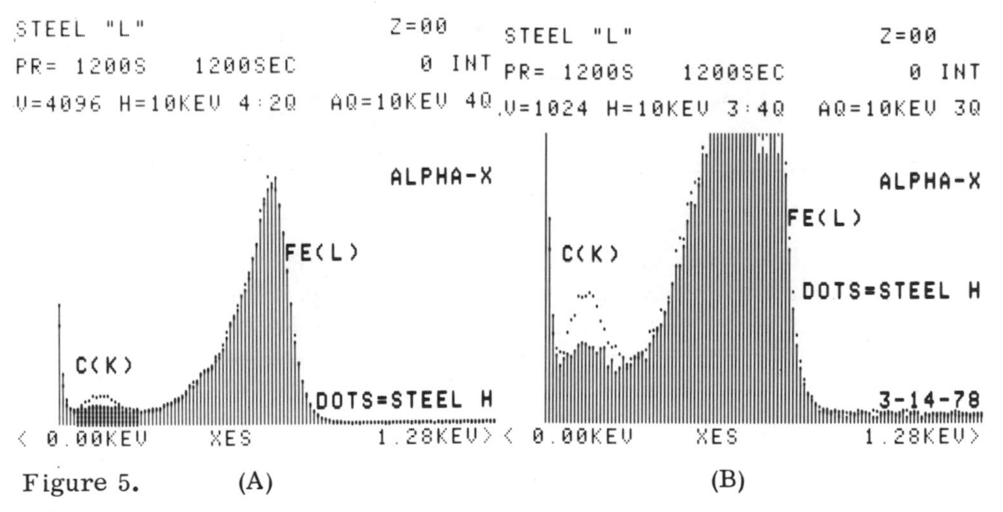

Figure 5. (A) (B)

Comparison of low-energy X-ray spectra of as-received Steel "H" and Steel "L". (A) 4096 counts full scale and (B) 1024 counts full scale.

except at the C(K) peak and at the hump corresponding to the O(K) peak of Steel "H". The sputtering removed both the surface carbon and oxygen from Steel "L". Analysis of the cleaned Steel "L" did not reveal any statistically meaningful carbon; thus, all the carbon detected on the as-received Steel "L" was surface carbon. Since the bulk carbon level was specified to be about 0.05 w/o carbon, this result indicated a minimum detection limit for carbon in steel of >0.05 w/o. Analysis of Steel "H" revealed a carbon level about 40 percent less than first measured. This suggested that a volatile component of the carbonaceous layer was sublimed during the 17-day vacuum storage interval between the first and second measurements.

The third analysis of Steel "H" was made ten days after the second and consisted of a series of six measurements over a twenty hour period of continuous alpha-particle bombardment. These measurements showed that the carbon concentration was not changed significantly by either the additional ten-day vacuum storage or the 20-hour alpha-particle irradiation (total fluence $\approx 5 \times 10^{12}/cm^2$).

Excluding the first measurement, the average of the other seven measurements on Steel "H" with the stable carbonaceous layer was 0.59 ± 0.07 μg carbon/cm^2 or 26 Å of carbon. The three-sigma minimum detection limit was determined to be 0.1 μg carbon/cm^2 (equivalent to $\simeq$5 Å of carbon) for a twenty minute analysis using

$$\text{MDL} = 3\sqrt{\text{Background counts}} \cdot \frac{(\text{Carbon surface density})}{(\text{Net C(K) counts})} .$$

Carbon in Steel

Fig. 7A-B exhibits the low-energy spectra of BCS Steel 407 before and after sputter-cleaning. Note the small-but-distinguishable differences at the energies of the C(K) and O(K) X-rays; this result is consistent with a thin oxide and light carbon contamination of the freshly polished surface. Since this steel was specified to contain 0.5 w/o carbon in the bulk, the MDL was determined to be 0.1 w/o bulk carbon for a twenty minute analysis. This result verifies the inability of ALPHA-X to detect the 0.05 w/o bulk carbon of Steel "L".

CONCLUSIONS

The ALPHA-X version of the alpha-induced X-ray emission technique has been shown to provide quantitative, non-destructive determinations of

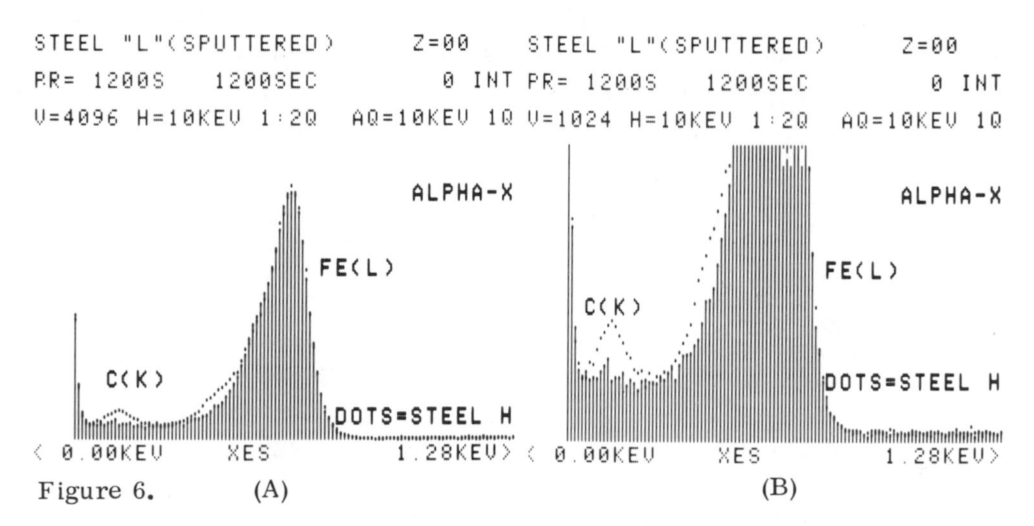

Figure 6. (A) (B)

Comparison of low-energy X-ray spectra of sputter-cleaned Steel "L" and Steel "H". (A) 4096 counts full scale and (B) 1024 counts full scale.

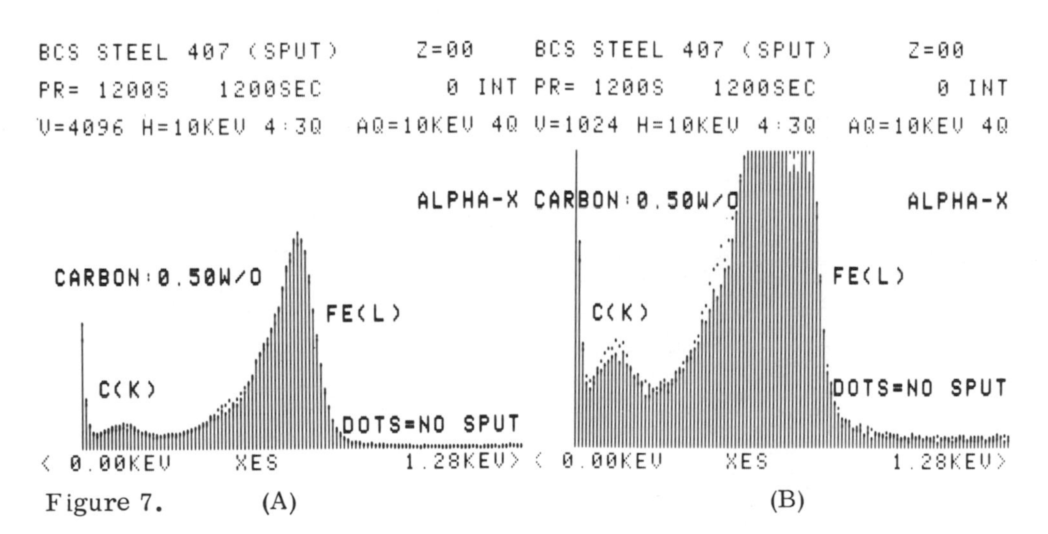

Figure 7. (A) (B)

Comparison of low-energy X-ray spectra of BCS Steel 407 before and after sputter-cleaning. (A) 4096 counts full scale and (B) 1024 counts full scale.

the carbon surface and bulk densities for steels with three-sigma detection limits of 0.1 μg carbon/cm^2 (i.e., ≃5 Å carbon) and 0.1 w/o carbon, respectively. Usefulness of ALPHA-X for such determinations is a consequence of the optimum interaction physics, ease of high-precision measurement, and direct data interpretation and analysis.

ACKNOWLEDGMENT

It is a pleasure to acknowledge the technical assistance of S. W. Taatjes.

REFERENCES

1. W. J. Veigele, Atomic Data Tables 5, 51 (1973).

2. C. J. Powell, Rev. Modern Physics 48, 33 (1976).

3. J. D. Garcia, Phys. Rev. A 4, 955 (1971).

4. F. Coppola, F. Maurice, and J. Ruste, Paper 152A, 8th International Conference on X-Ray Optics and Microanalysis and 12th Meeting of Microbeam Analysis Society, Boston, August, 1977.

5. Rigaku, Industrial Research, p. 96, February (1977).

6. R. G. Musket, Nucl. Instr. and Meth. 117, 385 (1974).

7. R. G. Musket, Research/Development 28 (10), 26 (1977).

8. J. F. Ziegler, Helium: Stopping Powers and Ranges in All Elemental Matter (Pergamon, NY, 1977).

9. Atomic Energy of Canada Limited, Ottawa, Canada.

10. Kevex Corporation, Foster City, CA 94404.

11. Physical Electronics Industries, Eden Prairie, Minn. 55343.

12. L. Maissel, in Handbook of Thin Film Technology, L. Maissel and R. Glang, eds (McGraw Hill, NY, 1970), Chapter 4.

DETERMINATION OF SOLIDS CONTENT IN SLURRIES BY X-RAY SCATTERING

J. Parus, J. Kierzek, T. Zółtowski, G. Kuc, and W. Ratyński

Institute of Nuclear Research, Warsaw, Poland

ABSTRACT

Using corrections applied to the XRF analysis of copper ore flotation materials, two methods of determining the ratio of solid to liquid content in slurries are described. Both are based on the use of coherently scattered radiation. In the first method the intensity of Cu, Pb and Fe is normalized using a coefficient defined as a ratio of scattered radiation from the sample and pure diluent. In the second, the regression equations are applied. Satisfactory results have been obtained for solid contents ranging from 10 to 40%.

INTRODUCTION

X-ray fluorescence analysis of ore processing slurries requires some means of solid content monitoring. It can be done in the simplest way by density measurement, but in plant conditions it may be inconvenient. This also increases installation and maintenance costs. On the other hand it complicates the whole system of analysis if many streams are to be analysed in sequence. It has been known for some time that coherently or elastically scattered radiation from the analysed stream can be used as a relatively accurate method of measuring solids content.

SIMPLE DENSITY CORRECTION FACTORS

Following this path we have studied in detail this problem using samples of copper concentrates, flotation heads and flotation tails from Polish copper ore processing; 18 samples of each material have been used. Copper, lead and iron were determined. The experimental set which has been described previously (1) consisted of 100 mCi Pu-238 annular source and Si(Li) spectrometer. The spectra and analytical results were evaluated with a PDP 11/45 computer. Because it is rather difficult to obtain homogenous water slurries in laboratory conditions, the slurries were simulated by mixing the powdered samples with different proportions of boric acid and pressed into pellets of 50 mm diameter. The "solid" content with respect to boric was: 10, 15, 20, 25, 30, 35 and 40%. All samples were measured in duplicate, so altogether 756 pellets have been evaluated.

As can be seen on the typical spectrum in Fig. 1, the intense peaks of coherently and incoherently scattered radiation appear. Incoherently scattered peaks were used for elimination of matrix effects (1). The intensity of copper Kα lines and coherently scattered ULβ lines is shown in Fig. 2 as a function of increasing solids content for a head sample containing about 2% Cu.

The change of both intensities goes in opposite directions as a function of dilution, so that their product should be less

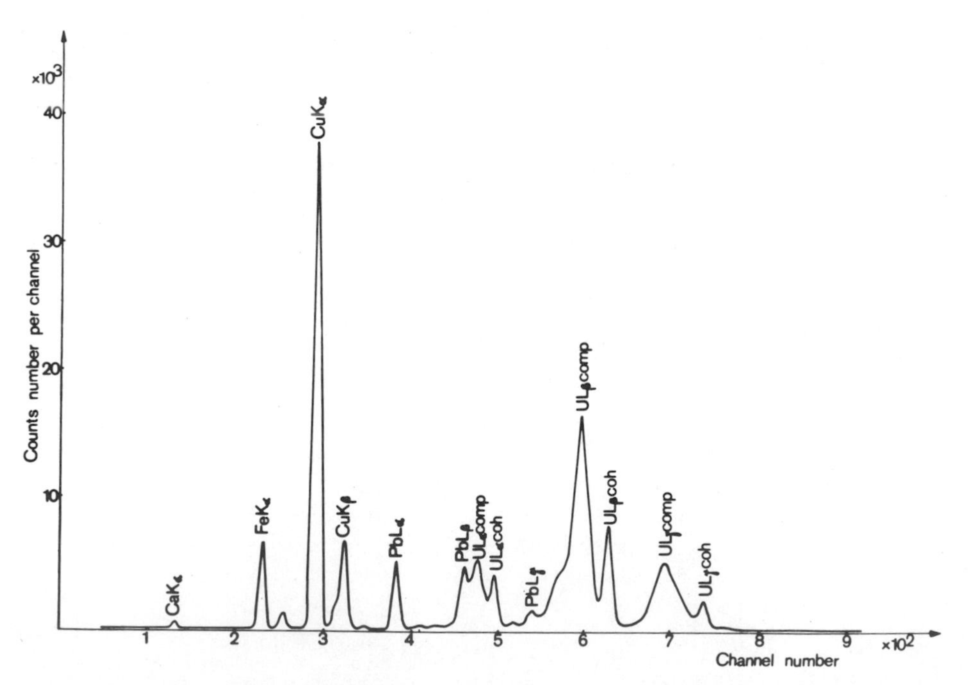

Figure 1. A typical XFA spectrum of copper ore processing materials.

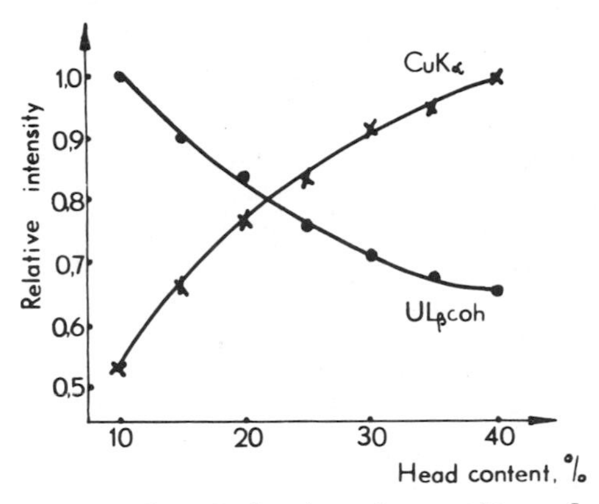

Figure 2. Relative intensity of copper Kα lines and coherently scattered ULβ lines vs. solids content for a head sample

sensitive to dilution and may be independent of it. To illustrate in detail the developed procedure the determination of copper in concentrates with variable dilution will be described. In Fig. 3 CuKα intensity is presented as a function of concentration for different dilutions. The straight line relationship between intensity and concentration has been assumed and the lines plotted according to their regression equations. To combine both intensities we introduced a coefficient B which is defined as:

$$B = \frac{\text{Intensity of coherently scattered UL}\beta\text{ for sample}}{\text{Intensity of coherently scattered UL}\beta\text{ for } H_3BO_3}$$

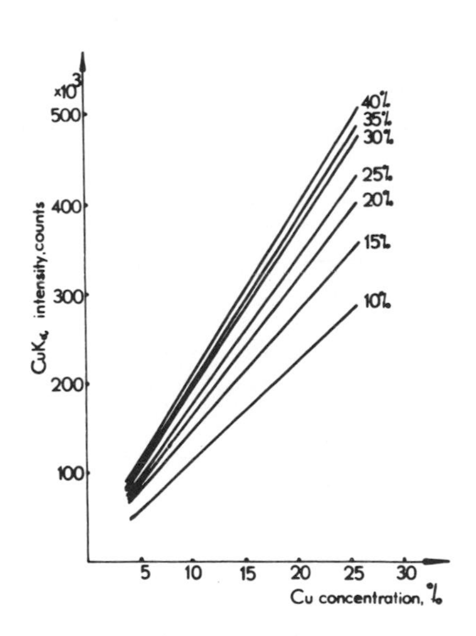

Figure 3. Cu Kα intensity vs. Cu concentration

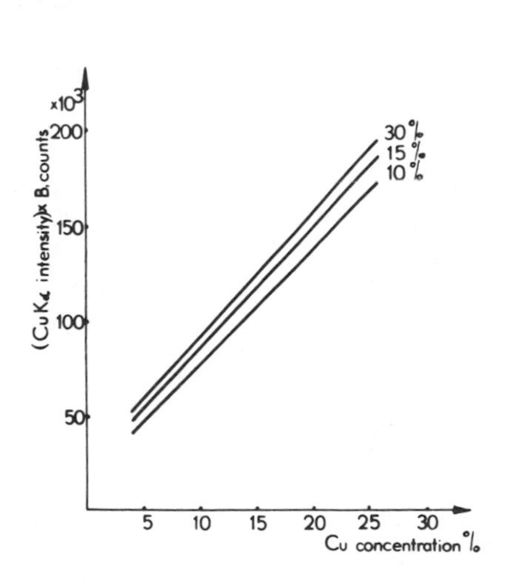

Figure 4. Cu Kα intensity x B vs. Cu concentration

The ULγ lines can also be used. Then CuKα intensities have been multiplied by suitable B coefficients and the equations of corrected regression lines calculated. The slopes a_i and intercepts a_o are shown in Table 1 and the lines in Fig. 4. Regression lines for 20, 25, 35 and 40% lie between 15 and 30% lines in Fig. 4 and are not shown. The analysis error caused by variable dilution does not exceed 1% Cu absolute except in the 10% range. Similar analysis of results has been carried out for iron and lead determination in concentrates and for all three elements in flotation heads and flotation tails. All results show usefulness of this approach, but the efficiency of this simple correction procedure for heads and tails was not as good as for concentrates.

Table 1. Parameters of regression lines

"solid" content %	intercept a_o	slope a_1
10	16793	6010
15	23187	6433
20	26377	6505
25	26378	6315
30	27710	6531
35	27614	6495
40	27238	6510

DENSITY CORRECTION WITH REGRESSION MODELS

It appears that changes in correction factors with dilution are too small to be fully effective. Therefore it was decided to check the effectiveness of the multiple regression models.

The parameters of regression models have been calculated using the program for function fitting to the experimental data based on a modified algorithm of Newton-Raphson. These models do not have linear parameters, so it is necessary to apply iteration methods. The program, however, can be executed with PDP 11/45 machine having 64 k memory.

A series of 12 models was selected. The first nine and the criteria of their selection have been described elsewhere (1). As can be seen in Fig. 5, they take into consideration the influence of other determined components, and Compton scattered radiation is used for matrix effect elimination. The influence of sample dilution is assumed to be linear against the coherently scattered radiation intensity. For calculation of model parameters all samples except the 25% dilution sample were treated as standards. The results obtained with models 10 and 12 are equally good;

Model 10S

$$Y_i = (\Theta_1+\Theta_2 I_{Cu}+\Theta_3 I_{Fe}+\Theta_4 I_{Pb}+\Theta_5 \frac{I_i}{I_{com_{i,j}}})\ \Theta_6 I_{coh_{i,j}}$$

Model 11S

$$Y_i = \left[\Theta_1+\Theta_2 I_{Cu}+\Theta_3 I_{Fe}+\Theta_4 I_{Pb}+\Theta_5 \frac{I_i}{I_{com_{i,j}}} + \Theta_6\left(\frac{I_i}{I_{com_{i,j}}}\right)^2\right]\ \Theta_7 I_{coh_{i,j}}$$

Model 12S

$$Y_i = (\Theta_1+\Theta_2 I_{Cu}+\Theta_3 I_{Fe}+\Theta_4 I_{Pb}+\Theta_5 \frac{I_i}{I_{com_\beta}+I_{com_\gamma}})\ \Theta_6(I_{coh_\beta}+I_{coh_\gamma})$$

where: $I_i = I_{Cu}$ or I_{Fe} or I_{Pb}

$I_{com_{i,j}} = I_{com_\beta}$ or I_{com_γ}

$I_{coh_{i,j}} = I_{coh_\beta}$ or I_{coh_γ}

Figure 5. Multiple regression correction models.

however, model 11 gives slightly inferior results because of its slow convergence, which is due to the quadratic term in the equation. The results of fitting for Cu, Pb and Fe in concentrates and heads are presented in Table 2. Results with ULβ lines of the source are significantly better than with ULγ lines.

The RMS values are for most cases between 10 and 15%, which is satisfactory provided they are only about twice the values for powdered, undiluted samples. S.D. est. is standard deviation of the residues, residue range gives the maximal difference of calculated and actual content, and the last two columns indicate the range of concentration for which the fitting was done. The results could be significantly improved by omitting samples with 10% solid content which differ very much from 25%, which is a mean value for the plant streams, as large dilution is rarely experienced in plant conditions.

To illustrate the validity of the fitting, scatter diagrams of the residues as a function of calculated values have been prepared. As an example, such a diagram for copper concentrates is shown in Fig. 6. Most of the points are in the range ± 1σ, the shape of scatter is random and does not show any correlation with

Table 2. Results of fitting by the correction models

		S.D. est.	RMS %	Residue Range %		Range of Fitted Concentrations %	
M 10S Concentrates	Cu_β	1.49	10.3	3.0	-3.51	25.3	0.56
	Cu_γ	1.92	11.7	6.0	-4.50	27.3	0.56
	Fe_β	0.59	12.1	1.29	-1.54	7.77	1.20
	Fe_γ	0.68	13.1	1.34	-1.55	7.68	1.21
	Pb_β	0.22	14.5	0.51	-0.51	3.24	0.18
	Pb_γ	0.24	14.8	0.53	-0.52	3.25	0.19
M 12S Concentrates	Cu	1.52	10.3	3.40	-3.32	25.9	0.49
	Fe	0.59	12.3	1.36	-1.56	7.62	1.21
	Pb	0.22	14.4	0.52	-0.51	3.27	0.18
M 10S Heads	Cu_β	0.35	12.1	1.22	-1.24	7.00	0.48
	Cu_γ	0.49	14.4	1.27	-1.70	7.01	0.54
	Fe_β	0.24	13.1	0.68	-0.57	3.70	0.25
	Fe_γ	0.35	17.2	0.72	-0.70	3.72	0.23
	Pb_β	0.05	17.3	0.12	-0.14	0.67	0.04
	Pb_γ	0.42	20.0	0.18	-0.26	0.93	0.04
M 12S Heads	Cu	0.36	11.6	0.95	-0.84	5.82	0.52
	Fe	0.24	13.0	0.67	-0.57	3.70	0.25
	Pb	0.50	17.6	0.11	-0.14	0.71	0.04

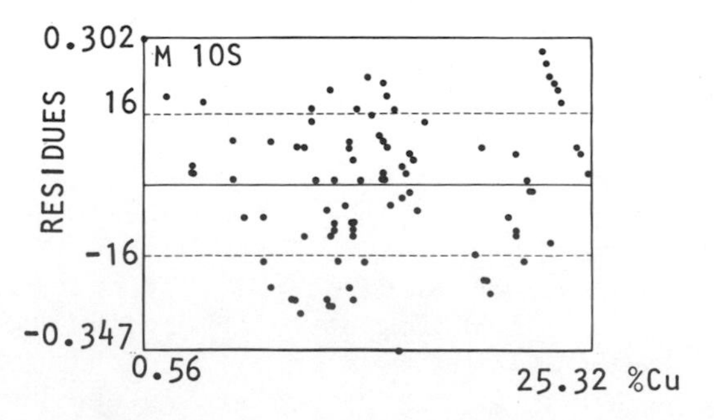

Figure 6. Residues *vs.* fitted copper concentrations

content of copper. The correlation between calculated and actual results is also typically random, independent of the actual value of dilution.

Finally some tests have been made with the same materials in the form of water slurries. The errors, *i.e.* mean standard deviation and RMS, were slightly higher, but RMS values were less than 20%. Only small correction was needed to fit the data to equations with coefficients calculated for boric acid samples.

CONCLUSION

In conclusion it can be said that, using coherently scattered radiation for correcting the changes in the solids to liquid ratio, it is possible in a relatively simple way to achieve results which are quite satisfactory for plant conditions.

REFERENCE

1. J. L. Parus, W. Ratyński, T. Zółtowski, J. Tys, J. Kierzek, E. Cieślak, and J. Kownacki, "Data Evaluation in Nondispersive X-Ray Fluorescence Analysis of Copper Ore Processing Materials," J. Radioanal. Chem. 44, 137-144 (1978).

DEVELOPMENTS IN THE USE OF X-RAYS IN BODY COMPOSITION ANALYSIS

L. E. Preuss and F. P. Bolin

Edsel B. Ford Institute for Medical Research
2799 West Grand Blvd.
Detroit, MI 48202

ABSTRACT

Quantitative assay of body components in the living human is a subject of keen interest. Measurement of such fractions on superficial view seems a simple analysis problem. In fact, the opposite is true. The 'specimen' is not amenable to the usual destructive analytical procedures. Due to its importance, a continued effort has been mounted to assay the most basic of such components. This paper and bibliography traces the recent applications of penetrating radiation for such analysis.

Some of the fractions whose measurement has been attempted are; bone mineral, adiposity (lipid to lean ratio), body hydration, tissue density, voids and spaces (i.e. lung). The importance of quantitative analysis of such gross components is apparent when taking mineralization as example. The common loss of bone apatite crystal results in collapsed vertebrae, broken femurs, crippling and morbidity in staggering numbers. Such assay thus can provide needed data on the important quantities of state of health and course of therapy.

In such measurements we are dealing with an analytical 'specimen' which imposes certain stringencies. The method must be nondestructive, rapid, simple in principle and operation, accurate, reproducible, adaptable and with a low cost-benefit ratio. The specimen's very nature also introduces certain unique problems. These are variables in; anatomical geometry, movement, positioning, specimen change with time, shadowing by overlying tissue or bone, specimen complexity (elemental through the molecular, structural and morphological levels). An assay system must take these into account.

It is instructive to consider earlier attempts at measurement; underwater weighing (body density and lean tissue mass), Lange caliper (adiposity), x-ray photodensitometry (bone mineralization and fat-lean ratio), dilution methods, i.e., deuterium oxide (hydration), fast neutron activation, (lean tissue mass), N_2 solution and dilution (muscle to fat ratio), ^{40}K (lean mass), ^{49}Ca by TNAA (bone mass), krypton absorption (lipids), Young's modulus, i.e., vibratory methods (bone mineral). Currently, some researchers are trying to remedy some of the problems of x-ray machine analysis by utilization of innovative techniques (Chan, Krokowski, Waggener, Wilson). To comment on only two: the caliper is notoriously inaccurate, and the Archimedian test is traumatic and involves substantial assumptions. Similar objections exist for the other systems. Clearly new analytical systems were needed.

The original analysis that utilized penetrating radiation from an isotopic source was that by Gershon-Cohen at the Albert Einstein Center. In 1958 he and his co-workers published a paper on a "gamma gauge" in which he passed photons from an ^{192}Ir source through tissue and bone; deriving soft tissue density, while obtaining bone thickness from a 'grid'. He claimed to have measured bone density quantitatively. His basic premise was good and he is credited as the catalyst who initiated research and development in this field. But his system and the source had flaws. As a result his ideas lay fallow for about five years.

In 1963 J. Cameron at the University of Wisconsin reported on the single beam absorptiometry using ^{125}I photons (27 keV) in a measurement on the arm bones (radius and ulna). Cameron solved the geometry problem with tissue equivalent absorbers. His contributions were; a monoenergetic source, softer radiation, stepping scans and correction for the geometry. He is credited with the advance that initiated quantitative bone assay. His quantification was in terms of mass per unit length. He did not measure density. His is a useful, accurate and precise analytical system in wide use (Zimmerman et al). Some have objected to the index of gm/cm and that the single energy does not correct for marrow lipid. Errors of the single beam bone mineral assay have been thoroughly explored (Cameron, Judy, Mazess, Sanrick, Watt, Wooton, Zeitz).

Single beam absorptiometry measures one unkown (apatite with one monochromatic photon) but a need exists to measure two components (i.e., adipose and lean tissues). A logical extension is a two-photon system solving for two unknowns. In the late sixties, workers employed isotopic sources (^{109}Cd, ^{241}Am, ^{125}I) with two photons in the 20 keV to 90 keV region. This sytem was applied to soft tissue and bone in the U.S., England, Germany, Sweden and Finland (Bolin, Burch, Gustafsson, Kan, Mazess, Pelc, Preuss, Price). It has been used for body hydration (Demling). Its accuracy, error and precision have been thoroughly established (Bolin, Watt, Mazess).

A challenging extension of dual beam absorptiometry is the approach with three or more beams and three or more tissue fractions. This is a difficult problem, but attempted by some centers with three photons (Hudepohl, Jacobson, Kairinto, Waggener). Workers at Vanderbilt University and Karolinska Institute used isotopic and x-ray tube sources respectively, but have encountered damaging errors in the absorption tables and from counting statistics. Three component analysis is in the research stage, but, if perfected, promises substantial benefit.

As early as 1955 workers at the Karolinska Institute and at AECL suggested using scattering principles for tissue analysis, but this remained in limbo until the seventies. Workers at Carleton and McMaster Universities in Canada, and then others reported on density measurement of trabecular bone using isotopic sources (Clarke, Garnett, Piper, Webber). Density can be a critical index of the state of mineralization, Clarke's and Garnett's system involves an ingenious 180° specimen rotation, eliminating the surround's, absorption and geometry effects. This is a promising analytical system since density of a specific bone region is a sought after goal.

Workers at the University of California, Newcastle, McMaster University and Siemens (Clarke, Farmer, Kauffman and Reiss), have studied Compton scatter with a single isotope and also x-ray tube sources. The isotope source is moved through two positions in Kauffman's system and density is obtained using expressions equivalent to the 180° transposition system of Clarke. Density of bone, liver and lung tissues have been obtained by single source Compton methods.

In 1975 Finnish workers combined Rayleigh and Compton scattering using 60 keV radiation from ^{241}Am (Olkkonen, Pummalanien). Pummalanien pointed out that as bone mineralization changes, the higher Z elements change. This in turn has a large effect on the coherent to incoherent scatter ratio, since it is sensitive to the third power of Z. The system thus far has been studied only by the Finns and they have applied it to bone and liver tissue density measurements.

Currently the use of penetrating radiation in the x-ray region for quantitation of specific tissue is undergoing a resurgence. A substantial number of laboratories are developing both isotopes and x-ray tubes as sources. The group at the University of Wisconsin remains an important center for these developments. Other groups at McMaster University, Toronto University, Vanderbilt University and the University of California are continuing developments in absorptiometry and scattering. Studies continue in England, Germany and Finland. Thus far, the analytical use of x-rays in single and dual beam absorptiometry in bone and soft tissue may be categorized as well established systems. The single beam principle has commerical instrumentation available. The Compton and the Rayleigh scattering systems are presently emerging as viable analytical tools, with the Compton principle currently being used routinely at a few centers. The scatter instrumentation still is assembled and studied on a research basis, but nevertheless is a principle begining to complete with absorptiometry.

For the future, it is possible that tomography using monochromatic isotopic sources of x-rays may be able to provide both accurate tissue densities as well as geometric mapping of that tissue. Thermal neutron activation analysis and proton activation analysis both have been proposed as another in-vivo analytical system, particularly in bone analysis. These latter methodologies are in the research stage.

BIBLIOGRAPHY

F.P. Bolin, L.E. Preuss, K.M. Glibert and C.K. Bugenis "Errors in Dual X-Ray Beam Differential Absorptiometry," Advances in X-Ray Analysis (Plenum Press, NY 1978) Vol. 21 p 155.

W. Burch and P. Bloch, "Two Wavelength Technique for the Measurement of Bone Mineral Content In Vivo," Proceedings of Bone Measurement Conference, Conf. 700515, 263 (1970).

J.R. Cameron and J.A. Sorenson, "Measurement of Bone Mineral In Vivo: an Improved Method," Science, 142, 230 (1963).

J.R. Cameron and J.A. Sorenson, "A Reliable Measurement of Bone Mineral Content In Vivo," (Abst.) Journal of Nuclear Medicine 8, 268 (1967).

J.R. Cameron, R.B. Mazess, and J.A. Sorenson, "Precision and Accuracy of Bone Mineral Determination by Direct Photon Absorptiometry," AEC Report COO-1422-32, (1968).

J.L.H. Chan, R.E. Alvarez and A. Macovski, "Measurement of Soft Tissue Overlying Bone Utilizing Broad Band Energy Spectrum Techniques," Nuclear Science, NS-13, 551 (1976).

R.L. Clarke and G. Van Dyk, "A New Method for Measurement of Bone Mineral Content Using both Transmitted and Scattered Beams of Gamma-Rays," Physics in Medicine and Biology, 18, 532 (1973).

R.H. Demling, R.B. Mazess, R.M. Witt, and W.H. Wolberg, "The Study of Burn Wound Edema Using Dichromatic Absorptiometry," Journal of Trauma 19 124 (1978).

W. Dohring, K.H. Reiss and H. Fabel, "Compton Scatter for Local In Vivo Assessment of Density in the Lung," Pneumonolgie 150, 345 (1974).

A.L. Evans, M. Davison, J. Kennedy, J.G. Shimmins, and G.R. Sutherland, "A Gamma Ray Densitometer for the Investigation of Pulmonary Function," Physics in Medicine and Biology 20 261 (1975).

F.T. Farmer and M.P. Collins, "A New Approach to the Determination of Anatomical Cross-Sections of the Body by Compton Scattering of Gamma-Rays," Physics in Medicine and Biology 16, 577 (1971).

E.S. Garnett, T.J. Kennett, D.B. Kenyon and C.E. Webber, "A Photon Scattering Technique for the Measurement of Absolute Bone Density in Man," Radiology 106, 209 (1973).

E.S. Garnett, C.E. Webber, G. Coates, W.P. Cockshott, C. Nahmias and N. Lassen, " Lung Density: Clinical Method for Quantitation of Pulmonary Congestion and Edema," CMA Journal 116, 153 (1977).

J. Gershon-Cohen, N.H. Cherry and M. Boehnke, "Bone Density Studies with a Gamma Gauge," Radiation Research 8, 509 (1958).

L. Gustafsson, B. Jacobson and L. Kusoffsky, "X Ray Spectrophotometry for Bone-Mineral Determinations," Medical and Biological Engineering p 113 (Jan 1974).

R. Hudepohl, D. Kedem, R.R. Price, J. Wagner and A.B. Brill, "Three Photon Absorption Techniques," Physics in Canada 32, 15.5 (1976).

B. Jacobson and Bjorn Lindberg, "X-Ray Spectrophotometer for Simultaneous Analysis of Several Elements," Review of Scientific Instruments 35, 1316 (1964).

P.F. Judy, J.R. Cameron and J.M. Vogel, "A Method to Estimate the Error Caused by Adipose Tissue in the Absorptiometric Measurement of Bone Mineral Mass," AEC Report COO-1422-141 (1973).

A.L. Kairento and E. Spring, "Measurement of Bone Mineral and Body Composition by the Attenuation of Three Low-Energy Radiations from ^{241}Am," Annals of Clinical Research 6, 80 (1974).

W.C. Kan, C.R. Wilson, R.M. Witt and R.B. Mazess, "Direct Readout of Bone Mineral Content with Dichromatic Absorptiometry," International Conference on Bone Mineral Measurement, (NIH) 75-683 66 1973.

L. Kaufman, O. Gamsu, C.H. Savoca, and S. Swann, "Three Dimensional Lung Densitometer Using CdTe Detectors for Diagnosis and Evaluation of the Progress of Pulmonary Edema," Revue de Physique Appliquee 12, 369 (1977).

E. Krokowski, "Calcium Determination in the Skeleton by Means of X-Ray Beams of Different Energies," Symposium Ossium, Editors A.M. Jelliffe, and B. Strickland (E & S Livingstone, Edinburgh and London, (1970) p 154.

R.B. Mazess, J.R. Cameron, J.A. Sorenson, "Determining Body Composition by Radiation Absorption Spectrometry, "Nature 228, 771 (1970).

R.B. Mazess, J.R. Cameron, R. O'Connor and D. Knutzen, "Accuracy of Bone Mineral Measurement," Science 145, 388 (1964).

H. Olkkonen, P. Puumalainen, P. Karjalainen, E.M. Alhava, Measurement of Bone Mineral Density Using Coherent and Compton Scattering," American Journal of Roentgenology, 126: 1279 (1976).

N. Pelc, "Body Composition Measurements Using ^{109}Cd," AEC Report COO-1422-185 (1974).

D.G. Piper, L.E. Preuss, "Absolute Bone Density Measurement

Using Compton Scattered Radiation," American Journal of Roentgenology, 126: 1279 (1976).

L.E. Preuss and W.G. Schmonsees, "^{109}Cd for Compositional Analysis of Soft Tissue," International Journal of Applied Radiation and Isotopes, 24, 9 (1972).

L.E. Preuss and F.P. Bolin, "In Vivo Analysis of Lipid-Protein Ratios in Human Muscle by Differential X-Ray Absorption Using ^{109}Cd Photons," Advances in X-Ray Analysis 16, 111 (1972).

L.E. Preuss and F.P. Bolin, "In Vivo Tissue Analysis with ^{109}Cd," Isotopes and Radiation Technology 9, 501 (1972).

L.E. Preuss and D.M. Leachy, "Low-Energy Photons in the Life Sciences," Proceedings of Symposium on Low-Energy X- and Gamma Sources and Applications, ORNL-11C-10 Vol. 1, 197 (1965).

R.R. Price, J. Wagner, K. Larsen, J. Patton, A.B. Brill, "Regional and Whole-Body Bone Mineral Content Measurement with a Rectilinear Scanner," American Journal of Roentgenology, 126: 1277 (1976).

P. Puumalainen, H. Olkkonen and P. Sikanen, "Assessment of Fat Content of Liver by a Photon Scattering Technique," International Journal of Applied Radiation and Isotopes 28, 785 (1977).

K. H. Reiss and B. Steinle, "Medical Application of the Compton Effect," Siemens Forsch.-Entwickl-Ber Bd. 2 (1973).

K. H. Reiss and W. Schuster, "Quantitative Measurements of Lung Function in Children by Means of Compton Backscatter," Radiology 102, 613 (1972).

J.M. Sandrik and P.F. Judy, "Effects of the Polyenergetic Character of the Spectrum of ^{125}I on the Measurement of Bone Mineral Content," Investigative Radiology 8, 143 (1973).

R.G. Waggener, L.F. Rogers and P. Zanca, "A Polychromatic X-Ray Beam to Represent n Monoenergetic Photon Sources," Proceedings of Bone Mineral Conference Conf. 700515, 314 (1970).

D.E. Watt, "Optimum Photon Energies for the Measurement of Bone Mineral and Fat Fractions," British Journal of Radiology 48, 265 (1975).

C.E. Webber and T.J. Kennett, "Bone Density Measured by Photon Scattering I. A System for Clinical Use," Physics in Medicine and Biology, 21, 760 (1976).

C.R. Wilson, J.R. Cameron, E.L. Ritman, E.R. Sturm and R.A. Robb, "Video-Roentgen Absorptiometry for the Measurement of Bone Mineral Mass," AEC Report COO-1422-152 (1973).

W.W. Wooten, P.F. Judy and M.A. Greenfield, "Analysis of the Effects of Adipose Tissue on the Absorptiometric Measurement of Bone Mineral Mass," Investigative Radiology 8, 84 (1973).

L. Zeitz, "Effect of Subcutaneous Fat on Bone Mineral Content Measurement, with the 'Single-Energy' Photon Absorptiometry Technique," ACTA Radiologica 11, 401 (1972).

R.E. Zimmerman, H.J. Griffiths and C. D'Orsi, "Bone Mineral Measurement by Means of Photon Absorption,"Radiology 106 561 (1973).

USING DEC OPERATING SYSTEM FOR X-RAY DIFFRACTION AND X-RAY FLUORESCENCE ANALYSIS

B. E. Artz, E. C. Kao and M. A. Short

Ford Motor Company, Engineering and Research Staff

Dearborn, Michigan 48121

ABSTRACT

The Digital Equipment Corporation (DEC) operating system RSX-11M has recently been installed on a DEC PDP 11/34 computer which is used for the control of, and to acquire and process data from, three X-ray diffractometers, one X-ray fluorescence analysis unit and an electron microprobe. The RSX-11M system replaced the modified DEC 1 - 8 User BASIC previously employed, thus replacing an operating system which was known in detail only to its writer by a system which is supported by the computer manufacturer. There are three major advantages in the use of RSX-11M over 1 - 8 User BASIC: an improved handling of program scheduling, the integration of the software driver for the computer - X-ray electronics interface into the operating system without a major modification of the latter, and the ability of RSX-11M to undertake concurrent execution of instrument control, data acquisition, and data reduction. The 11/34 - RSX-11M system has been implemented to use FORTRAN; a BASIC interpreter has, however, been added which allows users to interact on-line with the computer. A command interpreter which can accept a command line from a terminal has been included.

INTRODUCTION

In recent years both the hardware and the software used for the computer controlled automation of scientific instrumentation has become increasingly sophisticated. Indeed, it was not much more than ten years ago when X-ray diffraction, X-ray fluorescence, and electron microprobe instrumentation - with which this paper is concerned - were frequently supplied by equipment manufacturers without even hardwired automation facilities. In individual laboratories, however, hardwired and computer controlled automation for many years have been added on to equipment purchased from manufacturers. In 1963, for example, Cole, Okaya, and Chambers (1) described a fully automated four-axis single crystal X-ray diffractometer controlled by an IBM 1620 computer.

In little more than ten years, commercially available X-ray equipment has progressed from instrumentation with manually controlled goniometer motors, scaler-timers, and chart recorders to hardwired controllers which would carry out a very limited number of functions, such as an integrated scan on a diffractometer. These hardwired controllers were replaced by microprocessor controllers which were designed to carry out a considerable number of useful functions and could be communicated with by means of an input/output terminal such as a Teletype. To these, full computer control combined with appropriate data reduction capabilities have recently been added.

Most, if not all, commercially available instrumentation utilizes a dedicated computer whether it be for X-ray diffraction, X-ray fluorescence analysis, or for electron microprobe analysis. These dedicated computers are used for instrument control, data acquisition and data reduction. The computers may use languages and operating systems designed by the X-ray manufacturers for specific analytical instruments, for example FLEXTRAN (Tracor-Northern/energy dispersive analysis), BLISS (Applied Research Laboratories/electron microprobe analysis), CLASS (Canberra/X-ray analysis) or more conventional languages such as BASIC or FORTRAN in conjunction with - for those instruments which use a DEC computer - RT-11 as an operating system. Diano, for example, uses RT-11/FORTRAN; Siemens uses RT-11/BASIC.

Despite the continual decrease in the cost of computers, the principle of having a dedicated computer for each analytical instrument may become both expensive and duplicative for large laboratories with multiple X-ray systems. Consequently, many such laboratories are setting up single computer/multiple instrument systems. In some cases each instrument may have a dedicated microprocessor. It is obvious that such single computer/multiple instrument systems are not efficiently run under RT-11. In this paper the use of RSX-11M as an operating system is discussed.

HARDWARE

The computer system presently in use consists of a Digital Equipment Corporation PDP 11/34 with 80 Kwords of memory, two Plessey Microsystems dual hard disc drives giving a total of 10 megabytes of storage, two DEC dual DECtape drives, a DEC high speed reader punch, and DEC and Tektronix video terminals. Memory Management is an integral part of the PDP 11/34.

The computer system is directly connected to control and process data from one Philips X-ray diffractometer used for powder analysis, two Picker X-ray diffractometers used for the measurement of retained austenite and residual stress, and one Siemens SRS X-ray fluorescence analysis unit used for quantitative and qualitative elemental analysis. It is also directly connected to process data obtained on an Applied Research Laboratories EMX electron microprobe. A separate Teletype is associated with each instrument.

The interconnection of the axis positioners, scaler-timers, device actuators, terminal, computer interface, and computer has been described previously (2).

OPERATING SYSTEM

For a number of years a highly modified form of DEC's 1 - 8 User BASIC has been used (3). Because the modifications were known only to the writer of the software, it was found to be very difficult to modify the 1 - 8 User BASIC when need arose. It was therefore decided to change to a computer manufacturer supported operating system. The system design requirements called for an operating system which would support the following:

real time operation
independence of terminal and equipment
multiple users
multiple jobs for each user
simultaneous data collection and data reduction
asynchronous, interrupt driven operation
large range of data rates with no missed data points
error checking with recovery
high level programming languages
user choice of language
multiple access to data bases
upward compatibility to large computers.

When choosing an operating system that must interact with data collection equipment the over-riding concern in this choice should be the real time capabilities of the operating system. The computer must be able to respond to the equipment when the equipment needs the computer and not necessarily at some prescheduled time designated by the computer. The operating system must respond to the competing tasks in a prioritized, event driven manner. Whenever a significant event occurs, such as the presence of some data from the equipment, the operating system must interrupt the executing task and accept this data immediately. Operations such as this must have the highest possible priority while other jobs can be assigned a variety of lower priorities. This type of scheduling is best accomplished using an interrupt driven input/output method. Of the operating systems that DEC has available for PDP-11/34 computers (RT-11, RSX-11, RSTS) only RSX-11 satisfies all of the aforementioned requirements.

RSX-11M

The RSX-11M baseline operating system supports a variety of devices such as terminals, printers, discs, etc. In adapting the RSX-11M operating system for our use, we have added a driver which can control multiple instruments simultaneously through Canberra 6726 modules. The coding of the driver must observe all the programming protocols and conventions required by the system to insure system integrity. Diagrammatically, the operating system can be considered as consisting of "system directives" which are program modules to perform specific functions, "drivers" which are program modules that service all the devices present in the system, and the "executive dispatcher" which multiplexes and linearizes access to the system directives and device drivers. The data structure which is referenced by every element in the system is an aggregate of linked lists or tables. It is through the data structure that the system achieves process synchronization and communication. Our Canberra driver consists of a program module which interfaces the hardware to the operating system and a corresponding device data structure linked to the system data structure.

The program module of the Canberra driver has several functions of which the most obvious one is to service hardware interrupts. Under RSX-11M convention, there are three stages of interrupt processing. When an interrupt occurs, the computer is switched into the kernel or system state with all interrupts disabled, and the driver is entered at the appropriate point. In the first stage the driver identifies the unit number of the device and immediately thereafter re-enables higher priority interrupts. This first stage lasts for only a few microseconds and is the only time that the system is unable to respond to another high level interrupt. It is the very short duration of this stage that gives RSX-11M its real time capability. The driver then enters a second stage at a lower priority and continues to process the interrupt such as mapping data between program and instrument. If more processing is required, the driver enters into a third stage and is put into a first-in first-out queue with all interrupt priorities enabled and awaits processing completion. A distinct advantage of the 3-stage interrupt processing scheme is that the driver does not need to be written in re-entrant code. This considerably simplifies the coding of the driver. Other functions of the program module are input-output initiation, powerfail recovery, time-out and terminate I/O processing. These functions are defined by the system and must be included to achieve compatibility.

The device driver data structure consists of various linked lists. The lists contain information regarding static and dynamic characteristics of the device. The accuracy of the information is of vital importance, since based on these data, the executuve dispatcher would verify and queue each input-output request to the driver. The manner that the lists are linked determines whether multiple devices can operate in parallel or sequentially.

The user requests services of the driver by issuing system directives in his application program. The dispatcher receives and processes the directives. The following FORTRAN program illustrates some directive calls.

```
      IMPLICIT INTEGER (A-Z)

      DIMENSION PRL(6), ISB(2)

            .
            .
            .

10    CALL ANSLUN (LUN, NAME, UNIT, IDS)

            .
            .
            .

50    CALL WTQIO (LODE, LUN, FLG, IBS, PRL, IDS)
```

In statement 10, a call to the system directive ASNLUN is made. This is usually the first step during program initiation, and the purpose is to associate a physical device with the logical unit number used in FORTRAN I/O calls. The parameter LUN contains the logical unit number, and the parameters NAME and UNIT refer to a particular device to be referred to by that logical unit number. The parameter IDS is used by the system to return a condition code regarding the directive call. Statement 50 is an example of the most primitive I/O call, WTQIO,which can be issued by the user under RSX-11M. The parameter CODE contains an integer number indicating an I/O function such as read or write. The parameter LUN, defined in statement 10, indicates the targeted device. The buffer count and the address of the buffer to which the driver would map the data are contained in the arrays PRL. The condition codes of the directive call and the I/O operation are returned in IBS and IDS.

X-RAY APPLICATIONS

The new system has been used for a number of X-ray analysis procedures which were previously running on the modified 1 - 8 User BASIC. These include the X-ray fluorescence analysis of automobile emissions catalysts (2) and of aerosols on filter papers, the X-ray diffraction analysis of retained austenite and residual stress (4), and the reduction of data obtained on an electron microprobe. It is anticipated that programs for qualitative and quantitative X-ray diffractometry will be completed in the near future.

FURTHER SYSTEM DEVELOPMENTS

Although FORTRAN has been found to be an excellent language for instrument control and data reduction, it is by no means ideal. We are proposing to evaluate other high level languages such as PASCAL and FORTH to ascertain whether or not either of these would be an improvement on FORTRAN. Both PASCAL and FORTH belong to the class of structural languages. Programs written in structural language are typically in a highly modular form which can be easily understood and maintained. In addition, FORTH is an interactive language like BASIC but without many of the limitations of the latter such as speed and program size. It is anticipated that the adoption of either FORTH or PASCAL would result in reduced program development time.

We are also considering the possible networking of the PDP 11/34 with some or all of the other computer controlled instrumentation in our laboratory. This includes a 3M ion scattering spectrometer linked to a PDP 11/03, a Physical Electronics scanning Auger microprobe linked to a PDP 11/20, and a JEOL scanning electron microscope linked to a PDP 11/40. These three computers are currently operating with DEC's RT-11 operating system. A computer network would provide an opportunity to use distributive processing and resource sharing. The networking might be accomplished either with the use of DEC's DECnet software or by developing software locally. The DECnet software is a powerful communications package which, under different versions, will permit interconnection of all of Digital's operating system families. The penalty for this flexibility is in the cost of both software and hardware and in the vast amount of processing overhead, especially in cases where some of the DECnet functions (e.g., dial-up capability) are not required. The objective of locally developed software would be a simple system which would allow communication between the RT-11 and RSX-11M operating systems over an asynchronous serial line. This system would implement parts of DECnet's protocol and functions such as file transfer and resource access.

ACKNOWLEDGMENTS

We would like to thank M. F. Elgart for his help with the planning and installation of the computer system and P. M. Blass for his assistance in the development of the software.

REFERENCES

1. H. Cole, Y. Okaya, and F. W. Chambers, "Computer Controlled Diffractometer", Rev. Sci. Inst., 34, 872-876 (1963).

2. B. E. Artz, Carol J. Kelly, and M. A. Short, "A Computer Control for an X-Ray Fluorescence Analysis Unit", in W. L. Pickles et al., Editors, Advances in X-Ray Analysis, Vol. 18, pp. 309-316, Plenum Press (1975).

3. Carol J. Kelly and Carl A. Gagliardi, "Interactive X-Ray Laboratory Automation Utilizing Multiuser Basic", Proc. IEEE, 63, 1426-1431 (1975).

4. Carol J. Kelly and E. Eichen, "Computer Controlled X-Ray Diffraction Measurement of Residual Stress", in L. S. Birks et al., Editors, Advances in X-Ray Analysis, Vol. 16, pp. 344-353, Plenum Press (1973).

AUTOMATED DETERMINATION OF OPTIMUM EXCITATION CONDITIONS FOR SINGLE AND MULTIELEMENT ANALYSIS WITH ENERGY DISPERSIVE X-RAY FLUORESCENCE SPECTROMETRY

Wolfhard Wegscheider[1], Bruce B. Jablonski and Donald E. Leyden

Department of Chemistry, University of Denver

Denver, Colorado 80208

ABSTRACT

The determination of optimal excitation conditions for energy dispersive x-ray fluorescence is particularly critical for multi-element analysis covering a wide range (viz. 10 or 20 keV) of the spectrum. Functions that quantitatively describe the spectral quality are used as objective functions in pattern search algorithms. It is shown that the filters can be arranged in a definite order, at least with respect to the energy of the K-absorption edge of the tube and can therefore be employed as a dimension in the optimization process. Of the algorithms that were compared, the Nelder-Mead and Routh-Swartz-Denton versions of the sequential simplex search gave the best results if the excitation voltage and the current could be controlled in small increments. If the optimization includes dimensions with a few discrete stages (e.g. filters) the fixed size simplex proved to be of greatest value. The functions can be weighted to reflect special interest in one or more elements. Conditions for increasing the counting time and terminating the search are discussed.

INTRODUCTION

Energy dispersive x-ray fluorescence analysis (EDXRF) is a powerful tool for the simultaneous multielement analysis of a wide variety of samples. Elemental constituents can be determined

[1] On leave from: Institute for General Chemistry, Micro- and Radiochemistry, Technical University, Graz, Austria.

from the trace level to the percentage level with good precision and accuracy. A major factor contributing to the success of this method is the high degree of automation in the control of the instrument and in the subsequent data reduction. These features made EDXRF particularly attractive to new users of x-ray spectroscopy and a good many instruments are now used in medical, pathological and forensic laboratories. An important step in the course of a complete x-ray analysis that to date has not been automated is the selection of optimal excitation conditions for single and multielement analysis. One reason for this is the variability of conditions between element sets, instruments and samples. Another reason may be the lack of quantitative criteria or figures of merit to describe the quality of a multielement spectrum. It is believed that a procedure that copes successfully with all of the above mentioned problems will be of considerable help to technicians as well as expert spectroscopists, setting them free of the manual selection of excitation conditions. It is the goal of this communication to show that:

a) objective functions can be devised that closely resemble the qualities intrinsically sought by x-ray spectroscopists,

b) a certain class of optimization algorithms performs well on the hypersurfaces defined by these objective functions, and that

c) the time needed for the entire optimization process is on the order of 2-15 minutes for most cases depending on the resolution of the power supply, the dimensionality of the optimization problem and the desired level of precision.

EXPERIMENTAL

Instrument

A Nuclear Semiconductor NS-440 x-ray spectrometer with an Ag x-ray tube from Watkins-Johnson was used throughout this study. The tube was operated in pulsed mode for maximum throughput and the spectrometer had the capability of selecting from 5 different source filters: a Whatman 41 filter paper, an Al-filter (0.032 mm) an Ag-filter (0.025 mm), an Ag-filter (0.127 mm) and a Cu-filter (0.38 mm). The data acquisition was done employing an NS-880 analyzer from Tracor-Northern. The associated PDP-11/05 computer system was used for data reduction and quantitative evaluation of the figures of merit. All computations were performed in Flextran, a special purpose language provided by the vendor. The amplifier time constant was set at 4 microseconds which gives a maximum throughput of about 6×10^3 cps.

Samples

A mineral mix used as poultry feed premix was ground twice in a rotary mill (Dayton Electric) and a 500 mg portion was mixed with 500 mg of Somar Blend and pressed at 20,000 psi to give a pellet of 1.25 in. diameter. The premix contains manganese oxide, ferrous sulfate, ferrous carbonate, cobalt carbonate, calcium iodate and ground limestone. Another sample studied was an AISI Type 317 standard steel (United States Steel Corporation) with Si, Cr, Mn, Fe, Ni and Mo as major constituents. This sample is referred to as DDL-standard. A third type of sample consisted of a KBr/CdTe (1:3) pellet that was prepared similarly to the mineral premix sample.

PRINCIPLES OF OPTIMIZATION

Until recently, analytical procedures in general and spectroscopic techniques in particular have been published and used without considering the optimum of the response.(1) After the first application of an optimization in analytical chemistry,(2) it has been the distinct accomplishment of S.N. Deming and his co-workers to systematically explore the merits and limitations of experimental optimization procedures in analytical chemistry.(3-8) Mainly through their work, it is now generally accepted that classical optimization procedures which make use of the calculus to locate the optimal vector of the variables (experimental parameters) $\vec{X}$ in maximizing a function $f(\vec{X})$ exhibit a number of serious drawbacks. Generally speaking, the classical methods require the continuity or differentiability of $f(\vec{X})$ or both. Even though it is conceivable that such a function can be devised for x-ray processes, it would greatly rely on the exact determination of instrument parameters similar to fundamental parameter methods for data reduction. In most cases in which nothing is known about the functional relationship of response and experimental variables, one must resort to search methods.(9) In contrast to the classical methods, the only requirement that the objective function must satisfy is that it be computable.

In this study, all the algorithms used belong to the class of sequential methods. The most common of these, the univariate search technique, however, has been ruled out as it is well known that it fails in presence of factor interactions.(10,11) Among the algorithms evaluated are 1) the Sequential Simplex(12) with modifications by Nelder and Mead,(13) Routh and co-workers,(14) and Parkinson and Hutchinson,(15) and 2) the multivariate Grid Search.(16) All of these algorithms have been described exhaustively in the cited literature and also compared with respect to performance(17,18) for different kinds of numerical problems.

Although the results of these comparison studies cannot easily be projected onto the current problem, similar criteria can be used for their evaluation as described by Himmelblau.(18) These are the number of evaluations needed to arrive at an optimum, the total time needed for the optimization process, and the reliability of convergence to the correct optimum.

At this point it is appropriate to discuss the major differences between numerical optimization and experimental optimization, especially with respect to the problem of optimizing x-ray excitation conditions. Experimental optimization always suffers from random error. The simplex algorithm is largely self-correcting for moderate experimental error.(2,3,12) It is a special property of counted data that this error can be influenced by the counting up to the point where instrument fluctuations and drift become significant. The occurence of absorption edges in the space of experimental factors (viz. kV setting) causes a non-monotonic behavior of the hypersurfaces. Most optimization algorithms have neither been created nor tested for non-monotonic functions. In most other fields of analytical chemistry the surfaces are well behaved and smooth.(19,20) Another complication constitutes the limited resolution of the space of variables which is typically 1 kV in excitation voltage and 0.01 mA in current for low wattage x-ray generators. Conceptually, filters and secondary target emitters span the third and fourth dimensions of the variable space,(21) for these the typical resolution on commercial instruments is on the order of 1/5 or 1/6, corresponding to 5 or 6 automatic selections. This limitation of resolution leads to a significant "apparent" ruggedness of the surfaces in addition to the discontinuities introduced by x-ray physical processes.

RESULTS AND DISCUSSION

All methods of experimental optimization rely primarily on the unambiguous and quantitative definition of an objective function that must be evaluated after each trial. For this reason, considerable effort has been made to select functions which reflect the line of thought of experienced x-ray spectroscopists in the selection of operating parameters.(22) A well suited function derived from the detection limit(23) was found to be

$$Y_1 = \frac{1}{t_r} \prod_{i=1}^{n} \frac{I_{i,p} - I_{i,b}}{c_i(2.71 + 4.65\sqrt{I_{i,b}})}$$

for $(I_{i,p} - I_{i,b}) > (2.71 + 4.65\sqrt{Y_{i,b}})$ and $Y_1=0$ otherwise where t_r = real time needed to acquire a spectrum

n = number of elements of interest

$I_{i,p}$ = gross intensity of element i

$I_{i,b}$ = background of element i, and

c_i = the concentration of element i

Another function that gives an overall measure of precision is given by

$$Y_2 = \frac{1}{t_r} \prod_{i=1}^{n} \frac{I_{i,p} - I_{i,b}}{c_i\sqrt{I_{i,p} + I_{i,b}}} \quad (2)$$

for $(I_{i,p} - I_{i,b}) > (2.71 + 4.65\sqrt{I_{i,b}})$ and $Y_2=0$ otherwise.

Similarly, it is possible to derive expressions for the information rate.(24) The information rate of the qualitative x-ray analysis can be expressed as

$$Y_3 = \frac{1}{t_r} \log_2 \prod_{i=1}^{n} \frac{I_{i,p} - I_{i,b}}{c_i(2.71 + 4.65\sqrt{I_{i,b}})} \quad (3)$$

For a quantitative analysis, the respective formula is

$$Y_4 = \frac{1}{t_r} \log_2 \prod_{i=1}^{n} \frac{I_{i,p} - I_{i,b}}{c_i\sqrt{I_{i,p} + I_{i,b}}} \quad (4)$$

It is suggested that equations (3) and (4) should again only be defined for all signals of interest rising above the detection limit(22) and set to zero otherwise.

In addition to the above four equations, the probability of detecting all elements of interest simultaneously in a given unit of time(21,25) served as fifth objective function in this study.

$$P = \frac{1}{t_r} \prod_{i=1}^{n} \frac{I_{i,p} - I_{i,b} - (2.71 + 4.65\sqrt{I_{i,b}})}{I_{i,p} - I_{i,b}} \quad (5)$$

for all $(I_{i,p} - I_{i,b}) > (2.71 + 4.65\sqrt{I_{i,b}})$ and P=0 otherwise.

For the simultaneous analysis of Ca and Zn in a mineral premix sample, the response surfaces as defined by equations (1) and (2) are given in Figure 2. It can be seen that for elements emitting

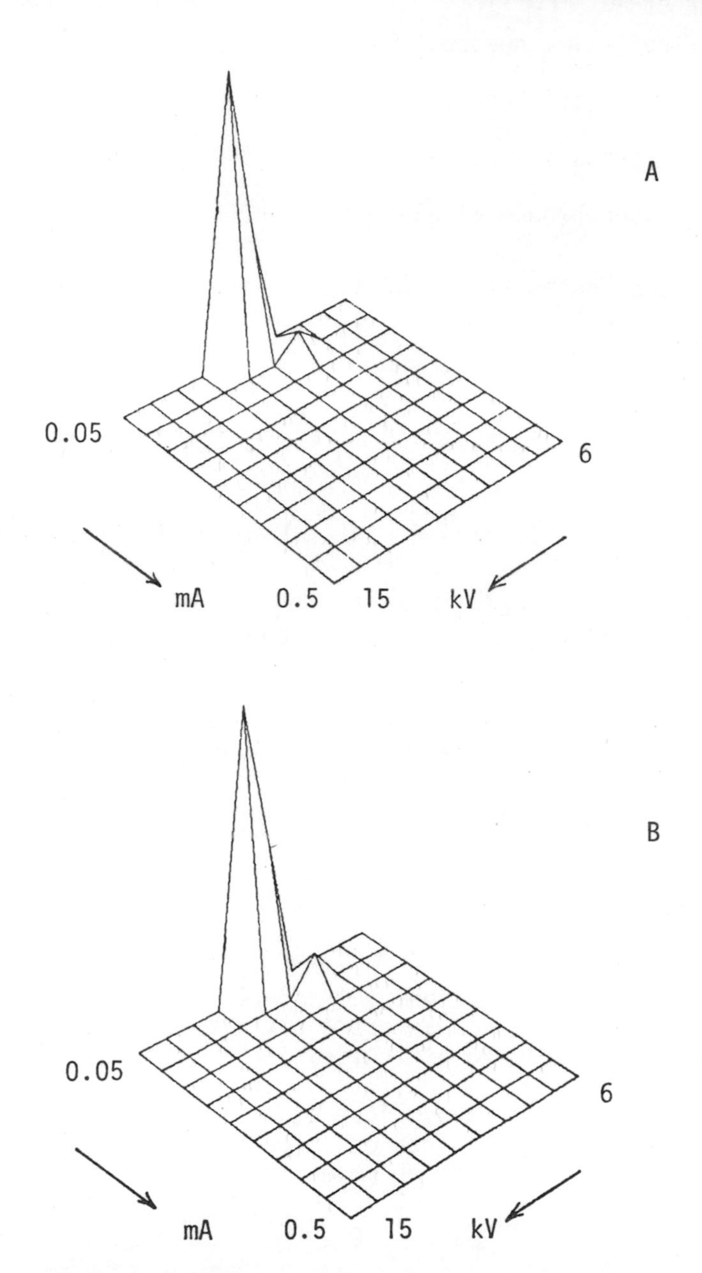

Figure 1. Response surfaces for the simultaneous determination of Ca and Zn in a mineral premix sample

A...equation 1
B...equation 2

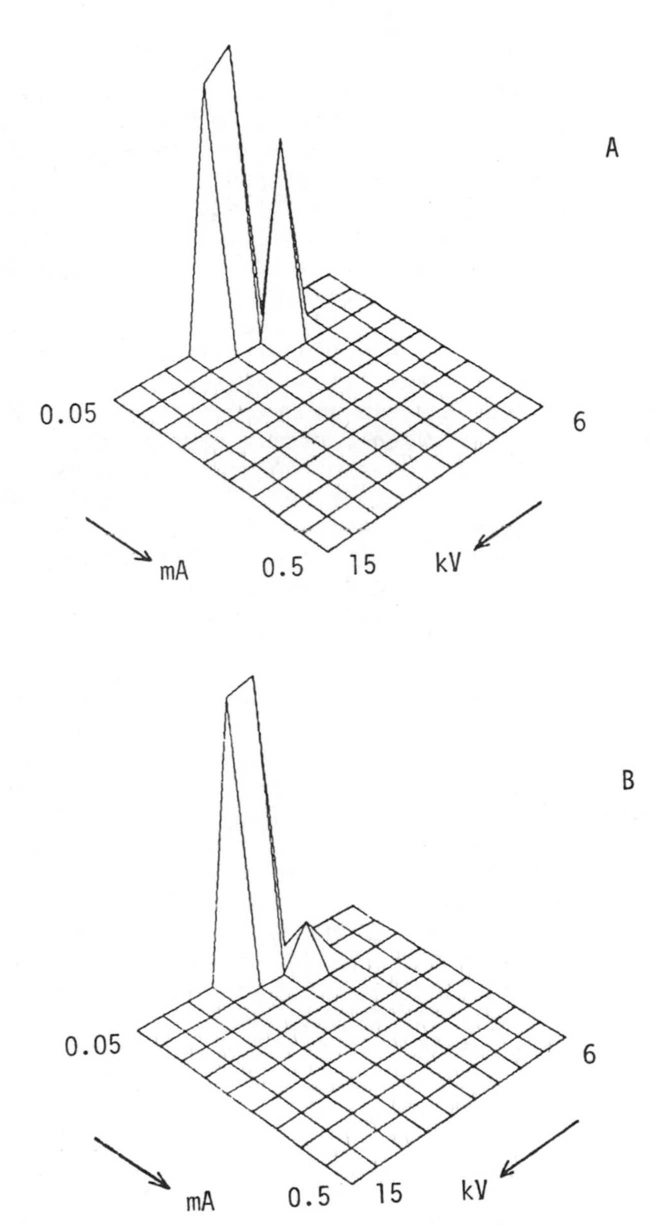

Figure 2. Response surfaces for the simultaneous determination of Ca and Zn in a mineral premix sample.

A...equation 3
B...equation 5

x-rays which are separated by a fairly wide energy range, the choice of excitation conditions appears to be particularly critical. In these and the following graphs, the functions have been restricted to the region below 50% system deadtime and the functional values in regions giving more than 50% deadtime have been set to zero as those conditions are unacceptable because of signal degradation.(26) Because it was desired to collect the data needed to evaluate these functions, it was essential to adopt Currie's definition of the detection limit for counted data(23) because the measured intensity constitutes a poor approximation of variance for a few counts. The two graphs in Figure 3 are a graphical representation of the same set of data once with the familiar definition of the detection limit(27) and once with Currie's used as a cutoff and it can be seen that the latter definition removes a considerable amount of noise. For this combination of elements in the mineral premix sample, the analytically useful region of the space of experimental variables is again extremely narrow and the optimum is well defined; this was found to be true of many multi-element analytical problems extending over fairly wide energy range. In these cases the minute determination of optimum excitation conditions may have to be done at a better resolution than generally provided by commercial automatic power supplies.

A basic assumption and necessary condition for the application of all sequential search algorithms is that the space has metric properties. This condition is obviously met for the excitation voltage and the tube current and also for secondary target emitters which can be ordered according to the atomic number and the fluorescence efficiency. For the dimension of the space that is spanned by the filters, the metric properties cannot be relied on a priori. Their usefulness lies in the filtering effect they have for low energy scatter (cellulose filter), low energy Compton and Raleigh scatter from the tube (Al-filter for Ag L-lines), for reducing the Bremsstrahlung (Ag 0.025 mm and Ag 0.127 mm filters) and removing the Ag K-lines (Cu 0.38 mm filters). At least the Ag-filters exhibit a qualitatively different property than the others as they affect primarily the Bremsstrahlung. The ordering of the Ag-filter and the relatively thick Cu-filter was especially in question because, in ascending order of K-absorption edges, the Cu-filter would come first, while taking into account its mass (and its use) it would be assigned a higher rank than both Ag-filters. To study the properties of the space in the dimension of the filters, these three filters (Ag 0.025 mm, Ag 0.127 mm and Cu 0.38 mm) were used to map the kV and mA space for several sets of elements. The response was scaled to the best response of all three filters to give an indication of the relative usefulness of each of the filters. For Cr, Mn, Fe and Ni in DDL-steel the optimum excitation conditions as defined by equation (2) are obtained

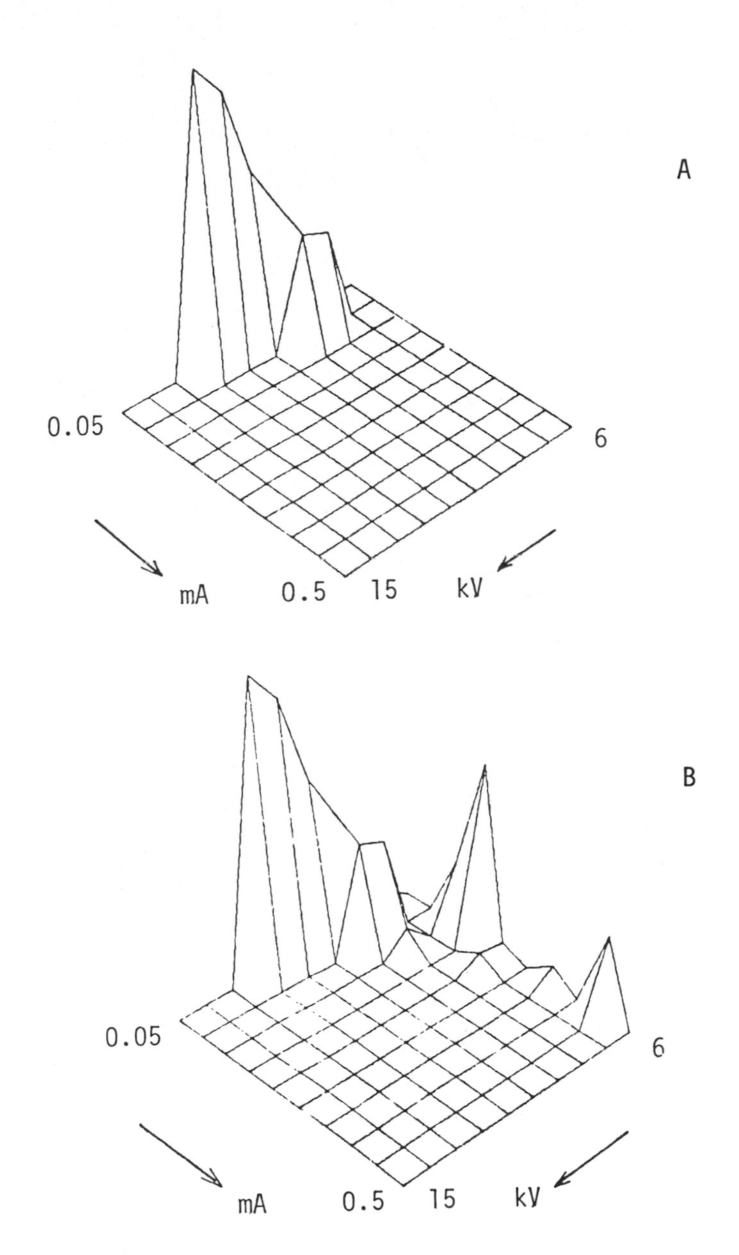

Figure 3. Response surfaces (equation 1) using different definitions of the detection limit.

A...Currie's definition(23)
B...Jenkins' definition(27)

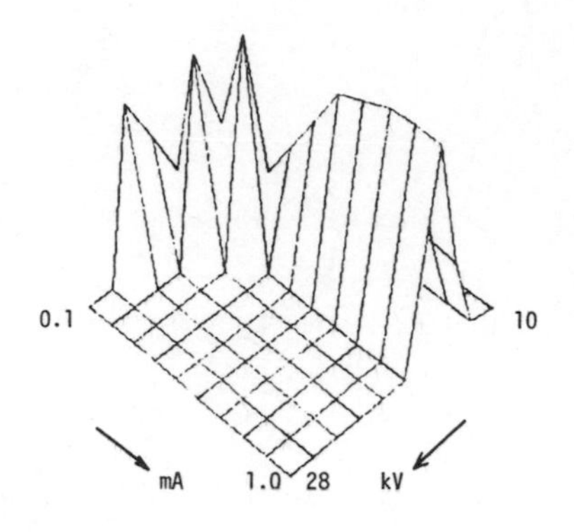

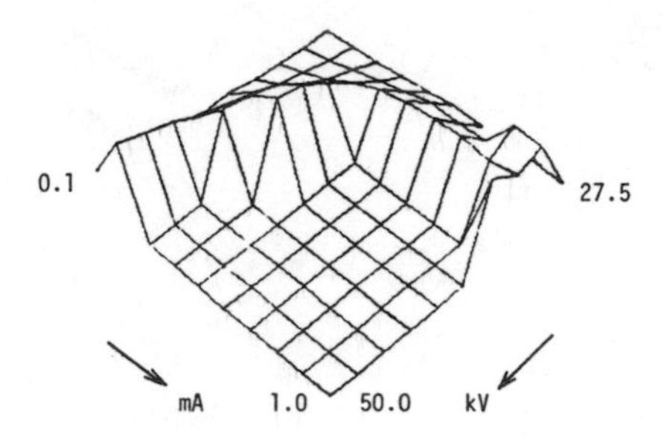

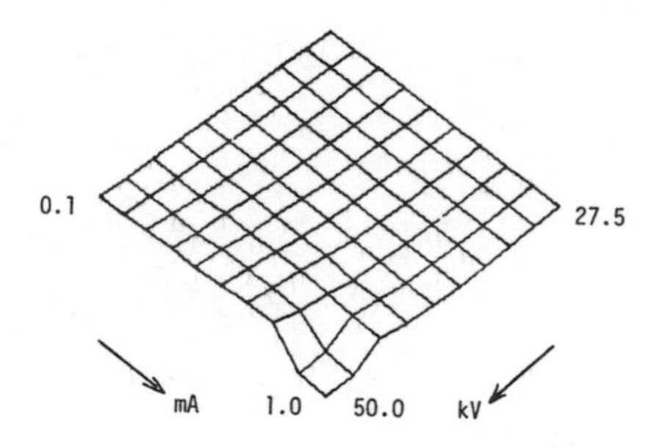

Figure 4. Response surfaces for the simultaneous determination of Cr, Mn, Fe and Ni in a DDL-steel standard sample.

A...0.025 mm Ag filter; optimum relative response 100%
B...0.127 mm Ag filter; optimum relative response 34%
C...0.38 mm Cu filter; optimum relative response 12%

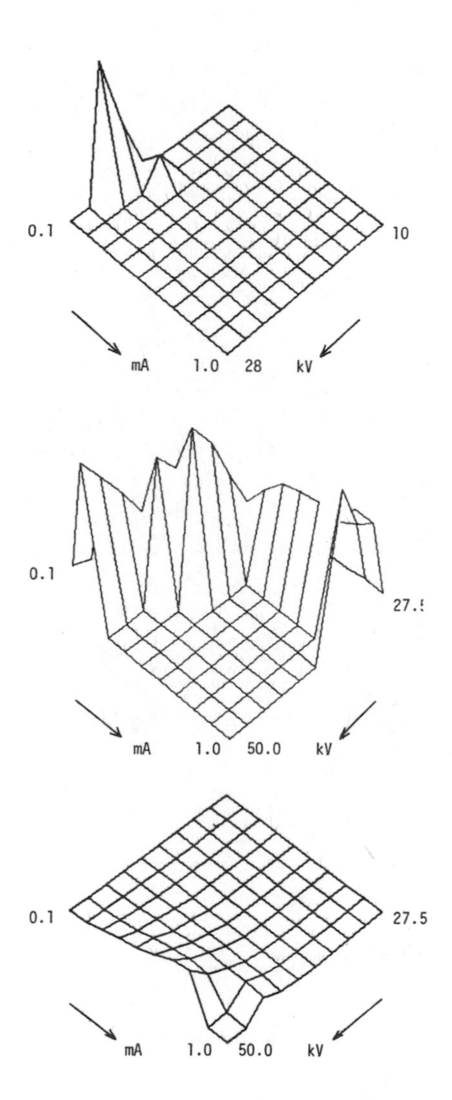

Figure 5. Response surfaces for the simultaneous determination of Cr, Mn, Fe, Ni and Mo in a DDL-steel standard sample.

A...0.025 mm Ag filter; optimum relative response 78%
B...0.127 mm Ag filter; optimum relative response 100%
C...0.38 mm Cu filter; optimum relative response 26%

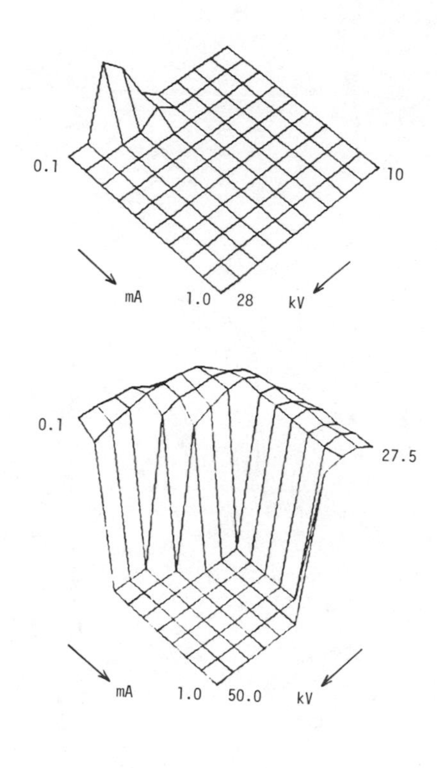

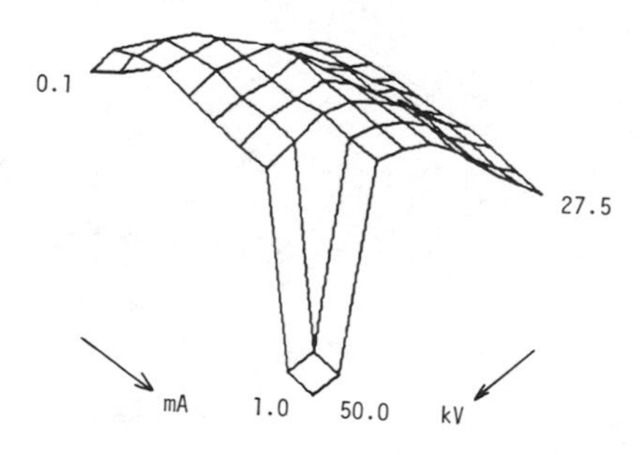

Figure 6. Response surfaces for the determination of Mo in a DDL-steel standard sample.

A...0.025 mm Ag filter; optimum relative response 42%
B...0.127 mm Ag filter; optimum relative response 100%
C...0.38 mm Cu filter; optimum relative response 82%

with the 0.025 mm Ag filter at about 18 kV and 0.3 mA (Figure 4A); this choice of filter is in agreement with the choice of experienced spectroscopists.(28) If Mo is added to the elements of interest, the optimum is shifted towards higher kV and now the 0.127 mm Ag-filter should be used (Figure 5B). Assuming now that Mo is the only element of interest in this steel sample, the optimum is again found if the 0.127 mm Ag filter is used even though the Cu-filter demonstrates a much smaller difference (Figure 6). It has been invariably found that only elements that are severely overlapped by the coherent and incoherent tube radiation should be determined using the Cu-filter. This now allows an establishment of an invariant ranking of all filters and establishes the metric properties of the entire space of optimization.

For the initialization of the algorithms it is necessary to choose a starting point; here an experienced operator is given the possibility to make a good initial guess. However, for the purpose of studying the performance of the optimization procedure, an alternative route was chosen. Those algorithms that do not have the capability to adjust the step size to the topography of the surface(12,16) were initiated using a pseudo random number generator; the proposed starting point was accepted if the function gave a non-zero value, otherwise the random search was continued. The variable size algorithms(3,14) were started using a fixed size simplex as suggested by Yarbro and Deming(5) spanning over approximately 1/3-2/3 of the entire domain in each dimension. A typical run of each of the algorithms is given in Figures 7 and 8.

Frequently, the excitation voltage is limited by the rate of electronic pulse processing; in those cases, the optimum lies at the upper limit of acceptable system deadtime. To obtain reliable results, it was necessary to introduce a penalty function; in doing so it was assured that the deadtime is acceptable, yet the algorithms were not impeded by vanishing functions. Good results were obtained for the following penalty function:

$$Y_{i,p} = Y_i \left[\frac{1}{\Delta t}\right]^{H(\Delta t)\cdot n} \qquad (6)$$

where $Y_{i,p}$ is function i modified by the penalty

Y_i is the original function as defined by equations (1) through (5),

Δt is that portion of the time used to complete the data acquistion that was above the tolerated real time limit (in 10^{-1} s)

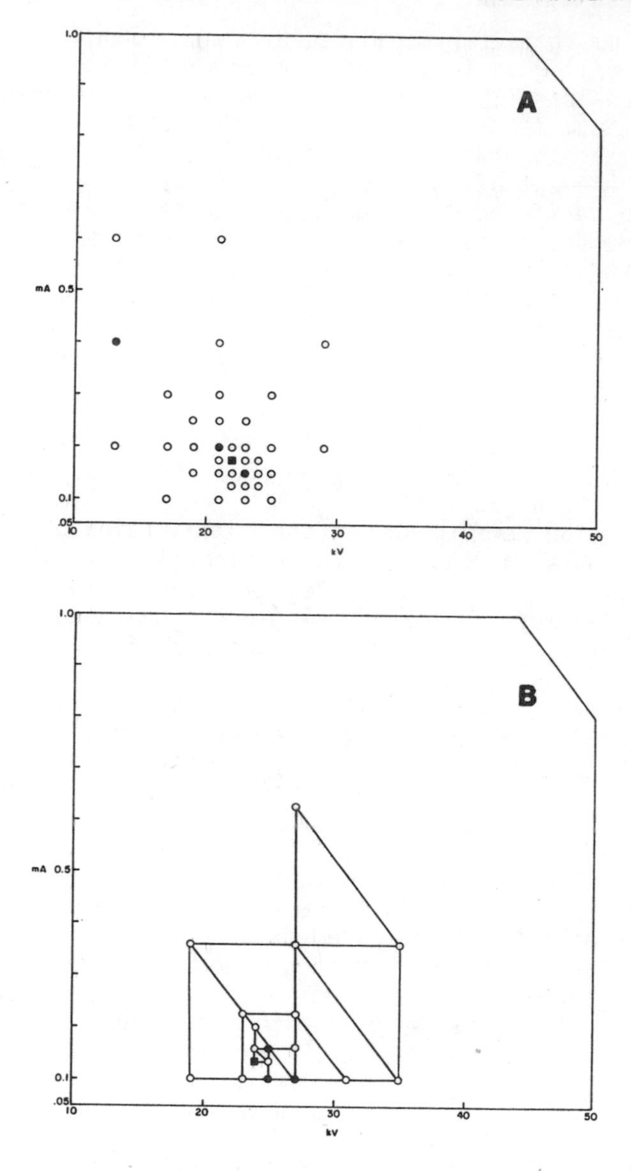

Figure 7. Progress of Multivariate Grid Search (A) and the Fixed Size Simplex Search (B) for the optimization of excitation conditions for Mn and Fe in the mineral premix. Experimental conditions: 0.025 mm Ag filter, pulsed tube, 4 s livetime per vertex. Function: Equation 1.

○ evaluated vertices
● vertices with reduction of grid size
■ optimum

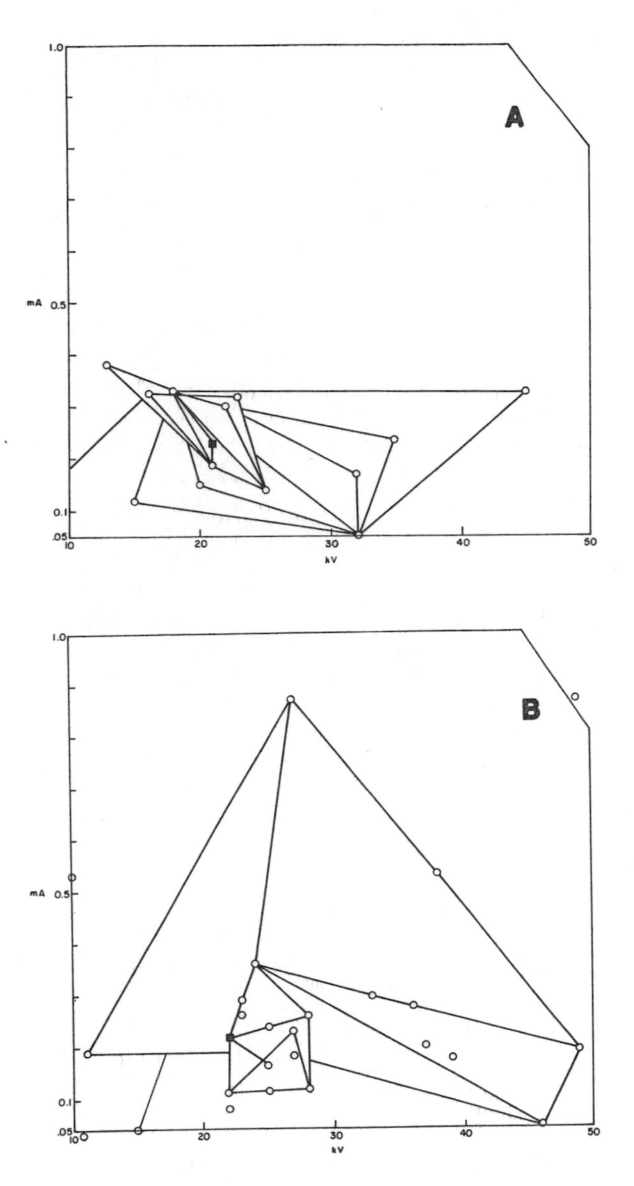

Figure 8. Progress of Nelder-Mead Simplex Search (A) and Routh-Swartz-Denton Simplex Search (B) for the optimization of excitation conditions for Mn and Fe in the mineral premix. Experimental conditions and function the same as in Figure 7.

o evaluated vertices
■ optimum

$$H(\Delta t) = 0 \text{ if } \Delta t \leq 10^{-1} s$$
$$= 1 \text{ if } \Delta t > 10^{-1} s$$

n is the number of elements of interest for which optimal excitation conditions are sought

If only the excitation voltage and the tube current were allowed to vary, the variable size algorithms gave consistently better results; even though all four algorithms were reliable, the total time needed for optimization was approximately 20% longer for the fixed size simplex and about 50% longer for the multivariate grid search. Generally, about 40-80% of the total optimization time was spent on counting. Provided that at least two of the initial vertices gave non-zero responses the Nelder-Mead version or the Routh-Swartz-Denton version of the variable size simplex gave equally good results. The actual time needed to complete the optimization depends strongly on the initial counting time and the final counting time. With large initial step sizes and random starting procedures it proved to be satisfactory for major constituents to start with 1 s livetime. In conducting a simplex search a vertex is reevaluated if it is retained as the best vertex in (n + 1) simplices. Generally, the two values, the original and the value from the reevaluation, were averaged and the search continued. At this point, use was made of the dependence of precision on counting time; the difference between the old and the new values was calculated and compared to the difference between the average of new and old and the next best value. If the latter difference was smaller, the counting time was increased by a factor of four. If the new counting time exceeded the maximum counting time, the search was discontinued. Similarly, the search was discontinued when the size of the simplex dropped below the resolution for all dimensions. The stepwise increase in counting time had to be bound to another criterion if the multivariate grid was used. Here, an additional evaluation of the original best grid point was done at the end of each search stage. Again, the criterion was based on the difference between recounting and the next best grid point. This added one additional evaluation at each stage, but the gain in precision and reduction of total optimization time offset this disadvantage by far.

Because of the very limited resolution in the dimension of the filters, it was found that variable size search algorithms give unpredictable results. This can best be understood in terms of rounding errors that amount to substantial portions of the entire domain. In this case, the fixed size algorithms gave by far more reliable results. With a completely randomized start, however, the chance of arriving at a local optimum is highly because of large discontinuities between the filters. This is an intrinsic problem for all search algorithms that can only be reduced by another search starting at a different region in space. With that restriction, the performance of both fixed size simplex and the multivariate grid search gave reliable results. The major difference is the vastly larger number of vertices that have to be evaluated for the multivariate grid. Here the total optimization time is approximately a factor of 5 less for the fixed size simplices, typically being on the order of 10-16 min for a maximal counting time of 16 s.

Frequently, in x-ray spectroscopy the operator wants to emphasize one or two elements while all the others may be measured with less precision or higher detection limits. It can be shown that weighting factors can easily be introduced in the objective functions(29) to express that selective interest. For the DDL-steel standard sample Cr was assigned the weight 5, while Mn through Mo were assigned the weight one. The resulting optimum for a two dimensional optimization was consistently found at 29-31 kV and 0.8-0.9 mA, while assigning equal weights to all elements gave the optimum at about 40 kV and 0.4 mA (Figure 5). Other element combinations in different samples gave equally encouraging results.

ACKNOWLEDGEMENTS

The authors greatly appreciate the valuable advice from S.N. Deming. This research is supported in part by the grant CHE-7618385 from the National Science Foundation. One of the authors (W.W.) expresses his gratitude to the Austrian-American Educational Commission for support under the Fulbright-Hays Act.

REFERENCES

1. S.L. Morgan and S.N. Deming, "Simplex Optimization of Analytical Chemical Methods", Anal. Chem., 46, 1170 (1974).

2. D.E. Long, "Simplex Optimization of the Response from Chemical Systems", Anal. Chim. Acta, 46, 193 (1969).

3. S.N. Deming and S.L. Morgan, "Simplex Optimization of Variables in Analytical Chemistry", Anal. Chem., 46, 278A (1973).

4. P.G. King and S.N. Deming, "Uniplex: Single-Factor Optimization of Response in the Presence of Error", Anal. Chem., 46, 1476, (1974).

5. L.A. Yarbro and S.N. Deming, "Selection and Preprocessing of Factors for Simplex Optimization", Anal. Chim. Acta, 73, 39 (1974).

6. S.L. Morgan and S.N. Deming, "Optimization Strategies for the Development of Gas-Liquid Chromatographic Methods", J. Chromat., 112, 267 (1975).

7. P.G. King, S.N. Deming and S.L. Morgan, "Difficulties in the Application of Simplex Optimization to Analytical Chemistry", Anal. Lett., 8, 369 (1975).

8. L.R. Parker, Jr., S.L. Morgan and S.N. Deming, "Simplex Optimization of Experimental Factors in Atomic Absorption Spectrometry", Appl. Spectr., 29, 429 (1975).

9. L. Cooper and D. Steinberg, Introduction to Methods of Optimization, p. 134 (W.B. Saunders, Philadelphia-London-Toronto: 1970).

10. G.E.P. Box, "The Exploration and Exploitation of Response Surfaces: Some General Considerations and Examples", Biometrics, 10, 16 (1954).

11. G. Wernimont, "Statistical Control of Measurement Process", in Validation of the Measurement Process, Vol. 63, ACS Symposium Series, J.R. DeVoe, Editor, p. 1, American Chemical Society (1977).

12. W. Spendley, G.R. Hext and F.R. Himsworth, "Sequential Application of Simplex Designs in Optimization and Evolutionary Operation", Technometrics, 4, 441 (1962).

13. J.A. Nelder and R. Mead, "A Simplex Method for Function Minimization", Comput. J., 7, 303 (1965).

14. M.W. Routh, P.A. Swartz and M.B. Denton, "Performance of the Super Modified Simplex", Anal. Chem., 49, 1422 (1977).

15. J.M. Parkinson and D. Hutchinson, "An Investigation into the Efficiency of Variants on the Simplex Method", in Numerical Methods for Non-Linear Optimization, F.A. Lootsma, Editor, p. 115 (Academic Press, London-New York: 1972).

16. L. Cooper and D. Steinberg, Introduction to Methods of Optimization, p. 156 (W.B. Saunders, Philadelphia-London-Toronto: 1970).

17. M.J. Box, "A Comparison of Several Current Optimization Methods and the Use of Transformations in Constraint Problems", Comput. J., 9, 67 (1966).

19. A.S. Olansky and S.N. Deming, "Optimization and Interpretation of Response in the Chromotropic Acid Determination of Formaldehyde", Anal. Chim. Acta, 83, 241 (1976).

20. S.N. Deming and S.L. Morgan, "Advances in the Application of Optimization Methodology in Chemistry", in Chemometrics: Theory and Application, Vol. 52, ACS Symposium Series, B.R. Kowalski, Editor, p. 1, American Chemical Society (1977).

21. W. Wegscheider, B.B. Jablonski and D.E. Leyden, "Development of an Automated Procedure for the Optimization of Multielement Analysis with Energy Dispersive X-ray Fluorescence Spectroscopy", Anal. Lett., All, 27 (1978).

22. W. Wegscheider, B.B. Jablonski and D.E. Leyden, "Response Functions for the Quantitative Evaluation of Compromise Conditions for Energy Dispersive X-ray Fluorescence, in preparation.

23. L.A. Currie, "Limits of Qualitative Detection and Quantitative Determination", Anal. Chem., 40, 586 (1968).

24. K. Eckschlager, "Informationsgehalt analytischer Ergebnisse", Z. Anal. Chem., 277, 1 (1975).

25. D.F. Brost, B. Malloy and K.W. Busch, "Determination of Optimum Compromise Flame Conditions in Simultaneous Multielement Flame Spectrometry", Anal. Chem., 49, 2280 (1977).

26. W. Wegscheider, H.C. Acree and D.E. Leyden, "Evaluation of a Pulsed Tube for Tube Excited X-ray Fluorescence Analysis", in preparation.

27. R. Jenkins, An Introduction to X-ray Spectrometry, p. 112 (Heyden, New York: 1974).

28. W. Wegscheider, B.B. Jablonski and D.E. Leyden, "Evaluation of Function Derived Optimization for Energy Dispersive X-ray Fluorescence Analysis", in preparation.

29. W. Wegscheider, B.B. Jablonski and D.E. Leyden, "Signal-Hyperflächen für die Multielement Analyse metallischer Proben mit energie-dispersiver Roentgenfluoreszenspektrometrie", Mikrochim. Acta, in press.

CODE	VOL	QTY	AUTHOR AND TITLE	
401630	22	1	ADVANCES IN X-Ray ANALYSIS	COMPLI

JR

IMPROVED X-RAY FLUORESCENCE CAPABILITIES BY EXCITATION WITH HIGH INTENSITY POLARIZED X-RAYS*

Richard W. Ryon and John D. Zahrt**

Lawrence Livermore Laboratory

P. O. Box 808, L-325, Livermore, California 94550

INTRODUCTION AND REVIEW

Energy dispersive x-ray fluorescence is an established and versatile tool for measuring major and trace elements in virtually any kind of solid or liquid specimen. The usefulness of the method could be extended even further if the time of analysis for multi-component samples could be reduced. In other words, we desire to analyze a wide range of elements with detection limits at least as good as obtained when the excitation conditions are optimized for a specific element or narrow range of elements. A major impediment to achieving this goal when analyzing bulk, low-Z materials is the scatter of source radiation into the detector by the specimen being analyzed. The adverse effects of the scattered radiation are its contribution to the background signal (i.e., "noise") and its overwhelming contribution to the limited counting rate of the system electronics.

---

*Work performed under the auspices of the U.S. Department of Energy by the Lawrence Livermore Laboratory under contract number W-7405-ENG-48.

**Present address: Northern Arizona University, Dept. of Chemistry Box 5698, Flagstaff, Arizona 86011

An excellent means of obtaining high sensitivity and low detection limits for a few elements is to excite the specimen with intense monochromatic radiation with an energy just above the absorption edges of the elements of interest. Nearly monochromatic radiation may be conveniently produced by using a secondary fluorescer. However, to obtain good sensitivity for all elements, several fluorescers must be used. For light elements, a separate fluorescer may be required for each individual element. Thus, the simultaneous multi-element capability of energy dispersive analysis is limited or even lost.

There is a means to reduce the foregoing limitation. A wide range of elements can be efficiently excited by using broad band (bremsstrahlung) radiation. The scatter of source radiation into the detector is reduced by first polarizing it. This is accomlished by simply replacing the secondary fluorescer by a suitable low-Z scatterer and using the proper geometry. We have previously demonstrated the usefulness of this approach (1). In our previous work, polarization was obtained by scattering a collimated beam of x-rays at an angle of 90° from boron carbide. Fluorescent radiation is detected along a line perpendicular to the plane defined by the incident and scattered beams at the polarizer. Ideally, the only radiation detected is due to fluorescence in the specimen. Count rate limitations and background due to scattered source radiation are eliminated.

Polarization is obtained only with a commensurate loss in intensity. A high degree of polarization requires very tightly collimated beams, so that intensity approaches zero as the theoretical polarization approaches 100%. Therefore, the ideal case is not realized. It has been shown, however, that in such a system the detection limits are proportional to $\bar{\omega}^{-2}$, where $\bar{\omega}$ is the average angular divergence about the system angles (i.e., $\pi/2 \pm \bar{\omega}$). Therefore, the lowest detection limits are obtained by opening the collimator apertures until the system maximum counting rate is obtained.

When a standard 2500 watt Mo anode x-ray tube is used, the maximum counting rate is obtained when the polarization is about 58% (polarization $\approx 1-2\bar{\omega}^2$). For example, with collimator tubes approximately 1 inch in length, the corresponding diameters are about 3/8 inch. Even though polarization is sacrificed, fluorescent radiation is enhanced over scattered radiation, as the latter is attenuated by the partial polarization factor ($\alpha \cos^2(\pi/2 \pm \bar{\omega}) \approx \bar{\omega}^2$). Using such a system, the detection limits for the elements K-Sr in NBS Standard Orchard Leaves range between about 3X to 1X lower as compared to the most favorable secondary exciter.

With the orthogonal arrangement described, a secondary fluorescer can always be used instead of the scatterer-polarizer for those cases where polarization would not be expected to offer any advantage. Such a case might be filter paper specimens, where there is little scatter of source radiation because of the low mass/unit area.

POLARIZER MATERIAL SELECTION

The particular materials selected for the polarizer-scatterer depend upon the energy range of interest. The first consideration is the need for a high ratio of scatter relative to photoelectric absorption. This need infers a low-Z material. However, the second consideration is that the interaction must take place over a limited region of space, namely the volume within the field of view of the collimators.

Equations of varying sophistication and accuracy have been derived to predict the scatter efficiencies of materials (2). One equation of intermediate complexity which proves to be quite useful is:

$$I_s = \frac{1}{4} I_o \left(\frac{\mu_s}{\mu_T}\right)\left(\frac{d\Omega}{2\pi}\right)\left(\frac{d\Omega}{4\pi}\right)\left\{1-e^{-2\sqrt{2}\mu_T\rho x} - \frac{1.273}{2\mu\rho d}\left[1-(1+2\sqrt{2}\mu\rho x)e^{-2\sqrt{2}\mu\rho x}\right]\right.$$
$$\left. + \frac{0.2122}{(2\mu\rho d)^3}\left[6-\left[6+12\sqrt{2}\mu\rho x+3(2\sqrt{2}\mu\rho x)^2+(2\sqrt{2}\mu\rho x)^3\right]e^{-2\sqrt{2}\mu\rho x}\right]\right\} \quad (1)$$

where I_s = intensity of scattered radiation

μ_s = mass scatter coefficient (cm^2/g) = $\mu_T - \mu_{P.E.}$

μ_T = total mass absorption coefficient

$\mu_{P.E.}$ = mass photoelectric coefficient

$\frac{d\Omega}{4\pi}$ = solid angle intercepted by the collimators $\approx \frac{1}{4}\bar{\omega}^2$

d = diameter of the collimators

x = scatterer thickness $\leq d/\sqrt{2}$

The terms within the brackets account for absorption within the field of view of the collimators, for the collimator diameters, and the thickness of the scatterer. The maximum value of I_s predicted is 25%, when the scatterer is "infinitely" thick to the x-rays and μ_s/μ_T approaches 1.

This equation is for monochromatic radiation. The geometry modeled is for circular collimators with angles of incidence and scatter = $\pi/4$ radians.

The degree of validity of Eq. 1 was experimentally evaluated at energies ranging from 3.7 to 76 keV. Fluorescent radiation from specimens following the scatterers was measured. Typical results are graphed in Figures 1 and 2 for CaKα and LuKα. The x-ray tube used was a W anode spectroscopy tube which may be operated up to 150 kV. The detector was high purity germanium. The intensities were calculated using equation 1; the interaction coefficients are those at the respective absorption edges.

The mathematical models can be used to guide the selection of scattering materials. Figures 3A and B illustrate the behavior of some typical materials as a function of energy using Eq. 1. It is seen the Be is an excellent material for low energy work, and that B_4C with its greater density is useful over a wide energy range. At the highest energies, medium-Z materials such as Fe are useful.

The goal stated at the outset is to obtain high sensitivity over a broad range of energies. This desire may be enhanced through the use of multi-layered scatter-polarizers. Successive layers of higher Z materials would efficiently scatter higher and higher energy x-rays, thereby providing broad-band polarized radiation to excite all elements simultaneously. The concept is illustrated in Figure 4.

CYLINDRICAL POLARIZER

We now have coming into operation a high intensity polarized x-ray source. The heart of this device is a cylindrical polarizer (3). The concept is illustrated in Figure 5. The polarizer gathers radiation emitted from the anode over approximately 240°, which is equivalent to 2.7 $\pi\omega$ steradians. (With a collimated beam, the area is only $\pi\omega^2$ steradians.) Because of the larger area, the maximum counting rate will occur at a polarization of greater than 95%. We therefore anticipate further significant reductions in detection limits.

The new device is designed with versatility in mind. The cylindrical polarizer has a radius of 1.25 in., and a variable height initially set at 3/8 in. The degree of polarization may be varied on either side of 95% by changing the focal spot size of the electron beam, the height of the polarizer, the polarizer thickness and the diameter of the detector collimator. The primary radiation may also be selected by changing anodes.

We anticipate that we will be able to fully exploit the promise of energy dispersive analysis with this device - that is, to analyze a large number of elements simultaneously with high sensitivity.

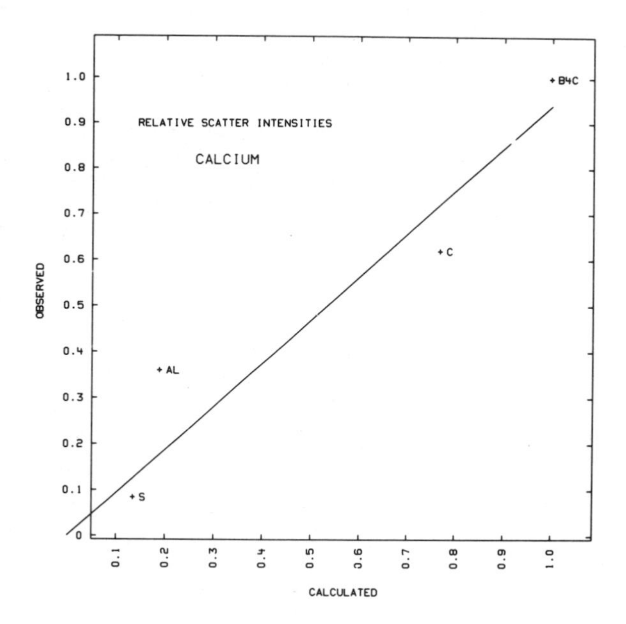

Figure 1

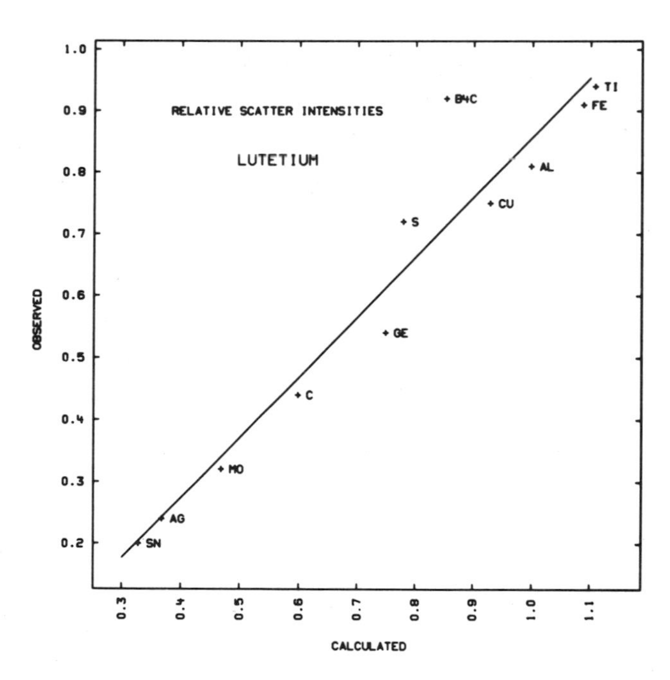

Figure 2

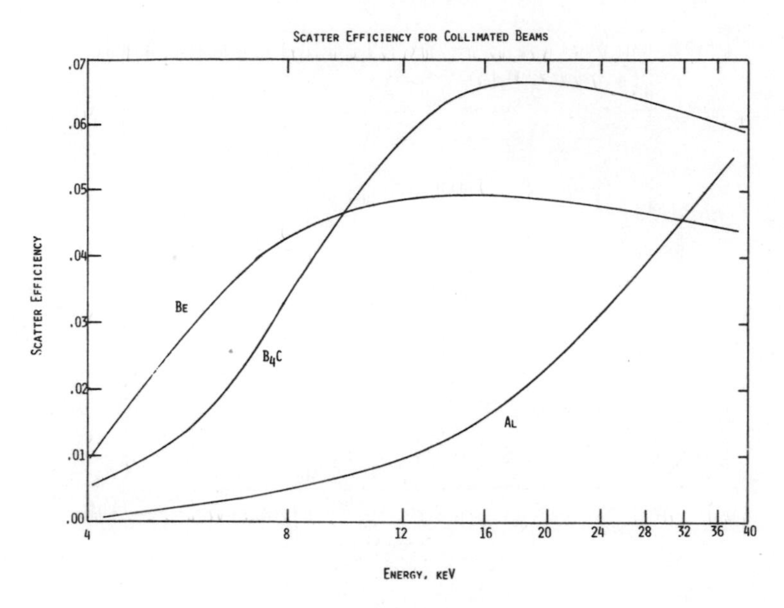

Figure 3A

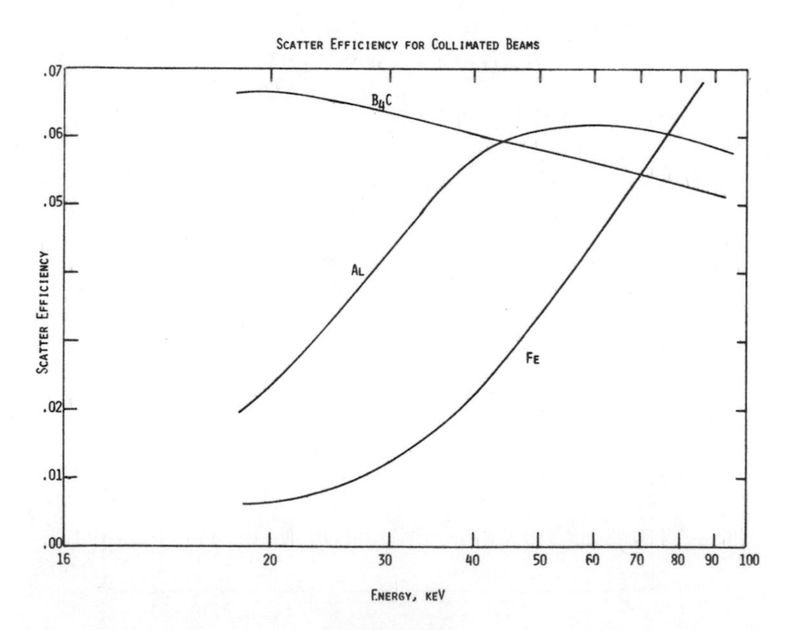

Figure 3B

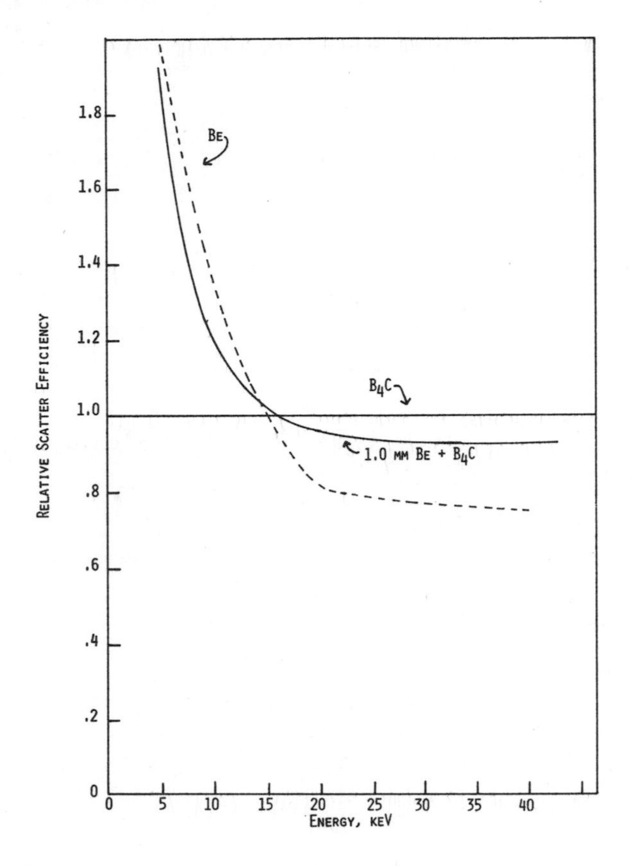

Figure 4

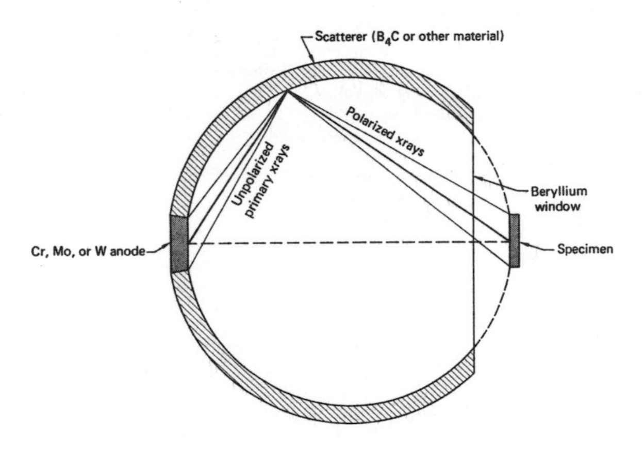

Figure 5

Further advances may be anticipated for a narrow band of elements by using a crystalline polarizer selected to diffract the anode characteristic line at 2θ = 90° (4). We would then have a high intensity, polarized, monochromatic source of x-rays, ideal for trace analysis. Work is continuing.

REFERENCES

1. R. W. Ryon, "Polarized Radiation Produced by Scatter for Energy Dispersive X-Ray Fluorescence Trace Analysis," in H. F. McMurdis, et al, Eds., Advances in X-Ray Analysis, Vol 20, p. 575-590, Plenum Press (1977).

2. J. D. Zahrt and R. W. Ryon, "Scatter Efficiencies for Polarized X-Ray Sources," to be published.

3. R. H. Howell, W. L. Pickles and J. L. Cate, Jr., "X-Ray Fluorescence Experiments with Polarized X-Rays," in W. H. Pickles, et al, Eds., Advances in X-Ray Analysis, Vol. 18, p. 265-277, Plenum Press (1974).

4. H. Aiginger, P. Wobrauschek, and C. Brauner, "Energy Dispersive Fluorescence analysis Using Bragg-Reflected Polarized X-Rays," Measurement, detection, and Control of Environmental Pollutants, International Atomic Energy Act, Vienna, 1976, pp. 197-212.

A NEW X-RAY SPECTROSCOPY CONCEPT-ROOM TEMPERATURE MERCURIC IODIDE WITH PELTIER-COOLED PREAMPLIFICATION(1)

G.C. Huth, A.J. Dabrowski(2), M. Singh, T.E. Economou*, A.L. Turkevich*

University of Southern California, School of Medicine, Medical Imaging Science Group, 4676 Admiralty Way, Marina del Rey, California 90291

*Enrico Fermi Institute, University of Chicago, Chicago, Illinois 60637

1. INTRODUCTION

Mercuric iodide (HgI_2) with high atomic numbers of 80 and 53 for the components, and wide bandgap (2.1 ev) has been considered a potentially useful material for gamma ray spectroscopy for the last number of years. Considerable effort by numerous groups has been directed toward purification and growth of large single crystals of this material. Even with continuing improvement, however, the material is still characterized by good electron transport properties and only modest to poor hole transport behavior. Interesting results nevertheless have been reported in mid-range gamma ray spectroscopy using up to millimeter thick sections of HgI_2. One question always asked is: has the fundamental of lattice scattering etc. been reached in increasing hole transport behavior in this material? There is interesting speculation at this time that this may not be so. The HgI_2 crystal with its lamellar nature (discussed later) may be sequestering impurities non-uniformly within lamella. If this is so

(1) Research supported by Department of Energy Funding under Contract EY-76-S-03-0113 and Planetology Program, Office of Space Science, NASA, under Contract NGR-14-001-135.

(2) On leave from Institute of Nuclear Research, Swierk, Poland.

the future may hold further improvements in hole transport properties.

Taking advantage of the high x-ray attenuation coefficient values of HgI_2, x-rays can be absorbed in very shallow regions. Effects of poor hole collection can then be minimized and this property, in contrast to higher energy gamma spectroscopy, is of less importance in x-ray spectrometry. Our work has focused on study of the use of thin sections (< 500 μm) of HgI_2 for spectrometry in the x-ray region. Initially we have concentrated in the very low energy x-ray region exploring the highest energy resolution attainable(1). The wide energy band of HgI_2 (which is, in fact, an insulator) results in extremely low values of room temperature detector leakage current and therefore detector noise. In terms of noise, the noise levels attainable with silicon or germanium can be approached. Those smaller band-gap materials require cooling to cryogenic temperature to reduce leakage current values. The chief advantage of HgI_2 x-ray spectrometry is considered to be the practicality etc. of room temperature operation and therefore any consideration of cryogenic cooling is ruled out. We describe herein the effect of electrical (Peltier) cooling of the noise limiting first stage of preamplification. This maintains the simplicity and considerable size advantage of the non-cryogenic HgI_2 x-ray spectrometry concept.

2. PROPERTIES OF MERCURIC IODIDE

The low temperature tetragonal crystalline structure is the basic form of mercuric iodide used. This is a lamellar crystal cleaving easily along the iodine-iodine planes. It is beautifully translucent ruby in color. This form of HgI_2 is stable up to $128^{\circ}C$ where it undergoes a destructive phase change to yellow β-HgI_2. Thus the upper operating temperature limit is of this order. The energy band-gap of HgI_2, as mentioned previously, is 2.13 ev ($300^{\circ}K$). It thus falls into the insulator or semi-insulator character. The ionization coefficient in ionizing radiation detection is 4.15 ev per electron-hole pair formed which places it at a slight disadvantage relative to silicon or germanium in potentially achievable energy resolution.

After purification of the starting material, the HgI_2 crystals are grown by vapor growth methods. This is usually accomplished in horizontal glass furnaces at low temperature. Large crystals weighing as much as 100 grams are common with the growth period being of several weeks duration.

Taking advantage of the lamellar nature of the crystal, "plates" are carefully cleaved to be used in the fabrication of x-ray detectors. These plates can be up to a centimeter in diameter reflecting the cross sectional area of the crystal. Subsequent x-ray detector fabrication is exceptionally simple. The detector plate is first etched to the required thickness (100-500 μm). The actual detector area is then defined by painting on contacts of emulsified carbon ("aquadag") to both sides of the crystal. Contact wires are embedded in the carbon contacts for remote attachment. Finally, the crystal is sprayed with a common commercial electronic encapsulant ("humiseal"). This effectively eliminates moisture or vacuum interaction with the HgI_2 surface and possible degradation of detector leakage current and noise behavior. In use, the detectors are biased in the 100-500 volt range corresponding to detector electric field values of 10^4 volts/cm which is sufficient for charge collection. Leakage current (I_d) values are very low. Less than 5 x 10^{-12} amperes are common and are required for high energy resolution.

In the x-ray energy region below 30 keV which encompasses most of the useful elements, an HgI_2 thickness of 400 μm suffices to provide high detection efficiency. This is indicated in figure 1. For the soft x-ray region below 10 keV even thinner sections suffice. At these thicknesses, (as noted in reference 1), effects of poor hole transport on attainable energy resolution are minimal. It is provocative that slightly thicker HgI_2 sections in the 500 μm to one millimeter range provide high detection efficiency over practically the entire x-ray range. Hole transport effects begin to intervene at this point however, and, although excellent energy resolution is possible, such factors as selection of crystal etc. must be considered. The reader is referred to the numerous papers on use of HgI_2 in higher energy gamma ray spectrometry. (2 , 3).

3. HgI_2 SPECTROMETER DEVELOPMENT

For x-ray energy resolution measurement the HgI_2 detector is coupled to a modified charge sensitive preamplifier. In essence, as will be described, the first stage of this preamplifier is extracted and modified in an external "probe" type arrangement containing the detector. Detailed description of the development and performance of this spectrometer is contained in reference (1) to which the interested reader is referred. We will summarize here the performance in terms of noise and energy resolution that have been obtained to date.

Figure 2 is a view of the first HgI_2 detector x-ray probe built for the accelerator measurements to be detailed in Section 4. The spectrometer is based on a modified Tennelec Model 161D preamplifier. Contained in the probe are the HgI_2 x-ray detector, input FET, feedback resistor, and a test capacitor. The input FET is thermally mounted to an electrical Peltier cooling element to reduce its temperature and thus noise. This probe also contained a vacuum mounting flange (visible in the photo) for accelerator mounting. No attempt was made to miniaturize this probe and it is in no sense optimized to that end.

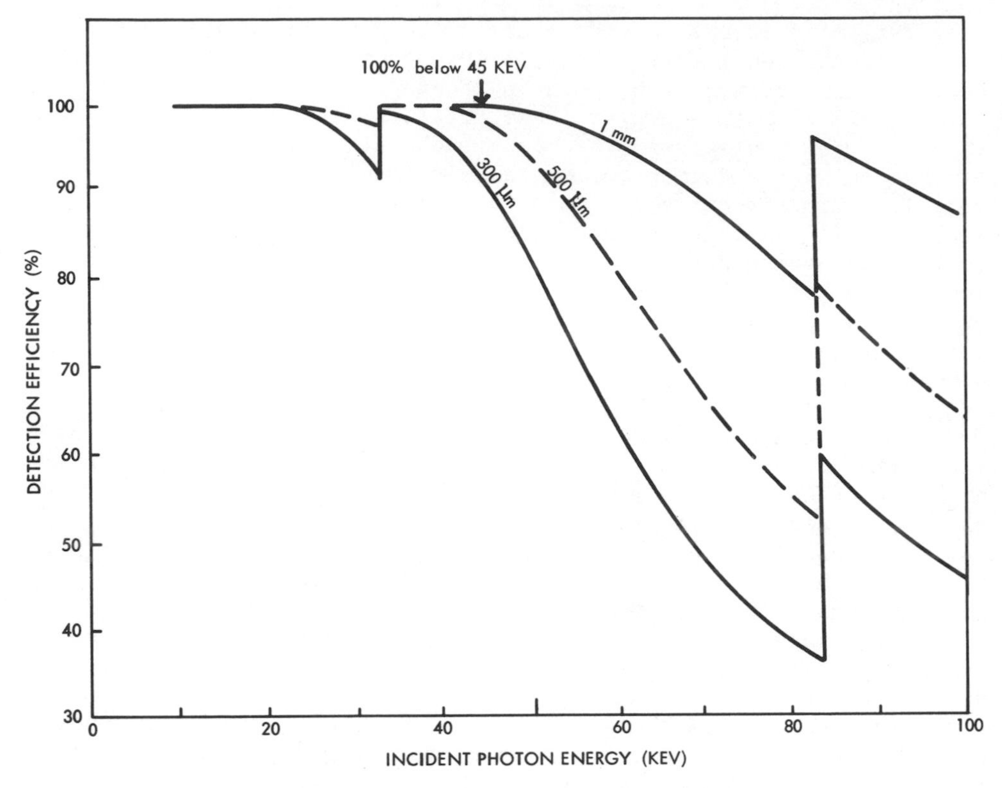

Figure 1: Photoelectric detection efficiency as a function of incident photon energy for mercuric iodide.

Figure 2: A photograph of the HgI_2 probe mounted inside the accelerator chamber for measuring alpha induced x-rays.

The detector cover shown in the view, although similar to the vacuum housing of a cryogenic spectrometer, was used only for this particular experiment. The noise and resolution remain constant in air or vacuum.

The x-ray spectrum obtained from a ^{55}Fe source using a ~9 mm^2 detector (~500 µm thick) is shown in figure 3. The energy resolution at 5.9 keV is 450 ev, and the electronic resolution (including detector) as measured by the pulser (shown at 663 ev in the figure) is 360 ev. The FET was thermoelectrically cooled to obtain these values.

Factors which influence the energy resolution of HgI_2 spectrometers have been studied theoretically and experimentally. Among various noise sources which degrade the energy spectrum, the statistical fluctuation in the creation of electron-hole pairs, trapping during the charge collection process and electronic noise of the detector-preamplifier system are most important (1).

Assuming a Fano factor of 0.46 for HgI_2 (1), statistical fluctuations in the charge generation process contribute 250 ev (fwhm) at 5.9 keV and 116 ev (fwhm) at 1.25 keV. Because of the relatively high value of the Fano factor for HgI_2, degradation due to statistical effects is not negligible.

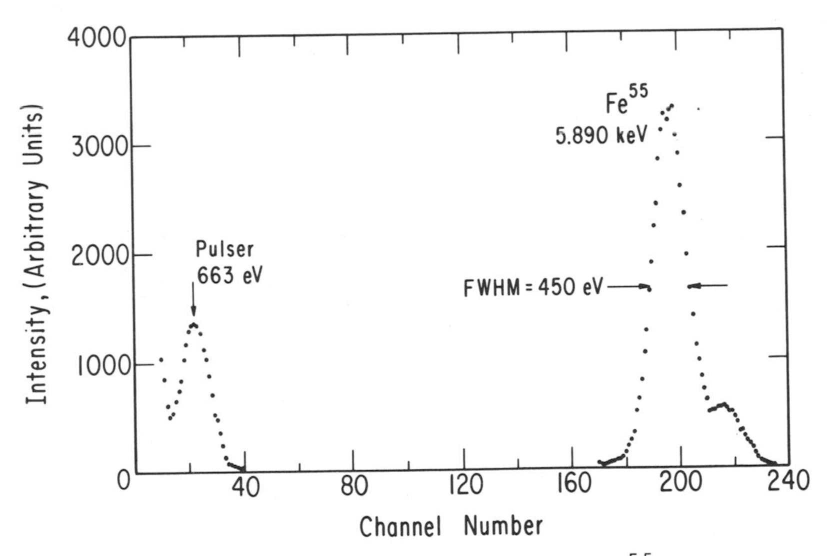

Figure 3: The x-ray spectrum obtained from a ^{55}Fe source using a 500 µm thick HgI_2 detector. Also shown is a pulser at a position corresponding to 663 ev, its fwhm is 360ev.

Table 1

Electronic Noise of the System for Three HgI_2 Detectors

DETECTOR			ELECTRONIC NOISE (ev)					
			Room Temperature FET (ΔE_p=360ev)			Electrically Cooled FET (ΔE_p=270ev)		
Active Area mm^2	Thickness μm	Leakage Current pA	Theoretical ΔE_d	Theoretical ΔE_n	Measured ΔE_n	Theoretical ΔE_d	Theoretical ΔE_n	Measured ΔE_n
7	400	3	218	421	430	206	340	355
9	500	4	228	426	435	216	346	360
12	500	6	274	452	470	261	376	400

Trapping during charge collection is not a limitation in low energy x-ray measurements because the value of mobility and trapping-time product for electrons is sufficiently high (10^{-4} -10^{-3} cm^2/V), and contributes only 100-150 ev (fwhm) to the line width.

The main factor which limits resolution of HgI_2 spectrometers has been found to be the electronic noise of the detector-preamplifier system (1). Following the discussion in reference 1 the electronic noise (ΔE_n) can be written as a square root of the sum of squares of the noise of the preamplifier without the detector (ΔE_p), and noise due to the detector (ΔE_d). In table 1 the predicted and measured values of electronic noise of the system for three HgI_2 detectors are shown, assuming the input capacitance of the preamplifier without the detector to be 5pF, equivalent series noise resistance of 140 Ω and shaping time constant 4 μs. It is seen that when the input FET is cooled to -30°C, the preamplifier noise diminishes considerably (from 360 ev to 270 ev) but is still the predominant factor contributing to the noise of the system.

Based on the values in table 1, we can estimate the resolution for a given system. Assuming for example a HgI_2 detector 7 mm^2 in area and 400 μm thick, the estimated value of energy resolution will be 500 ev (fwhm) at 5.9 keV and 440 ev (fwhm) at 1.25 keV for the FET at room temperature. With the FET cooled to -30°C, the corresponding values will be 435 ev (fwhm) and 370 ev (fwhm), respectively. These estimated values could be compared with the experimental results presented in figure 3, and those discussed in the next section.

4. LOW ENERGY X-RAY SPECTRAL MEASUREMENTS

An alpha particle beam obtained from a Van de Graaff accelerator at the California Institute of Technology was used to excite characteristic x-rays from several elements, which were then measured with the probe shown in figure 2. In addition to a general evaluation of the HgI_2 spectrometer, this experiment was used to assess the suitability of HgI_2 spectrometry for in-situ chemical analysis of extraterrestrial bodies. A schematic of the experimental arrangement is shown in figure 4. The α beam-sample-x-ray detector relationship for this measurement was the same as in the alpha particle instrument (4), with the HPGe detector being replaced by the HgI_2 detector. Thus, the probe containing the HgI_2 detector was designed to fit into the space taken by the bulky housing of the germanium detector, and is much larger than necessary.

The mercuric iodide detector used was 7 mm^2 in area and 310 μm thick, and was masked down to an exposed area of $3mm^2$ to avoid edge effects. Alpha beam energies of 2.0 and 6.1 Mev were used. The intensity of the beam was ~1 nA. Characteristic lines <10 keV from different elements [sodium, magnesium, aluminum, silicon, iron, nickle, gold (M line)] were studied.

The spectrum obtained from a magnesium target using a 6.0 Mev α-beam is shown in figure 5. The K_α line at 1.25 keV has a fwhm of 390 ev and is well separated from the noise. The spectrum recorded from

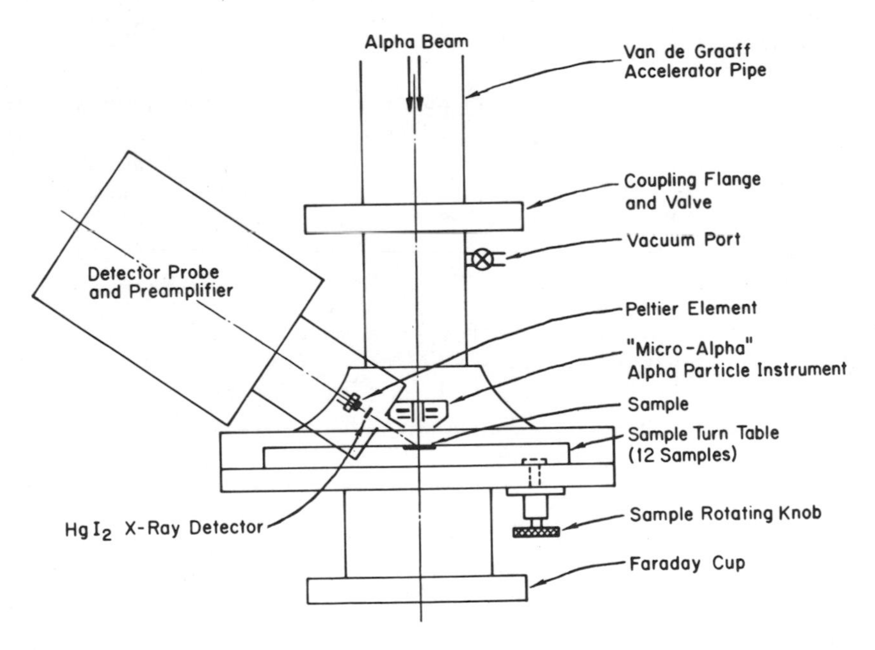

Figure 4: Experimental arrangement for measuring alpha-induced x-rays with a mercuric iodide probe mounted inside the accelerator chamber.

a silicon target is shown in figure 6. Although not shown here, the K_α line of sodium at 1.04 keV was also well separated from the noise (see reference 5 for the sodium spectrum and a detailed discussion). In general, all x-ray peaks obtained with HgI_2 detectors, down to the lowest energies, exhibit excellent symmetrical shape. This characteristic is in contrast to HPGe which shows serious tailing below 2.3 keV (6). Thus, HgI_2 seems promising for space applications.

5. X-RAY FLUORESCENCE APPLICATION

Analysis by x-ray fluorescence involves measurement of characteristic x-rays induced by radiation from a radioisotope or an x-ray tube. In practical applications, x-rays under 30 keV can be used to analyse a large range of elements, and almost 100% detection efficiency can be obtained with a ~500 μm thick detector (see figure 1). Absorption losses in the entrance window mainly arise from the aquadag coating. These are relatively small even for ultra-soft x-rays under 5 keV. Consequently, a high detection efficiency is obtained both for low and high energy x-rays. This characteristic of HgI_2 provides a distinct advantage over silicon detectors which, because of their low Z, show a reduction in detection efficiency above 15 keV for typical detectors, and over germanium detectors which are not suitable for energies under 2.3 keV because of a window located in the detector material (6).

We have measured x-ray fluorescent spectra from several metal, alloy and rock samples using ^{109}Cd and ^{55}Fe sources and a mercuric iodide detector. As an example, we shall discuss the spectra obtained from ores* containing uranium. A ^{109}Cd source was used to excite the L x-rays of uranium. The geometrical arrangement for making this measurement is shown in figure 7. The sample holder was made of teflon, with a thin polypropylene window. An angle of approximately 110° was formed between the exciting source radiation and the radiation being detected. The detector to sample distance was ~1 cm, and the source to sample distance was ~2 cm. The ^{109}Cd source (~10 mCi) was sealed in a special holder made out of tungsten alloy and high-purity silver, and was collimated as shown in figure 7. The detector was about 500 μm thick and had an active area of ~9 mm^2, which was masked down to ~3.2 mm^2 (to reduce edge effects) with a lead collimator. The energy resolution of this detector system was 450 ev for the 5.9 keV peak as measured with a collimated ^{55}Fe source. The source, sample holder, and detector including components for the first stage of preamplification were enclosed in a chamber which was evacuated by a mechanical pump.

---

*Uranium ore samples were supplied by Scintrex Corporation, Toronto, Canada and were chemically analyzed by New Brunswick Laboratory.

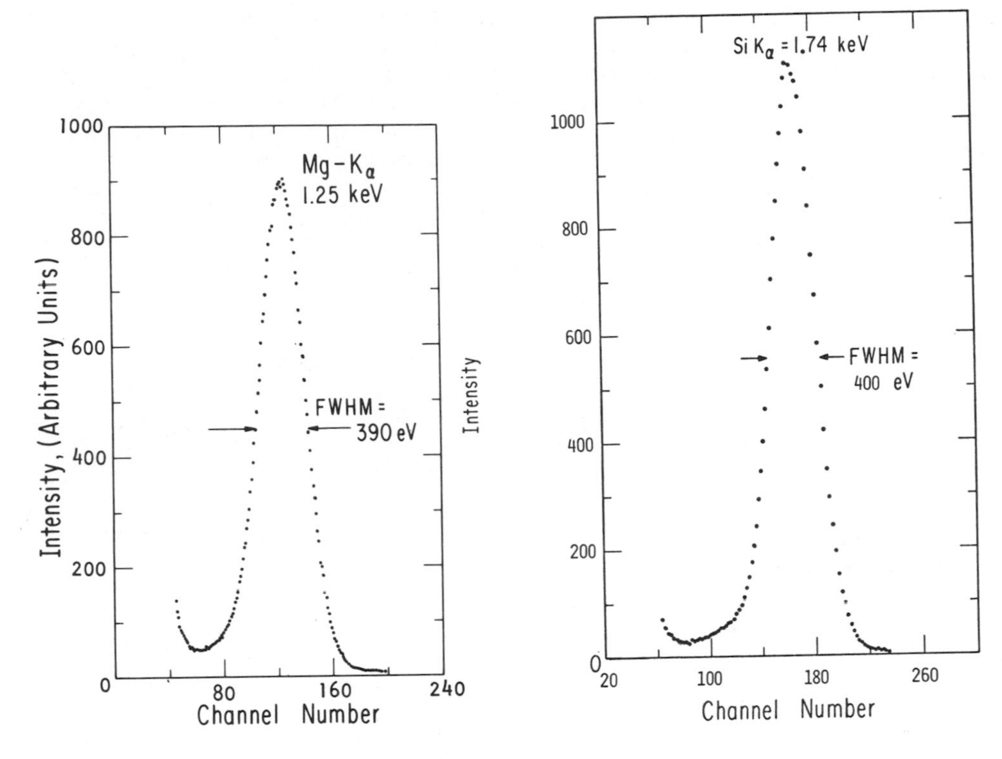

Figure 5· X-ray spectrum obtained from a magnesium target.

Figure 6: X-ray spectrum obtained from a silicon target.

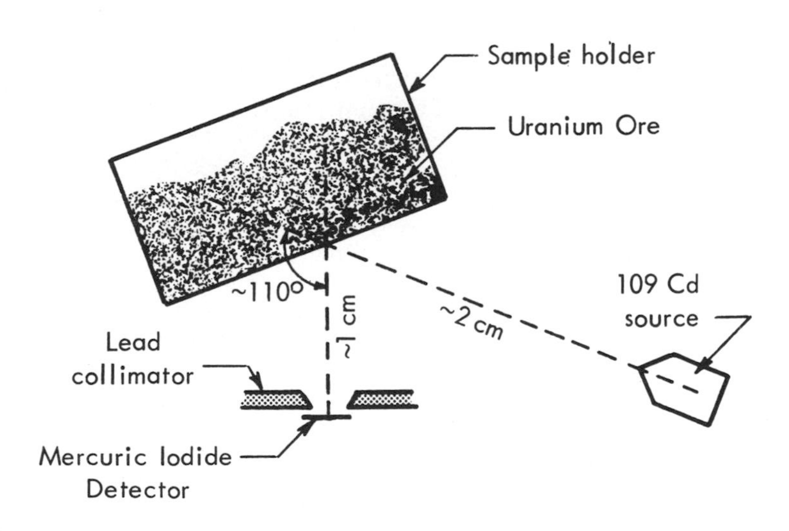

Figure 7: Geometrical arrangement for x-ray fluorescence measurements.

The spectra obtained from ores containing 4% and 0.05% concentrations of uranium (as determined by a prior chemical analysis), are shown in figures 8a) and 8b). In addition to characteristic x-rays, scattered source radiation from the sample into the detector produces several prominent peaks. Both coherent (elastic) and incoherent (compton) scattering contribute to this process, the relative strength depending on the composition of the sample matrix. The coherently scattered peaks are at 24.94 and 22.16 keV and the incoherently scattered peaks corresponding to an angle of 110° are at 23.4 and 20.9 keV, as predicted by the kinematics of compton scattering.

The characteristic L x-rays of uranium observed are the L_α peak at 13.6 keV, $L_{\beta 1}$ and $L_{\beta 2}$ at 17.22 and 16.4 keV, and $L\gamma_1$ at 20.16 keV. The energy resolution for the 13.6 keV L_α peak is 864 ev (fwhm) in this measurement. All of these peaks can be separated in the spectrum recorded from the ore sample containing 4% uranium. For lower concentration of uranium, the $L\gamma$ peak becomes obscured by the scattered source radiation. However, the L_α and L_β peaks remain visible down to 0.05% of uranium, as shown in figure 8b). Although not shown in the figure, the L_α x-ray from an ore containing 0.002% uranium was also visible above the background.

Another example of analysis by x-ray fluorescence using HgI_2 detectors is shown in figure 8c), which represents the spectrum obtained from a sample containing 50% iron and 50% nickel.** The geometrical arrangement was similar to that shown in figure 7, and the same ^{109}Cd source and detector were used. Although the iron K_β peak is obscured by the large K_α peak from nickel, the K_α peak of iron (6.4 keV) and the K_α and K_β peaks of nickel at 7.47 and 8.27 keV are well separated.

In addition to the scattered source radiation which produces large peaks, there are several mechanisms that produce background in the form of a continuum or small peaks in the spectra recorded by mercuric iodide detectors. The following are the main contributing factors. Similar mechanisms also occur in other semiconductor detectors (7). a) Production and subsequent escape of characteristic x-rays of mercury and iodine from the detector material, which results in small peaks in the spectra. The K absorption edges of mercury and iodine are at 83.1 keV and 33.17 keV respectively. Therefore K_α escape will be a factor only for high energy x-ray detection. For the region below 30 keV which covers a large range of elements since either K or L x-rays can be detected, escape peaks can arise from the L and M shells of mercury and the L shells of

---

**Details of these measurements made in collaboration with Dr. Benton Clark of Martin-Marietta Corporation, Denver, will be published separately.

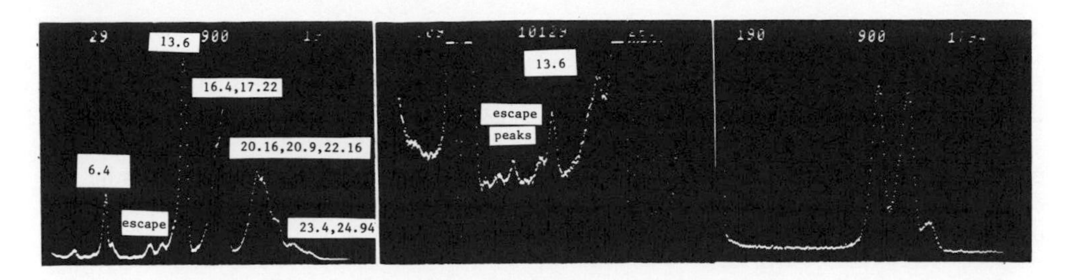

Figure 8: X-ray fluorescent spectra obtained with a mercuric iodide spectrometer and a ^{109}Cd source from samples containing a) 4% uranium, b) 0.05% uranium and c) 50% iron and 50% nickle.

iodine. In applications involving a ^{109}Cd source, escape of mercury L x-rays, mainly L_{α} which has an energy of 9.99 keV, are responsible for producing small peaks in the spectrum. These peaks have an intensity of about 1-2% of the parent peaks, and can be seen quite clearly in the uranium spectra. They are best visualized in figure 8b) on the left of the 13.6 keV uranium peak. Interference effects of escape peaks originating from the M shells of mercury and the L shells of iodine are relatively small. Using an ^{55}Fe source it was observed that these peaks had <1% intensity of the parent peak. In general, for a given detector, the strength and location of suspected escape peaks can be predicted quite accurately from an inspection of the parent peaks, and therefore their contribution to a spectrum can be subtracted for quantitative analysis. b) A continuum background can be produced from compton scatter of the incident photons within the detector, and subsequent escape of the scattered photons, wherefrom only a small amount of energy imparted to recoil electrons would be deposited in the detector. However, for the energy range of interest in x-ray spectroscopy (<100 keV) the photoelectric absorption cross-section is very high for mercuric iodide, and incident photons are predominately absorbed by the photoelectric process. Even when a compton interaction occurs, the scattered photon has a high probability of being absorbed photoelectrically, thereby resulting in a full-energy peak. We have studied this interaction mechanism in detail theoretically, and have computed the probability of upto two compton interactions, and subsequent escape of the scattered photon from the detector. Our results indicate that for typical mercuric iodide detectors the probability of obtaining such events which lead to a continuum background is $\leq$ 1% for incident energies upto 100 keV. Under similar conditions, the background generated by this process is higher for typical germanium and silicon detectors. Therefore, because of its high Z, HgI_2 has an advantage even in this respect. c) A continuum background can also be produced from an escape of the photoelectron (produced as a result of the photoelectric process) from the detector surface. The photoelectron loses energy along its track, and if it escapes, only a partial energy is deposited in the detector. Neglecting electron channeling effects, the background intensity from this process is calculated to be ~0.8% of the incident photon intensity for 20 keV photons. d) Another mechanism which causes background is incomplete charge collection arising from a distortion of the electric field on the surface and edges of the detector. This can be reduced by either masking the detector (as was done in these measurements) or by using a guard ring and logic electronics (8).

6. CONCLUSIONS

The present results demonstrate that mercuric iodide provides high resolution, energy dispersive x-ray spectrometry at room temperature. Decapsulation of the FET, more effective Peltier cooling, and a better selection of the components should further improve the energy resolution. Elimination of cryogenic reservoir and plumbing allow for an exceptional degree of miniaturization, and increased practicality for extra-laboratory use of x-ray spectrometry. Portable x-ray fluorescence elemental measurement systems should become more practical and useful. It should be feasible to place XRF systems in applications that were hitherto impossible. Small diameter, bore hole analysis comes to mind. Lastly, the capability may lead to more fundamental advances in such areas as x-ray microprobe analysis by providing almost a two order of magnitude geometrical advantage over conventional systems in detection of characteristic x-rays.

The authors thank the research group of E.G.&G. Santa Barbara, California, for HgI_2 detector fabrication, Professor T. Tombrello and the accelerator staff at Cal Tech, and Dr. Benton Clark of Martin-Marietta Corporation, Denver.

7. REFERENCES

1. A.J. Dabrowski, G.C. Huth: IEEE Trans. Nucl. Sci. NS-25(1), 205-211 (1978).
2. J. Beinglass, G. Dishon, A. Holzer et al: Appl. Phys. Lett. 30, 611-613 (1977).
3. L. van den Berg, R.C. Whited: IEEE Trans. Nucl. Sci. NS-25(1), 395-397 (1978).
4. T.E. Economou, A.L. Turkevich: Nucl. Instrum. Methods 134, 391 (1976).
5. A.J. Dabrowski, G.C. Huth, M. Singh et al: Appl. Phys. Lett. 33(2) 211-213 (1978).
6. J. Llacer, E.E. Maller, R.C. Cordi: IEEE Trans. Nucl. Sci. NS-24, (1977).
7. F.S. Goulding and J.M. Jaklevic: LBL-5367 (1976).
8. F.S. Goulding and J.M. Jaklevic: UCRL-20625 (1971).

SOME STUDIES OF CHLORINATED POLY(VINYL CHLORIDE) USING X-RAY PHOTOELECTRON SPECTROSCOPY

James Gianelos and Eric A. Grulke

The BFGoodrich Research & Development Center

Brecksville, Ohio 44141

Chlorinated poly(vinyl chloride) (CPVC) is similar in many ways to its thermoplastic parent poly(vinyl chloride) (PVC). Chlorination imparts superior high temperature properties, which make CPVC preferable for many and varied applications.

Our primary purpose in studying CPVC with x-ray photoelectron spectroscopy (XPS) was to see whether we could gain insight into how chlorine substitutes into the PVC molecule. We also hoped we would obtain insights into the chlorination reaction itself, and how variations would affect the final product.

All of the spectra in this paper were taken from polymers in their virgin powder resin form, as they come from a reactor. The average particle size was 150 micrometers (dry). All polymers were totally free of various compounding ingredients usually found in commercial versions, including stabilizers. The only exception was a commercially made poly(ethylene), which was used as a reference material.

All data were obtained using a computer operated Varian VIEE-15 XPS spectrometer equipped with a magnesium anode soft x-ray source. The vacuum system of this instrument has been improved permitting operation at 6×10^{-6} Pa (5×10^{-7} torr). Curve resolving was done with a DuPont Model 310 curve resolver.

Figure 1 shows the XPS spectrum obtained from a PVC. Initially we were surprised to see only one carbon peak. From its formula $(CH_2CHCl)_n$ we knew that half of the carbon atoms in PVC are bonded to a chlorine atom. Therefore, we expected to see two equal side by side carbon peaks, one corresponding to $-CH_2$ and the other to $-CHCl$.

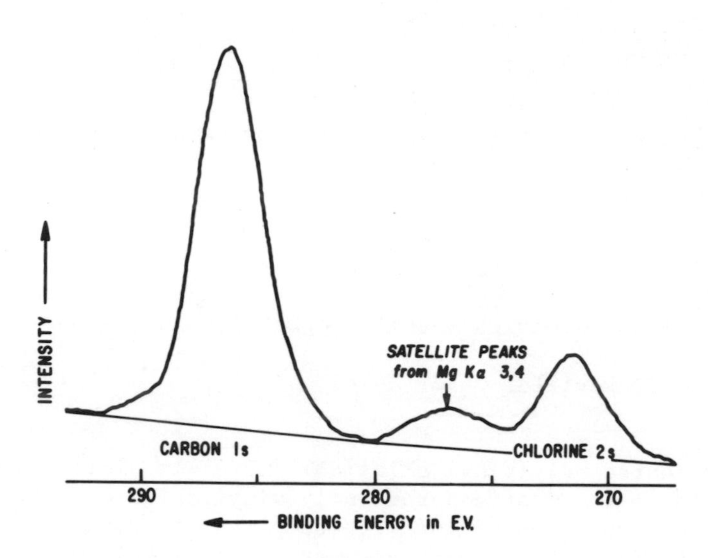

Figure 1. XPS spectrum of PVC

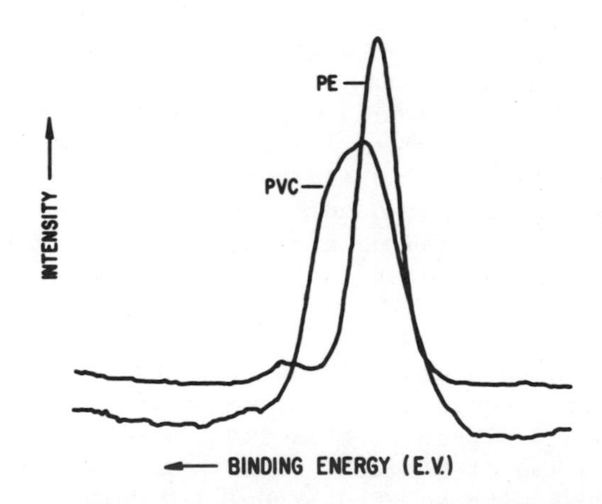

Figure 2. Superimposed carbon Is spectra of PVC and polyethylene

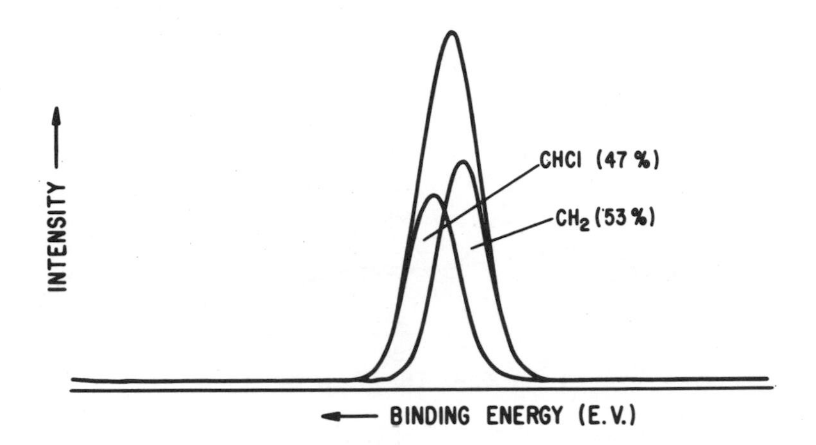

Figure 3. Curve resolved carbon Is spectrum of PVC

That two peaks do, in fact, exist can be seen in Figure 2. This figure shows just the carbon 1s portion (left peak) of the PVC spectrum. Superimposed upon it is the carbon 1s spectrum of poly(ethylene) which is essentially 100% $-CH_2$. From both the peak width and peak position of the $-CH_2$ peak, it is obvious that there must be a second peak occupying the left half of the PVC carbon 1s envelope. This is the -CHCl peak, and is shown deconvoluted in Figure 3. The peak centers are about 1.5 EV apart.

Figure 4 is an XPS spectrum of a CPVC. The chlorine 2s peak is considerably larger and the carbon 1s peak envelope is broadened in comparison to the PVC spectrum (Figure 1).

Figure 5 is just the carbon peak envelope (left half) of this CPVC spectrum. Superimposed over it is a PVC carbon peak envelope, which we know is an unresolved doublet. It can be seen that the CPVC spectrum is broadened to the left or higher binding energy, indicating the presence of a third peak corresponding to $-CCl_2$. This is shown curve resolved in Figure 6. The center of the $-CCl_2$ peak is shifted an additional 1.5 EV from the -CHCl peak, and 3.0 EV from the $-CH_2$ peak. These data are summarized in Table I.

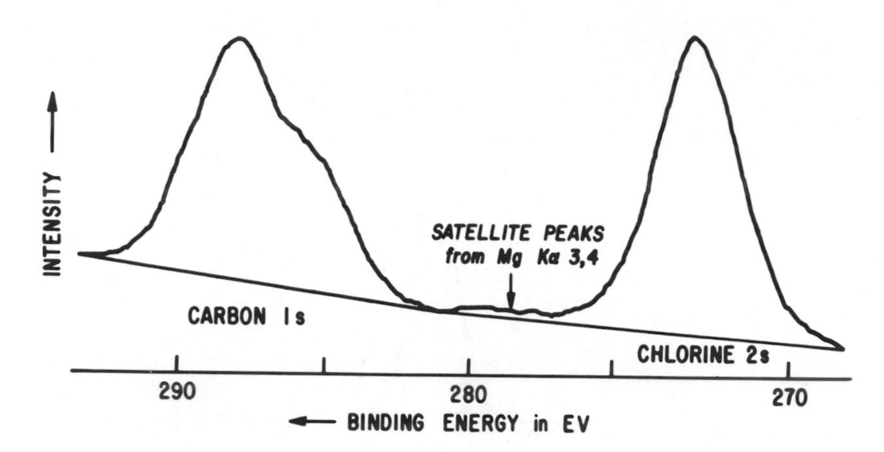

Figure 4. XPS spectrum of chlorinated poly(vinylchloride) CPVC

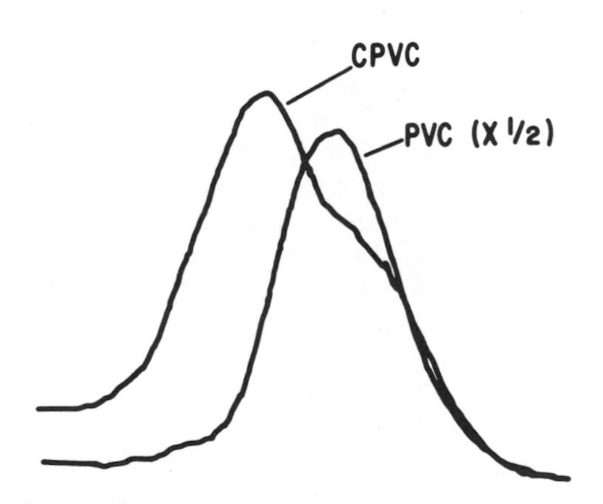

Figure 5. Superimposed carbon Is spectra of PVC and CPVC

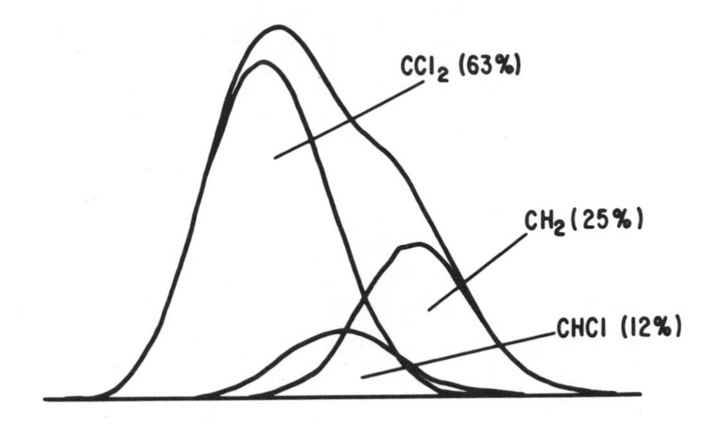

Figure 6. CPVC carbon Is deconvolved spectrum

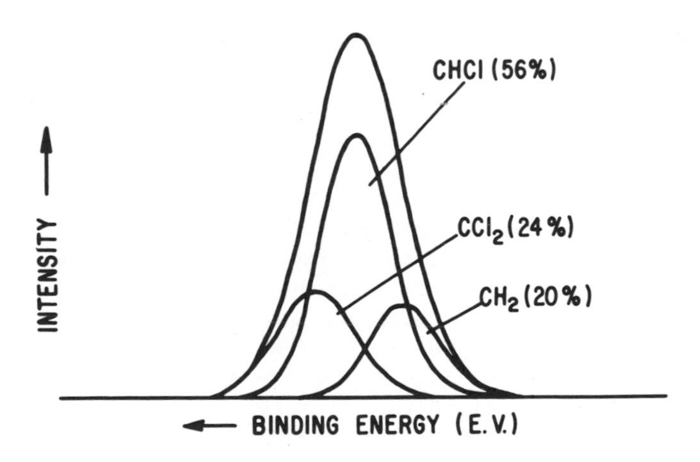

Figure 7. Carbon Is spectrum for a CPVC made under different reactor conditions

Table I

CPVC CARBON 1s BINDING ENERGIES

$-CH_2$	285.5 EV
$-CHCl$	287.0 EV
$-CCl_2$	288.5 EV

The CPVC spectrum in Figure 4 was selected in order to best show the position of the $-CCl_2$ component in the carbon peak envelope. Being the largest peak by far, its position is clear and unmistakable. However, in many CPVCs the $-CCl_2$ component is much smaller, and the ratios of the three peak components may be considerably different. For example, the carbon peak envelope for a CPVC produced under different chlorination conditions is shown in Figure 7. In this example, the -CHCl component is the largest; in the previous figure it was the smallest. These data show that chlorine can substitute on either the -CHCl carbon or the $-CH_2$ carbon. Or, both 1,1 and 1,2 chlorination can and do occur.

Since we have a three component carbon peak envelope, curve deconvolution requires precise knowledge of the individual peak positions. Otherwise, there would be a very large number of possible ways to fit three peaks within a given envelope. We feel our procedure has established peak positions with enough certainty to make curve deconvolution acceptably accurate, especially for large components. When a component is present at or below 10%, however, accuracy probably deteriorates greatly.

In obtaining CPVC spectra, care must be taken to keep x-ray exposures as short as possible in order to minimize two undesirable effects which can distort the spectrum obtained. First, surface dehydrohalogenation (HCl splitting off) can occur. Second, carbon contamination can build up on the surface, as has been documented by other Varian VIEE-15 users. (1) Figure 8 shows these two effects combined. The lower spectrum (normal) was taken during the first ten minutes of x-ray exposure. The upper spectrum is the same sample, but was recorded after the x-rays had been on for 60 minutes. The chlorine 2s peak is somewhat smaller in the upper spectrum, and a shoulder has developed on the carbon 1s peak envelope as well.

We took advantage of XPS's surface sensitivity to answer the question "Does chlorination occur simultaneously throughout the resin particles; or does the surface chlorinate first, followed by chlorination of the bulk interior?" The answer is revealed in Figure 9 which was prepared by removing many small samples from

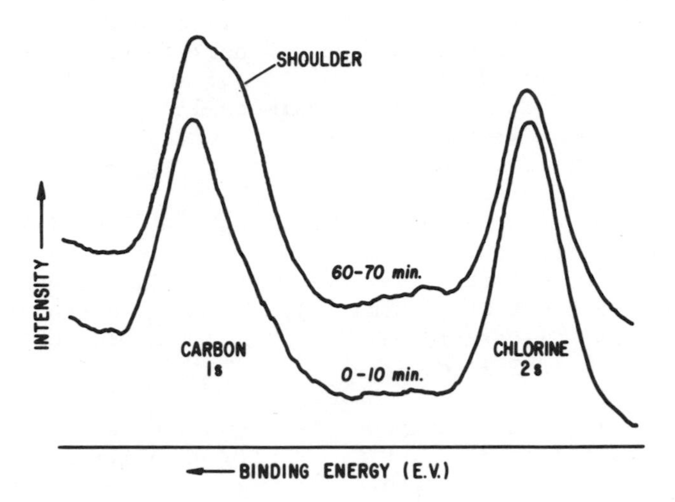

Figure 8. Effect of excessive X-ray irradiation on CPVC spectrum

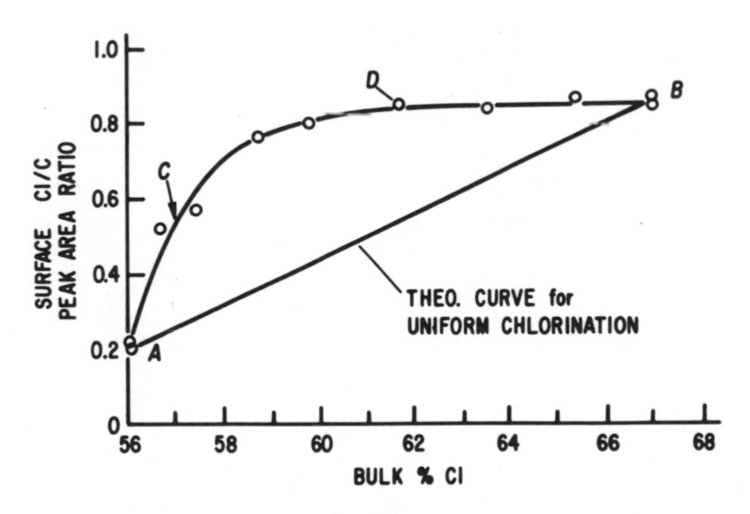

Figure 9. Comparison of experimental chlorine/carbon ratio with theoretically expected curve

the reactor during the course of a chlorination. Point A represents the starting PVC which contained 56% chlorine (by weight) and had a surface Cl/C peak area ratio of 0.22. Point B is the CPVC end product which contained 67% chlorine, and had a surface Cl/C ratio of 0.85. If the surface and bulk interior chlorinated simultaneously, we would expect the plot of surface Cl/C ratio versus bulk % Cl to follow and approximate line AB. However, it does not. Instead, the curve climbs very steeply from Point A indicating very rapid surface chlorination. It then flattens out as the chlorination progresses, and as the bulk interior of the particles becomes chlorinated. Point C on the curve (57% Cl, Cl/C = .57) indicates that the surface of the resin particles has already reached 55% of its final Cl/C ratio value, when the chlorination is only 9% complete. Similarly, at Point D the surface has reached its final Cl/C ratio value, and does not increase further, when chlorination is 53% complete.

In conclusion, we have shown that XPS is an excellent and practical means for studying CPVC, or related polymers. And when used with complementary structure probing techniques such as NMR, IR, SIMS, etc., a very full complete picture of CPVC molecular structure can be obtained.

REFERENCES

1. R. S. Swingle II, "Quantitative Surface Analysis by X-Ray Photoelectron Spectroscopy (ESCA)," Analytical Chemistry 47, 21-24 (1975).

CONTRIBUTOR INDEX

SUBJECT INDEX

C

D

E